Statistical Analysis and Optimization for VLSI: Timing and Power

Ashish Srivastava
Dennis Sylvester
David Blaauw
University of Michigan, Ann Arbor

Springer

Library of Congress Cataloging-in-Publication Data

A C.I.P. Catalogue record for this book is available from the Library of Congress.

ISBN-10: 0-387-25738-1
ISBN-13: 9780387257389

ISBN-10: 0-387-26528-7 (e-book)
ISBN-13: 9780387265285

Printed on acid-free paper.

Printed in the United States of America.

9 8 7 6 5 4 3 2 1 SPIN 11380634

springeronline.com

Dedicated to our families

Preface

Traditional deterministic computer-aided-design (CAD) tools no longer serve the needs of the integrated circuit (IC) designer. These tools rely on the use of corner case models which assume worst-case values for process parameters such as channel length, threshold voltage, and metal linewidth. However, process technologies today are pushed closer to the theoretical limits of the process equipment than ever before (sub-wavelength lithography is a prime example) – this leads to growing levels of uncertainty in these key parameters. With larger process spreads, corner case models become highly pessimistic forcing designers to overdesign products, particularly in an application-specific integrated circuit (ASIC) environment. This growing degree of guardbanding erodes profits, increases time to market, and generall will make it more difficult to maintain Moore's Law in the near future.

The concept of statistical CAD tools, where performance (commonly gate delay) is modeled as a distribution rather than a deterministic quantity, has gained favor in the past five years as a result of the aforementioned growing process spreads. By propagating expected delay distributions through a circuit and not a pessimistic worst-case delay value, we can arrive at a much more accurate estimation of actual circuit performance. The major tradeoff in taking this approach is computational efficiency. Therefore, we can only afford to use statistical CAD tools when their performance benefit is compelling. In earlier technologies this was not the case. However, many companies now feel that the levels of variability, and the stakes, are high enough that the day of statistical CAD has arrived. An inspection of current CAD conference technical programs reflect a large amount of interest from both academia and industry; the current year's Design Automation Conference (DAC) has at least a dozen papers on this topic, nearly 10% of the conference program. While a large fraction of this work has been in extending traditional deterministic static timing analysis (STA) to the statistical regime, power is also critical due to the exponential dependencies of leakage current on process parameters.

As a result of the above trends, the pace of progress, in the past few years in statistical timing and power analysis has been rapid. This book attempts to summarize recent research highlights in this evolving field. Due to the rapid pace of progress we have made every effort to include the very latest work in this book (e.g., at least five conference publications from the current year are included in the reference list). The goal is to provide a "snapshot" of the field circa mid-2005, allowing new researchers in the area to come up to speed quickly, as well as provide a handy reference for those already working in this field. Note that we do not discuss circuit techniques aimed at reducing the impact of variability or monitoring variability, although we feel these will play a key role in meeting timing, power, and yield constraints in future ICs. The focus here is on CAD approaches, algorithms, modeling techniques, etc.

On a final note, a key to the widespread adoption of statistical timing and power analysis/optimization tools is designer buy-in. This will only come about when there is open discussion of variability data, variation modeling approaches (e.g., Does a Quad-Tree model accurately capture the actual behavior of spatially correlated process parameters?), and related topics. We believe that the recent progress in algorithms for statistical analysis and optimization has brought us to the point where these practical issues, and not the underlying tool capabilities, are the limiting factor in commercial acceptance of the approaches described in this book.

This book is organized into six chapters. The first chapter provides an overview of process variability: types, sources, and trends. The second chapter sets the stage for the following four chapters by introducing different statistical modeling approaches, both generic (Monte Carlo, principal components) and specific to the topic of integrated circuit design (Quad-Tree). The third chapter summarizes recent work in statistical timing analysis, a ripe field of research in the past 4-5 years. Both block-based and path-based techniques are described in this chapter. Chapter 4 turns attention to power for the first time – both high-level and gate-level approaches to modeling variation in power are presented with emphasis on leakage variability. Chapter 5 combines ideas from the previous two chapters in examining parametric yield. This important performance metric may replace other more traditional metrics, such as delay or power, in future ICs as the primary objective function during the design phase. Finally, Chapter 6 describes current state-of-the-art in the statistical optimization area – the work to date is primarily aimed at timing yield optimization and ranges from sensitivity-based to dynamic programming and Lagrangian relaxation techniques.

The authors would like to thank Carl Harris of Springer Publishers for arranging for this book to be published and also for consistently pushing us to the finish line. We thank Sachin Sapatnekar for comments on the general content of the book and we also thank Amanda Brown and Paulette Ream for help in proofreading and generating figures.

Ann Arbor Michigan,
May 2005

Ashish Srivastava
Dennis Sylvester
David Blaauw

Contents

1

Introduction

The impact of process and environmental variations on performance has been increasing with each semiconductor technology generation. Traditional corner-model based analysis and design approaches provide guard-bands for parameter variations and are, therefore, prone to introducing pessimism in the design. Such pessimism can lead to increased design effort and a longer time to market, which ultimately may result in lost revenues. In some cases, a change in the original specifications might also be required while, unbeknownst to the designer performance is actually left on the table. Furthermore, traditional analysis is limited to verifying the functional correctness by simulating the design at a number of process corners. However, worst case conditions in a circuit may not always occur with all parameters at their worst or best process conditions. As an example, the worst case for a pipeline stage will occur when the wires within the logic are at their slowest process corner and the wires responsible for the clock delay or skew between the two stages is at the best case corner. However, a single corner file cannot simultaneously model best-case and worst-case process parameters for different interconnects in a single simulation. Hence, a traditional analysis requires that two parts of the design are simulated separately, resulting in a less unified, more cumbersome and less reliable analysis approach. The strength of statistical analysis is that the impact of parameter variation on all portions of a design are simultaneously captured in a single comprehensive analysis, allowing correlations and impact on yield to be properly understood.

As the magnitude of process variations have grown, there has been an increasing realization that traditional design methodologies (both for analysis and optimization) are no longer acceptable. The magnitude of variations in gate length, as an example, are predicted to increase from 35% in a 130 nm technology to almost 60% in a 70 nm technology. These variations are generally specified as the fraction $3\sigma/\mu$ (3σ is assumed to be the worst case shift in the parameter), where σ and μ are the standard deviation and mean of the process parameter, respectively. Thus a 60% variation in 70 nm technology implies that the standard deviation of the distribution of gate length across a large number of samples is 14 nm. With variations as large as these, it becomes extremely important that the designers treat these variation in a statistical manner rather than using gaurd-bands in deterministic analysis.

1.1 Sources of Variations

The traditional approach to ensuring acceptable yield is to estimate margins, while assuming worst-case process and environmental conditions. With increasing clock frequency and the growth of variations, these margins have become a larger fraction of the total clock cycle, making the traditional techniques hard to sustain. Part of this difficulty is that margins do not result from a single source of randomness. They are, in fact, used to capture a host of physical effects that are either truly statistical (and hence unknown at design time), or are hard to model while performing analysis.

The first step to consider the impact of variations during the design process is to understand the sources of variations and the impact they have on performance. We first characterize the variations based on their sources.

1.1.1 Process Variations

Process variations are fluctuations in the value of process parameters observed after fabrication. These variations result from a wide range of factors during the fabrication process which determine the ranges of variations. It is obvious that large variations in process parameters will lead to designs that deviate strongly from their specifications. These variations effect the performance characteristics of devices as well as interconnects. The resulting distribution for performance across a large set of fabricated samples leads to the definition of *parametric yield*, which is the fraction of manufactured samples that meet the performance constraints. Parametric yield should be contrasted to *manufacturing yield* that defines the fraction of samples manufactured without catastrophic manufacturing failures (such as wire shorts and opens) that render a given sample useless at any frequency.

For a given process technology, two different designs can have significantly different parametric yield. This results from the fact that the same variations in process parameters may influence two designs in very different manners. For example, we will see in Chap. 2 that designs with a large number of timing critical signals have an increased susceptibility to process variations. In this context, we define the so-called *timing yield* as the fraction of samples of a design that meet the timing constraint, and similarly we define the *power yield* as the fraction of samples that meet the power constraint.

1.1.2 Environmental Variations

These variations capture the variations in the surrounding environment in which a chip sits during its operation. This includes temperature variations, variation in the power supply and variations in switching activity (defined by the input vectors). A reduced power supply lowers the drive strengths of the devices and hence degrades performance. Similarly, an increased temperature results in performance degradation for both devices and interconnects. It is important to understand that these variations depend on the work-load of the processor and are hence time-dependent. Thus, the set of input vector combinations that result in a worst-case voltage supply drop can occur on any possible sample of the design but will, in all likelihood, occur only intermittently during its operational life time. Thus, power supply and temperature variations are generally not treated statistically, since every shipped chip is required to operate without failures over its entire operational life-time. Power supply drops and high temperatures are, therefore, assumed during the verification of a design. However, identifying specific worst-case conditions for temperature and power supply variation is extremely difficult. Therefore, designers often focus on minimizing temperature and supply variations as much as possible, such as ensuring that the voltage drop on a power grid is always within 5%–10% of the nominal supply voltage.

A particularly interesting situation occurs when process variations increases the current demands on the power supply grids. In older technologies, leakage power dissipation was a concern only in designs that spent a large fraction of their time in stand-by. With leakage power becoming a significant contributor to total power dissipation, leakage currents flowing through the power grid can result in significant supply voltage drops. Moreover, assuming that all devices are operating at their highest leakage will be extremely pessimistic. In this situation, it becomes important to estimate the mean and variance of voltage drops and temperature hot-spots based on variation in process parameters [50], [51], since worst-case leakage induced power-supply drops and hot-spots cannot be expected to occur on each sample of a design.

Leakage currents themselves also increase strongly with an increase in temperature, just as increasing leakage currents may result in a higher temperature. In certain cases, this positive feedback can be strong enough to cause *thermal runaway*, where the currents and temperature in the design continue to increase until failure. Thus, it is important that chip level leakage and temperature analysis are performed in a self-consistent manner [156].

1.1.3 Modeling Variations

These variations result from the fact that the power and delay models used to perform design analysis and optimization are inaccurate and do not perfectly capture device characteristics. These models, if conservative, will make it harder to meet design specifications, whereas aggressive models will result in yield loss. The sample-space of these variations is over design iterations, with different modeling errors at different design points. The tradeoff, in using smaller margins to capture modeling variations, involves the likelihood of tuning particular paths post-fabrication or going through the entire design process again. Thus, we typically want to be conservative while accounting for modeling variations, since it affects all fabricated samples of a design.

1.1.4 Other Sources of Variations

Though most variations are included within the previous three classes of variations, there are physical effects that result in a change in process parameter with time. These effects include phenomena such as hot electrons, negative bias temperature instability (NBTI) and electromigration. Hot electron and NBTI effects result in device degradation with time causing the threshold voltage of the device to rise. Electromigration may cause increased wire resistance due to a reduction in the width of a wire, which increases the resistance of the wire and increases propagation delay. In the worst case, it will result in wire opens and shorts causing functional failure. The impact of these variations depends strongly on process and environmental variations. A wire that has a smaller width to start-off (due to patterning) and is used to provide current to a hot section of the design that demands large currents is much more likely to fail due to electromigration. If these effects are not properly accounted during the design process, they may result in timing errors that become visible during operation or burn-in. The analysis of these variations is particularly difficult, since they become visible after a reasonable time of operation. Therefore, techniques such as burn-in, which are accelerated test techniques, are used. These testing techniques are used to stress the design to operate under worst-case conditions. However, these testing techniques are expensive and have a large application time.

1.2 Components of Variation

For the purpose of design analysis, it is beneficial to divide the variations into two categories: inter-die and intra-die variations. As we will see in later chapters, these components influence the performance of a design differently. Moreover, the influence of these components also depends on how well the design is optimized, which impacts the number of critical paths in a design.

1.2.1 Inter-die Variations

Inter-die variations refer to a parameter variation that has the same value across a single die, and hence captures variations that occur from die-to-die, wafer-to-wafer and lot-to-lot. Since these variations are independent, they are all represented using a single variational term for ease of analysis. These variations are thus represented by a single value for each die and represent a shift in the mean or expected value of the parameter distribution from the nominal value. These variations include gate-length variations due to fluctuations in the time of exposure during fabrication and metal thickness variations between different metal layers. Thus, considering inter-die variations for a process parameter, we can write the value of a parameter for a device as a random variable (RV).

$$P = P_{\text{nom}} + \Delta P_{\text{inter}} \tag{1.1}$$

where P_{nom} is the nominal value of the process parameter and P_{inter} is a zero mean RV that captures the inter-die variation. The RV P_{inter} has a single value for all components on the die. The inter-die variations are generally assumed to have a simple distribution, such as Gaussian, with a given variance. These variations may have systematic trends across dies that can be captured if the specific orientation and location of a die on the wafer is known. However, the designer typically has no control where his chip will be placed on a wafer. Moreover, this information is not available at design time and hence the impact of these factors on process parameters must be captured using a random variable.

Inter-die variations in a single process parameter are easily captured by corner models, which assume that all devices and interconnects on a given sample of the design have a value that is shifted away from the mean by a fixed value that degrades (improves) performance, for slow (fast) path analysis. However, when a number of process parameters are considered simultaneously it is important to consider the correlation between these process parameters. As discussed above, thickness of metal layers that are negatively correlated can result in timing failures when the logic is slower than nominal and clock is faster than nominal. The number of process corners at which a design needs to be simulated for functional correctness thus increase exponentially with the increase in process parameters.

1.2.2 Intra-die Variations

Intra-die variation is the component of variation that causes device parameters to vary across different locations within a single die. Thus, each device on a die requires a separate RV to represent its intra-die variation. Depending on the source of variations, intra-die variations may be spatially correlated or spatially uncorrelated. Though all variations are random, the accepted terminology is to use the term random variations specifically to refer to the uncorrelated component of intra-die variations.

It is obvious that intra-die variations result in a huge increase in the dimensionality of the problem by requiring an extra RV for each device. In addition, these RVs are correlated due to proximity-effects. Since, it is computationally very expensive to generate samples of correlated RVs of high dimensionality, traditional statistical analysis methodologies such as Monte Carlo become unsuitable in scenarios where intra-die variations are significant, whereas deterministic approaches fail to capture the effect of intra-die variations completely. Spatially correlated random variations can be handled by dividing the chip into regions that can be assumed to be perfectly correlated and using a correlation matrix to capture the correlation among these RVs. If the number of these perfectly correlated regions are small, they can be handled easily.

Now, considering both intra-die and inter-die variations for a process parameter, we can write the value of a process parameter as

$$\begin{aligned} P &= P_{\text{nom}} + \Delta P_{\text{inter}} + \Delta P_{\text{intra}}(x_i, y_i) \\ &= P_{\text{nom}} + \Delta P_{\text{inter}} + \Delta P_{\text{spatial}}(x_i, y_i) + \Delta P_{\text{random, i}} \end{aligned} \quad (1.2)$$

where $\Delta P_{\text{intra}}(x_i, y_i)$ represents intra-die variation that consists of a spatially correlated component $\Delta P_{\text{spatial}}$, which is a function of the location on the die and an independent or so-called random component $\Delta P_{\text{random, i}}$ that has no correlation with other devices and is represented as a separate RV for each device.

Intra-die variations can also be classified based on their origin as: wafer-level trends, layout dependent variations and statistical variations.

Wafer-level Variations

Wafer-level variation originate due to effects such as *lens aberrations* and result in *bowl-shaped* or other known distributions over the entire reticle, which results in a *slanted* profile of the process parameter across a single die. Again, the direction of slant varies depending on the orientation of the die on the wafer and cannot be ascertained a priori.

Layout Dependent Variations

Layout dependent variations result in different geometric dimensions due to lithographic and etching techniques that are used during fabrication. These include fabrication steps such as chemical mechanical polishing (CMP) and optical proximity correction (OPC). CMP results in variations in dimensions due to *dishing* (shown in Fig. 1.1) and *erosion*. Dishing arises from the fact that all excess copper must be removed from the wafer – to accomplish this goal, a wafer is typically over-polished, removing some of the copper that is supposed to remain. As copper etches much faster than the surrounding dielectric, the wire ends up being shorter than the oxide. Dishing is the vertical distance between the final oxide level and the lowest point in the copper wire. A substantial amount of dishing leads to increased resistance, worsened planarity, and overall process non-uniformity. Constraints are set on the processing equipment (including slurries and pads) to limit the amount of dishing in the widest wire expected in a given process. Oxide erosion is another problem – normally in this case CMP is applied to an array of dense lines. The oxide between wires in a dense array tends to be over-polished compared to nearby areas of wider insulators (that is, oxide between sparse features will be thicker

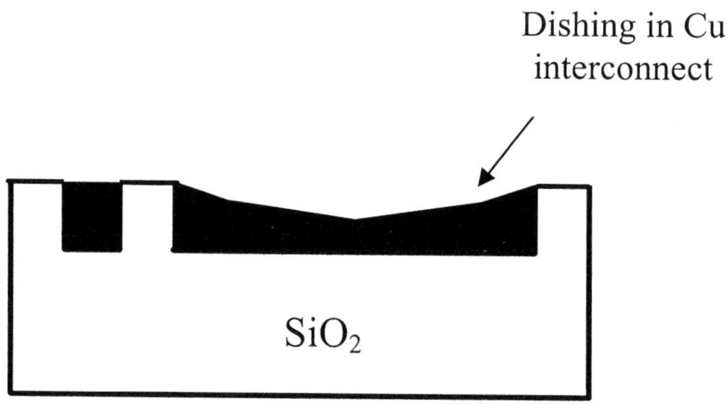

Fig. 1.1. Dishing results in smaller height of copper interconnects resulting in higher resistance, with wider wires having the largest impact.

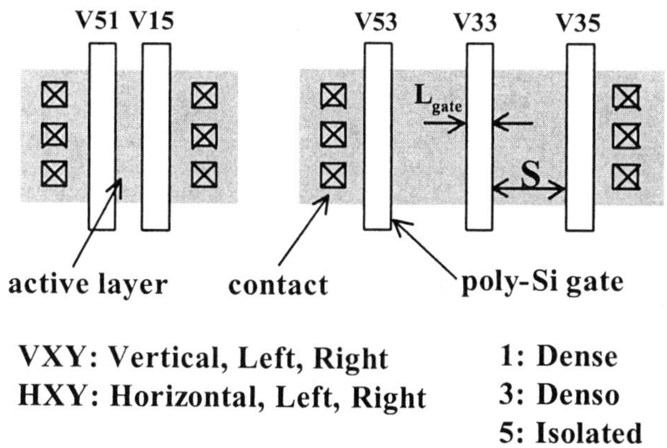

Fig. 1.2. Characterization of polysilicon lines based on their orientation and distance to nearby polysilicon lines [104]. (©2005 IEEE)

than that between dense features). Both dishing and oxide erosion are problematic in wide lines and dense arrays, respectively, and are therefore layout dependent. They lead to higher resistances and more surface non-uniformity.

The patterning of features smaller than the wavelength of light used in optical lithography results in distortions due to the diffraction of light referred to as optical proximity effects (OPE). Shorter wavelength lithography technology is too costly and unstable to be used in current technologies. Changes made to the mask layout to account for these distortions are known as optimal proximity corrections (OPC). Another technique that is used to improve the performance of sub-wavelength lithography is phase-shift masks (PSM), which exploits the phenomenon of interference to enable patterning of features with higher resolution. OPEs are also layout dependent and result in different CD variations depending on their environment (presence of neighboring lines) and orientation (vertical or horizontal). Figure 1.2 shows the classification of polysilicon lines based on their orientation and distance to the neighboring lines from the left and right edges. The edge is characterized as being *dense* if the next line is at the minimum possible distance, *denso* if the next line is at some intermediate distance, and *isolated* if the next line is further apart. Based on test-chip measurements, the work in [104] found that proximity CD variation is a strong function of both the orientation and the nearby environment. Controlling these variations has become extremely critical in current technologies and has resulted in an explosion in the number of design rules. Polysilicon routing in two orthogonal directions may no longer be allowed in certain technologies, so that better control can be achieved in one single direction. Since these variations are layout dependent, they are generally treated as spatially correlated intra-die variations.

Statistical Variations

Statistical quantization effects, such as random dopant variations, have also grown with scaling of process dimensions. The number of dopant atoms in the channel region of a device decreases as the critical dimension is scaled down. As the number of dopant atoms becomes less, small variation in their number result in a large variation in device performance. Moreover, the actual location of these atoms also plays a role in determining the threshold voltage of a device, further increasing the variability. These variations are true random variations with no correlation across devices and represent one source of intra-die random variations. Such random variations can result from a host of other sources as well, such as lithography, etching, CMP etc. Although their impact in current technologies is small, it is expected to grow as process parameters scale. Their impact on performance has been manageable since random intra-die variations have the well known *averaging effect*, and their impact on path delay decreases with increasing logic depth. However, they result in an increase in mean circuit delay. In addition, the trend to increase clock frequency of a design using aggressive pipelining has resulted in smaller logic depths, which increases the effect of these random intra-die variations.

These variations have a strong influence on leakage power as well, which has become a big cause for concern even in current technologies. As an example, increased V_{th} variability and lower V_{th} values (which result in a much higher leakage) can result in functional failures in dynamic logic designs. To counter worst-case leakage scenarios, a stronger *keeper device* is required which has a negative impact on both power and performance. Adaptive post-fabrication techniques such as [74], which turn on a subset of parallel keeper devices depending on the variations will become useful in these scenarios.

We have classified variations as being inter- and intra-die variations with intra-die variations having spatially correlated and random components. Another equivalent view is to divide variations as being spatially uncorrelated and correlated with the correlated variation further divided as being intra- or inter-die variations depending on their correlation distance [158]. However, we will work with the previous definition of variations throughout the remainder of this book.

1.3 Impact on Performance

In this section, we will discuss the impact of variation on performance parameters. However, first we need to establish the components of variations that dominate each of the device and interconnect parameters. Variation in gate-length is perhaps the most critical device variation and has significant components of both inter-die variation (resulting from variation in duration of exposure) and intra-die variation (resulting from lens aberration and other lithography effects) [158], [124]. The intra-die variations in gate length are also expected to have significant components of spatially correlated variation with a small amount of random variations.

Device threshold voltage presents an interesting picture, since it is dependent on a number of process parameters such as channel doping concentration and gate length. Variations in gate length result in a change in the Drain Induced Barrier Lowering (DIBL) coefficient which results in a change in the threshold voltage. Thus, it is beneficial to separate the variation of threshold voltage between gate length independent variation, resulting from channel doping variations which are random intra-die variations, and gate length dependent variation (which has equal components of inter-die and spatially correlated intra-die variations). In current technologies, most of the variation in threshold voltage is due to variation in gate length and is thus spatially correlated. However, in future technologies random dopant variations are expected to increase raising the level of random variations significantly. In terms of interconnect parameters variations, most of the variations are spatially correlated intra-die variations and inter-die variations.

The trends in the magnitude of process variations is shown in Fig. 1.3 based on the National Technology Roadmap of Semiconductors [99]. The figure shows the increase in the variability of interconnect parameters such as wire width W, wire thickness T, wire height H and resistivity ρ, along with device parameters such as gate-oxide thickness T_{ox} and threshold voltage V_T and environmental factors such as power supply voltage V_{dd}. It shows that variations in gate-length are expected to increase significantly as compared to other process parameters, with variability increasing in all parameters.

The impact of the variations on power and performance was highlighted in [20], which showed measured data over 1000 samples of a design manufactured in an 180 nm technology. The results showed a 20X variation in leakage current

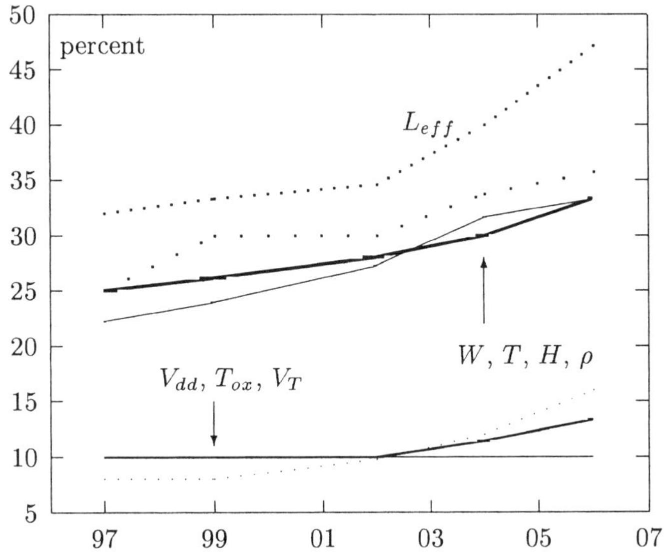

Fig. 1.3. Variability trends in key process parameters with scaling process technology. The x-axis is time with numbers representing the last two digits of the year and the y-axis represents variability in process parameters [99]. (©2005 IEEE)

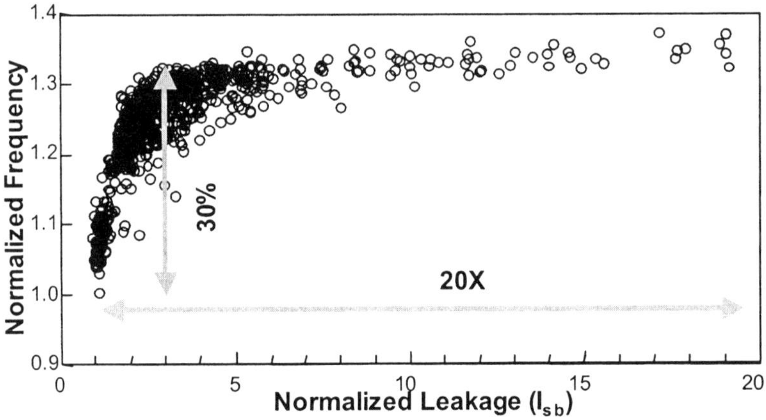

Fig. 1.4. Large variations in leakage power and performance are attributed to process variations [20]. (©2005 IEEE)

for a 1.3X variation in performance. The large variations in leakage result in a large fraction of samples that fail to meet the power constraint. Moreover, these samples are the high performance samples of a design and hence result in a two sided constraint on the region that represents samples that meet both the timing and power constraint.

Though the problem of variations seems to be growing tremendously, [124] recently showed that spatial correlated variations have been kept within manageable limits due to better polysilicon CD control. It was argued that the impact of inter-die variation can be kept within limits through better analysis and design techniques.

2

Statistical Models and Techniques

Traditionally, circuit performance has been modeled in the industry using worst-case models which are used to predict the performance of a design under worst-case process, temperature, and voltage conditions. However, with scaling process dimensions, the impact of process variations has grown, making traditional worst-case models extremely pessimistic. This results in the reduction of feasible regions for the design and increases design effort. Additionally, most of this effort is aimed at accounting for worst-case situations that will most likely not occur in actual designs. This has resulted in significant interest in statistical modeling techniques that can be used to enable statistical analysis and optimization.

Although the need for statistical modeling has been acknowledged to be critical, industry has been reluctant in adopting modeling techniques that can be used to replace traditional worst-case models. This stems from the fact that statistical models are expensive and difficult to construct, and unless analysis and optimizations tools are built on top of these modeling techniques, the utility and validity of these models will be questionable.

In this chapter, we will discuss key statistical techniques, such as principal component analysis, that have been extensively used in developing techniques for process variation modeling and analysis to simplify the problem of simultaneously considering different components of variations. We will also look at specialized modeling techniques to account for sources of variations as discussed in Chap. 1. Having developed the basic infrastructure to model process variation, we will then discuss performance modeling techniques using response surfaces. Then we will discuss statistical gate-delay models and interconnect-delay models that have seen substantial research activity in the past few years.

Before we discuss modeling techniques, let us spend some time understanding the basics of a crucial statistical technique known as Monte Carlo. This will serve as a benchmark against which all modeling and analysis techniques will be tested for accuracy. The need for techniques such as Monte Carlo becomes obvious as soon as we look at the scale of the problem at hand. We will show that the error in Monte Carlo techniques reduces with the number of samples n as $O(n^{-1/2})$. Hence, obtaining an accuracy improvement of two orders of magnitude requires that the number of samples be increased by four orders of magnitude. Thus, the number of simulations required to obtain reasonable accuracy using Monte Carlo is generally extremely large and using a Monte Carlo based analysis or optimization engine will be prohibitive. Even though this seems to be computationally demanding, this dependence is much better than non-statistical techniques where the error reduces as $O(n^{-1/d})$, where d is the dimensionality of the problem.

Therefore, Monte Carlo methods are used in almost all cases to evaluate the results obtained using newly developed analysis techniques. These techniques, which are, in general, orders of magnitudes faster than performing Monte Carlo simulations, lay the framework for the development of optimization engines that provide improvements in a reasonable amount of time. However, it is important to understand the basics of Monte Carlo simulations, so that they are used reasonably as golden models to test the accuracy of new techniques.

2.1 Monte Carlo Techniques

Numerical methods that make use of random numbers are known as *Monte Carlo* methods. One of the most important applications of Monte Carlo methods is in the evaluation of multi-dimensional integrals, and hence finds extensive application in areas such as yield estimation [154].

Non-statistical numerical techniques to estimate one dimensional definite integrals proceed by dividing the region, over which the integration needs to be performed, into a number of identical parts. Let us apply the technique to estimate the definite integral as shown in Fig. 2.1

$$I = \int_a^b f(x)\mathrm{d}x. \tag{2.1}$$

The interval $[a, b]$ is divided into n equal subintervals such that $a = x_0 < x_1 < x_2 < \cdots < x_n = b$. The integral (2.1) can then be approximated by

$$I = \int_a^b f(x)\mathrm{d}x \approx \sum_{i=0}^{i=n-1} f\left(\frac{x_i + x_{i+1}}{2}\right) h \tag{2.2}$$

where $h = (b - a)/n$. This method is known as the *midpoint* method, since it approximates the area under the curve $f(x)$ in a subinterval using the value of the function at the midpoint of the subinterval. If the function varies linearly within the subinterval, then the value estimated using the midpoint method is exact. Hence, in the general case, midpoint method incurs an $O(h^2)$ error in each subinterval of the integral. Since the total number of subintervals is

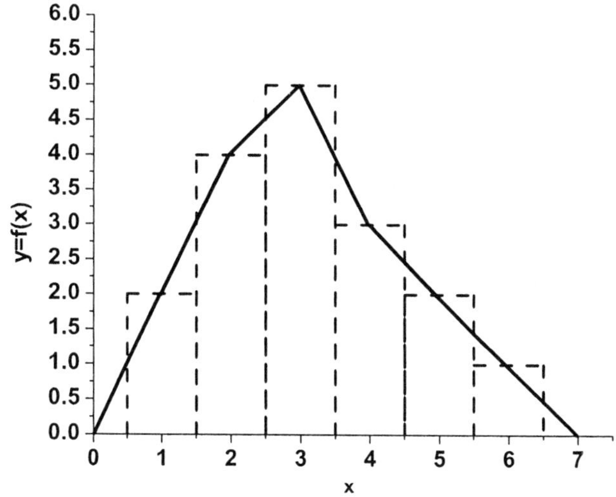

Fig. 2.1. Midpoint method to approximate the integral of $f(x)$, or the area under a curve.

inversely proportional to h, the overall error incurred in estimating the integral is $O(h)$. Thus, we can finally write

$$\begin{aligned} I = \int_a^b f(x)\mathrm{d}x &= \sum_{i=0}^{i=n-1} f\left(\frac{x_i + x_{i+1}}{2}\right) h + O(h) \\ &= \sum_{i=0}^{i=n-1} f\left(\frac{x_i + x_{i+1}}{2}\right) h + O(n^{-1}). \end{aligned} \tag{2.3}$$

The approach can be easily extended to two dimensional integrals. We now consider the case where the area enclosed by a curve is estimated as shown in Fig. 2.2. Using the ideas from the one dimensional case, the two dimensional surface is divided into a set of n equal sized squares with dimensions (h, h). If the midpoint of the square is enclosed by the curve, then the square contributes to the integral, otherwise not. Note that the square either contributes fully to the area or contributes nothing. The error in estimating the area of the square that actually contributes to the area of the curve is therefore $O(h^2)$. Since the number of squares that intersect the curve is $O(h)$, the overall error in estimating the area is again $O(h)$. However, the number of function evaluations required to estimate the area is now proportional to $1/h^2$, which results in an overall error in the integral of $O(n^{-1/2})$. Note that if this idea is extended to the evaluation of multi-dimensional integrals of dimension d, the

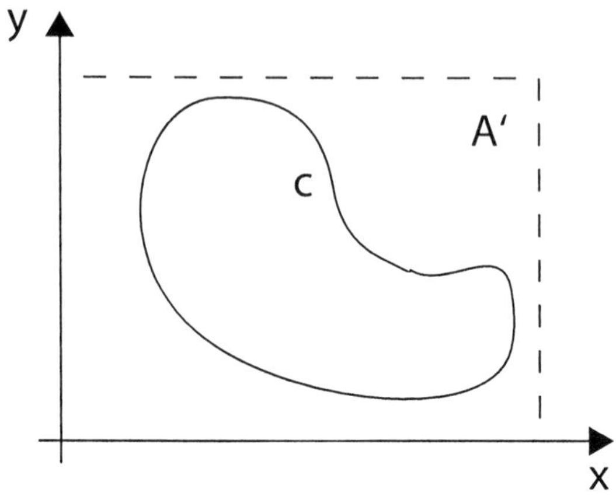

Fig. 2.2. Estimating the area enclosed by the curve C enclosed by a rectangular bounding box A'.

error falls off at a very slow rate of $O(n^{-1/d})$ as the number of samples in the d-dimensional space are increased. Thus we see that to maintain a reasonable accuracy, the number of function evaluations required by the midpoint method grows rapidly with the dimensionality of the integral.

Let us again estimate the area enclosed by a curve as shown in Fig. 2.2, now using a statistical technique. Instead of partitioning the entire region A', we generate n random points independently and assume that n_0 of these points lie within the region enclosed by the curve. Now we can approximate the area enclosed by the curve as

$$A_C \approx \widehat{A_C} = A_{A'} \frac{n_0}{n} \tag{2.4}$$

where $A_{A'}$ is the area of the region A' and A_C is the area enclosed by the curve C as shown in the figure. What is the advantage of this method compared to the midpoint method? To answer this question we need to estimate the error incurred in using approximation (2.4). The probability that a randomly generated point lies within the area enclosed by the curve is simply $A_C/A_{A'}$. If we generate n such samples, then the number of points found to be within C can be expressed as

$$n_0 = \sum_{i=1}^{n} x_i \tag{2.5}$$

where x_i is the result of the ith measurement of x, which is 1 if the randomly generated i^{th} point lies within C and 0 otherwise. The expected value of n_0 can then be expressed as

$$E[n_0] = E\left[\sum_{i=1}^{n} x_i\right] = \sum_{i=1}^{n} E[x] \tag{2.6}$$

where $E[x]$ is the expected value of x, which has a binomial distribution with n samples and a probability of success $A_C/A_{A'}$. The expected value of x can then expressed as

$$E[x] = 0 * \left(1 - \frac{A_C}{A_{A'}}\right) + 1 * \frac{A_C}{A_{A'}} = \frac{A_C}{A_{A'}}. \tag{2.7}$$

Substituting (2.6) and (2.7) into (2.4) and taking expectations we get

$$E[\widehat{A_C}] = A_{A'}\frac{E[n_0]}{n} = A_{A'}\frac{nA_C}{nA_{A'}} = A_C \tag{2.8}$$

and we find that on average the measurement of n_0 will result in an accurate estimate of the area enclosed by C. The class of estimators whose expected value of error is zero are known as *unbiased estimators*, therefore Monte Carlo provides an unbiased estimate of the area.

Let us now consider the variance of the estimate provided by Monte Carlo. We know from *Chebyshev's inequality* [109] that for a RV x

$$\mathcal{P}\left(|x - \eta| \geq \epsilon\right) \leq \frac{\sigma^2}{\epsilon^2} \tag{2.9}$$

where η and σ are the expected value and the standard deviation of x, respectively. Setting $\delta = \sigma^2/\epsilon^2$ we can rewrite (2.9) as

$$\mathcal{P}\left(|x - \eta| \geq \frac{\sigma}{\sqrt{\delta}}\right) \leq \delta. \tag{2.10}$$

Since the expected value of n_0 gives the exact value of A_C, using (2.10) allows us to estimate the error in the value of n_0 in terms of the number of samples for a fixed desired level of accuracy. First, let us calculate the variance of n_0:

$$\begin{aligned} Var[n_0] &= E\left[(n_0 - E[n_0])^2\right] \\ &= E\left[\left(\sum_{i=1}^{n} x_i - E\left[\sum_{i=1}^{n} x_i\right]\right)^2\right] \\ &= E\left[\left(\sum_{i=1}^{n} (x_i - E[x])\right)^2\right] \\ &= E\left[\sum_{i=1}^{n}(x_i - E[x])^2 + 2\sum_{i,j=1}^{n} (x_i - E[x])(x_j - E[x])\right]. \end{aligned} \tag{2.11}$$

Since different measurements of x are assumed to be independent, the second term on the right in (2.11) does not contribute to the expression and (2.11) can be simplified as

$$\begin{aligned} Var[n_0] &= E\left[\sum_{i=1}^{n}(x_i - E[x])^2\right] \\ &= nE[x^2 - 2x_iE[x] + E^2[x]] \\ &= n(E[x^2] - E^2[x]). \end{aligned} \tag{2.12}$$

Now

$$E[x^2] = 0^2 * \left(1 - \frac{A_C}{A_{A'}}\right) + 1^2 * \frac{A_C}{A_{A'}} = \frac{A_C}{A_{A'}} \tag{2.13}$$

therefore, the standard deviation σ of n_0 can be written as

$$\sigma_{n_0} = \sqrt{Var[n_0]} = \sqrt{n\left(\frac{A_C}{A_{A'}}\right)\left(1 - \frac{A_C}{A_{A'}}\right)}. \tag{2.14}$$

Since the estimate of the area enclosed by C is proportional to the ratio n_0/n, using (2.10) and (2.14), the error in the estimate is $O(n^{-1/2})$. Note that the estimation in error is independent of the dimensionality of the problem. This gives us the very interesting and important result that the *error incurred by Monte Carlo methods does not depend on the dimensionality of the problem.* Note that the error in Monte Carlo is fundamentally of a different nature. The error in the midpoint method was due to the inability of the linear approximation to fit the actual integrand, whereas in Monte Carlo methods, the error has a probabilistic origin. Additionally, for one dimensional integrals the midpoint method is more accurate since the error is $O(n^{-1})$ whereas Monte Carlo methods provide an accuracy which is $O(n^{-1/2})$. For two dimensional integrals both the methods provide similar accuracy, and for higher dimensions Monte Carlo methods are always more accurate. The disparity between the accuracy of both the methods increases with the dimensionality of the problem, since the inaccuracy of the midpoint method increases rapidly.

Note that to improve the accuracy of the integral by a factor of two while using Monte Carlo would always require an increase in the number of samples by a factor of four. On the other hand, analytical methods such as the midpoint method require an increase in the number of samples by a factor $2^{D/2}$, where D is the dimensionality of the integral. If $D > 4$, then Monte Carlo methods fare better in this respect as well as compared to analytical midpoint methods.

For our purposes, we will use Monte Carlo methods to estimate the moments of physical or performance parameters. The main goal will be to estimate the quantity

$$E[g(X)] = \int_{\Re} g(x)f(x)\mathrm{dx} \tag{2.15}$$

where X is a RV with probability density function $f(x)$, $g(x)$ is a function of the RV X, and $\Re$ is the region of interest. If we can generate samples of the RV X, then the integral can be estimated as an average of the values of $g(x)$ at these sample points. This approach shows better convergence properties and reduces the runtime of Monte Carlo based techniques.

2.1.1 Sampling Probability Distributions

Monte Carlo methods rely on sampling the space of interest using random samples by generating uniform statistically independent values in the region. As it turns out, it is very difficult to generate truly random numbers using computers. Specialized pieces of hardware are used in certain applications to generate random numbers that amplify the thermal noise of a resistor or a diode and then sample it using a *Schmitt trigger.* If these samples are taken at sufficient intervals of time, we obtain a series of random bits. However, in software, random numbers have to be modeled using *pseudo-random* number generators. Pseudo-random numbers, as the name suggests, are not truly random and are typically generated using a mathematical formula. Most computer languages use *linear congruential* generators. These generators are defined by three positive integers a (multiplier), b (increment), and m (modulus) and given an initial seed (the first pseudo-random number r_0), generates pseudo-random numbers in the following fashion:

$$r_{k+1} = ar_k + b(\mathrm{mod\,m}). \tag{2.16}$$

If desired, the random numbers generated can be mapped to a given range by dividing the numbers obtained using the above generator by m. Note that the r_k's can only take one of the m values. Hence, in all practical implementations m is a very large number (eg. 2^{32}). Also, the choice of a is critical to the randomness of the number generated. More details regarding pseudo-random generators can be found in [75].

We will now review some of the general techniques used to sample arbitrary probability distributions and algorithms to generate samples of some of the pertinent RVs that we will deal with throughout this book.

Inverse Transform Method

Let us assume that the probability distribution function (pdf) of a RV X that we want to sample is given by $f(x)$. The cumulative probability distribution (cdf) $F(x)$, which gives the probability that $X \leq x$, is then given by

$$F(x) = \int_{-\infty}^{x} f(x)\mathrm{d}x. \tag{2.17}$$

Let us take samples of X, which will have a probability density of $f(x)$. Now we will use these samples of X to obtain samples of F. Consider a small region $x < X < x + \mathrm{d}x$ on the x-axis of the cdf. The number of sample points in this region will be proportional to the integral of the pdf in this range. Note that this is equal to the change in the value of the cdf. Hence, the number of sampling points within a range is equal to the length of the region sampled as well. Therefore, these samples of $F(x)$ will be uniformly distributed in the range [0,1].

Using this idea we can write

$$\begin{aligned} u &= F(x) \\ x &= F^{-1}(u) \end{aligned} \tag{2.18}$$

where u represents samples of a uniformly distributed random variable, and F^{-1} is the inverse of F. Hence, if we can find the inverse of F we can use this technique to generate random numbers distributed according to the probability distribution $f(x)$.

Transformation Method

Now let us consider two RVs, X and Y, which are related such that $Y = f(X)$, where f is a monotonic function (inverse of f is well defined). Let the pdf of X and Y be $f_x(x)$ and $f_y(y)$, respectively. Then from the conservation of probability it follows that

$$|\mathcal{P}_x(x)\mathrm{d}x| = |\mathcal{P}_y(y)\mathrm{d}y| \tag{2.19}$$

which states that the probability of finding X between x and $x + \mathrm{d}x$ is the same as the probability of finding Y between $y = f(x)$ and $y + \mathrm{dy} = \mathrm{f(x + dx)}$ as illustrated in Fig. 2.3. From (2.19) it follows that

$$f_y(y) = \frac{f_x(x)}{|f'(x)|}. \tag{2.20}$$

When f is non-monotonic, the left hand side in (2.19) is replaced by a summation of the ranges of x that correspond to the given range of y on the right hand side in (2.19). An equivalent for (2.20) can then be immediately constructed [109]. Therefore, to generate samples of a RV Y we need to find a RV X whose samples can be easily obtained such that X and Y satisfy (2.20).

Consider the case where we want to generate samples of a *Poisson distribution.* The pdf of the Poisson distribution is expressed as

$$f_y(y) = \begin{cases} \mathrm{e}^{-y} & \text{if } \ 0 \leq \mathrm{y} \leq \infty \\ 0 & \text{o.w.} \end{cases} \tag{2.21}$$

then choosing $y = -\ln x$ we get

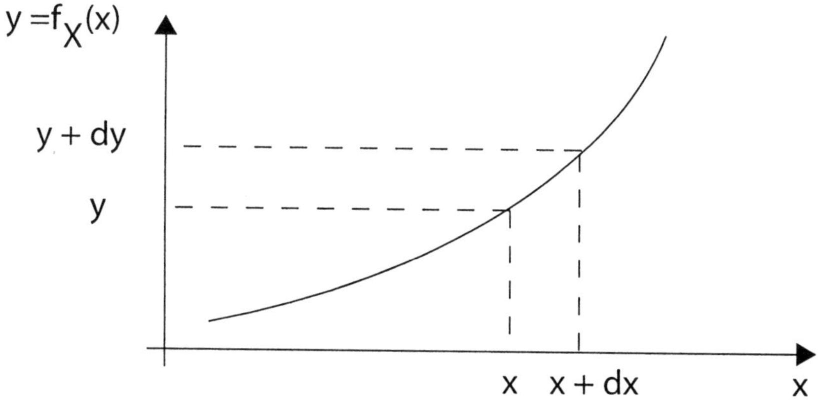

Fig. 2.3. The probability that $x \leq< X <\leq x + dx$ is equal to the probability that $y \leq< Y <\leq y + dy$ for the case when Y varies monotonically with X.

$$f_x(x) = \begin{cases} 1 & \text{if } 0 \leq \mathrm{x} \leq 1 \\ 0 & \text{o.w.} \end{cases} \tag{2.22}$$

hence the pdf of Y and X satisfy (2.20). Therefore, if we generate uniform samples in the range [0,1], then the negative natural log of these samples will have a Poisson distribution. This method requires a differentiable pdf, which is a restriction particularly when dealing with discrete RVs.

Acceptance-Rejection Method

If both the above methods are inapplicable due to the restrictions imposed on the pdf of the RV then the acceptance-rejection method may be used. Let us consider the case where we want to generate samples of a RV X whose pdf is as shown in Fig. 2.4. The acceptance-rejection method consists of the following steps. First, generate uniform samples in the range $[x_{min}, x_{max}]$. For each sample x_i evaluate the value of $f_x(x)$. Next, generate another random sample a in the range $[0, \max f_x(x)]$. If $x_i \geq a$, then accept the sample x_i, otherwise reject it. The accepted samples are then distributed according to the pdf f_x.

To generate samples of a Gaussian RV using this approach, we must truncate the pdf of the RV. Since most of the values of a Gaussian RV are concentrated around its mean, a $\pm 4\sigma$ range around the mean is sufficient to capture the behavior of the Gaussian RV. The steps outlined can then be applied to this *truncated* Gaussian RV to generate the desired random samples.

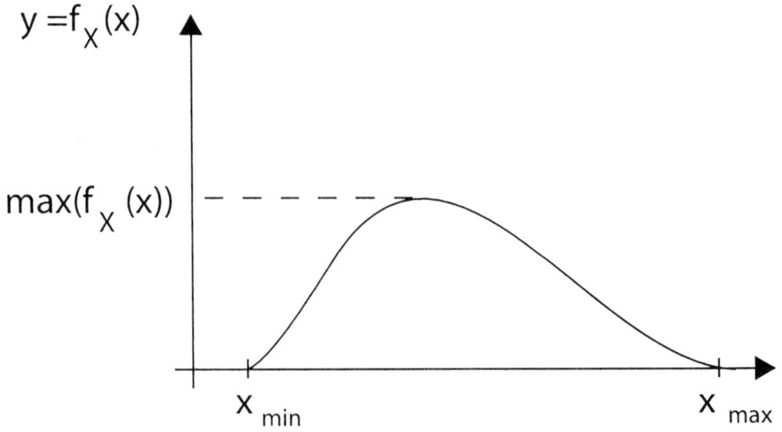

Fig. 2.4. The acceptance-rejection method to generate samples of a RV with a given distribution function.

Generating Multivariate Gaussian RVs

Now let us look at techniques that may be used to generate multivariate Gaussian RVs. We will use the transformation method to generate samples of a one dimensional Gaussian RV. If u_1 and u_2 are independent uniform RVs in the range [0,1], then

$$\begin{aligned} y_1 &= \sin 2\pi u_1 \sqrt{-2 \ln u_2} \\ y_2 &= \cos 2\pi u_1 \sqrt{-2 \ln u_2} \end{aligned} \tag{2.23}$$

are two independent Gaussian RVs with zero mean and unit variance. The Gaussian random numbers generated using the above transformation, also known as the *Box-Muller* transformation, can then be used to generate samples of a Gaussian RV with an arbitrary mean and variance. To obtain the desired mean and variance for the Gaussian RV, we use the fact that given two Gaussian RVs that are related as $Y = aX + b$

$$\begin{aligned} E[Y] &= aE[X] + b \\ Var[Y] &= E[Y^2] - E^2[Y] = a^2 Var[X]. \end{aligned} \tag{2.24}$$

To generate an n-dimensional multivariate random variable with a covariance matrix $\mathbf{\Sigma}$ and mean $\mathbf{\Delta}$, the first step is to generate n independent random variables with zero mean and unit variance. Then, take a sample of these RVs ($\mathbf{X}$), and generate a new sample $\mathbf{X}'$ from $\mathbf{X}$ such that

$$\mathbf{X}' = \mathbf{\Delta} + \mathbf{L}\mathbf{X} \tag{2.25}$$

which gives

$$\begin{aligned} E[\mathbf{X}'] &= \mathbf{\Delta} \\ Cov[\mathbf{X}'] &= E[\mathbf{X}'\mathbf{X}'^T] - E[\mathbf{X}']E[\mathbf{X}'^T] = \mathbf{L}\mathbf{L}^T. \end{aligned} \tag{2.26}$$

Hence, if $\mathbf{L}\mathbf{L}^T = \mathbf{\Sigma}$ we have the samples of the desired multivariate Gaussian RV. The evaluation of $\mathbf{L}$ from $\mathbf{\Sigma}$ is a popular technique in matrix computation, known as *Cholesky decomposition* [55], for symmetric positive-definite matrices. Cholesky decomposition factorizes a symmetric positive-definite matrix into a product of a lower and upper triangular matrix, which are the transpose of each other. Hence, $\mathbf{L}$ is a lower triangular matrix in (2.26). The covariance matrix of a set of RVs is known to be symmetric and positive-definite. Therefore, Cholesky factors can be obtained for $\mathbf{\Sigma}$. Writing out (2.26) in components, we can solve for the elements of $\mathbf{L}$ and readily obtain

$$\begin{aligned} L_{ii} &= \left(\Sigma_{ii} - \sum_{k=1}^{i-1} L_{ik}^2\right)^{1/2} \\ L_{ji} &= \frac{1}{L_{ii}}\left(\Sigma_{ij} - \sum_{k=1}^{i-1} L_{ik}L_{jk}\right) \quad j = i+1, i+2, \ldots, N. \end{aligned} \tag{2.27}$$

Using these techniques, samples of the required RV to perform Monte Carlo analysis can be generated. For most purposes, variations in VLSI designs are assumed to be Gaussian. Consequently, while analyzing intra-die variations, we need to generate samples of a multi-normal RV.

Though we have described the theoretical foundation of Monte Carlo based simulations, there are a number of practical issues that must be kept in mind. It must be ensured that the result of a Monte Carlo based simulation has converged and that further increase in the number of samples will not result in a large change in the value of the target parameter. A number of issues arising in Monte Carlo simulations in VLSI designs were highlighted in [126]. One of the most important computational issues involved in generating the required RVs is that, in most cases, correlated RVs are required to sample the space. This requires Cholesky decomposition of the correlation matrix, which has a computational complexity of $O(n^3)$ for an $n \times n$ matrix. This becomes prohibitive as soon as the number of RVs being considered increases beyond a few thousand (this may be the case if intra-die variations are considered). Additionally, techniques that aim to reduce the complexity of matrix operations based on sparsity are not applicable. This is due to the fact that some process parameters, such as threshold voltage, have significant correlations across the chip. Since threshold voltage variations have a strong component of random dopant variations as well, these RVs cannot be assumed to be perfectly correlated as well. Spatial correlation models such as PCA and Quad-Tree based modeling, which we will discuss in the next section, can be used in these situations to reduce the computational complexity.

2.2 Process Variation Modeling

Process variation can be considered to operate at two different levels; at the chip level, which we call inter-die variation, and at the transistor level, which we refer to as intra-die variation. As discussed in Chap. 1, the demarcation between these two components is not very strict since variations that need to be modeled as intra-die variations may have a correlated component across a chip as well. However, the impact of inter-die variations and intra-die variations on circuit performance is very different. Additionally, as we saw earlier, the number of RVs that we deal with increases rapidly when intra-die variations are considered, which increases computational costs (particularly when intra-die variations are spatially correlated). In this section, we will discuss techniques that have been proposed to simplify modeling and analysis techniques when dealing simultaneously with both correlated and independent sources of variations. We also discuss models that have been developed to specifically understand the impact of certain physical phenomena, such as random dopant effects, which exhibit themselves as variations (process or time-dependent) on circuit performance. However, we first discuss Pelgrom's model, which has been widely used to understand the mismatch in devices resulting from random and correlated sources of variations.

2.2.1 Pelgrom's Model

Pelgrom's model [111] has been the most widely used modeling technique to capture the mismatch in transistors arising due to variations in process parameters. The approach is based on analyzing the impact of variations (both random and correlated) in the frequency domain and abstracting key features of both intra-die and inter-die variation.

Let us consider a parameter P that varies over the surface of a die in the x-y plane due to process variations. Variations in P for different values of coordinates (x, y) result in mismatch of transistors, which have been designed to have the same characteristics. The overall mismatch between two regions (R_1) and (R_2) corresponding to the points (x_1, y_1) and (x_2, y_2), respectively, which have an area A_0 can be expressed as

$$\Delta P = \frac{1}{A_0}\left(\iint_{R_1} P(x,y)\mathrm{d}x\,\mathrm{d}y - \iint_{R_2} P(x,y)\mathrm{d}x\,\mathrm{d}y\right). \tag{2.28}$$

The integral in (2.28) can be viewed as a convolution of the function describing P $(P(x,y))$ in the x-y plane and a function $f_g(x,y)$ which describes the geometry of the problem and can be expressed in this case as

$$f_g(x,y) = \begin{cases} +1/A_0 & \text{if } (x,y) \in R_1 \\ -1/A_0 & \text{if } (x,y) \in R_2 \\ 0 & \text{o.w.} \end{cases} \tag{2.29}$$

Thus, we can rewrite (2.28) as

$$\Delta P(x,y) = (P * f_g)(x,y) = \int_{-\infty}^{\infty}\int_{-\infty}^{\infty} f_g(x',y')P(x-x',y-y')\mathrm{d}x'\mathrm{d}y' \tag{2.30}$$

where $*$ is the convolution operator. If we take the *Fourier transform* of (2.30), the convolution can be written as a product of Fourier transforms of the two functions, which effectively separates the process and the geometry dependent terms of the mismatch. The equation in the frequency domain can be written as

$$\Delta P(\omega_x,\omega_y) = \mathcal{F}(f_g(x,y))\ \mathcal{F}(P(x,y)) = \mathcal{F}_g(\omega_x,\omega_y)\ \mathcal{P}(\omega_x,\omega_y) \tag{2.31}$$

where the operator $\mathcal{F}$ represents the two-dimensional Fourier transform. Let us consider the specific case where the transistors have a nominal device width W, gate length L, are separated by a distance d_x and are laid out as shown in Fig. 2.5. In this case, the Fourier transform of the geometry dependent part takes the form

$$\begin{aligned}\mathcal{F}_g(\omega_x,\omega_y) &= \frac{1}{WL}\int\int_{\Re^2} f_g(x,y)\mathrm{e}^{-\mathrm{i}\omega_x x}\mathrm{e}^{-\mathrm{i}\omega_y y}\mathrm{d}x\,\mathrm{d}y \\ &= \frac{1}{WL}\int_{-d_x/2-L/2}^{-d_x/2+L/2}\int_{-W/2}^{W/2} \mathrm{e}^{-\mathrm{i}(\omega_x x+\omega_y y)}\mathrm{d}x\,\mathrm{d}y \\ &\quad - \frac{1}{WL}\int_{d_x/2-L/2}^{d_x/2+L/2}\int_{-W/2}^{W/2} \mathrm{e}^{-\mathrm{i}(\omega_x x+\omega_y y)}\mathrm{d}x\,\mathrm{d}y = I_1 - I_2.\end{aligned} \tag{2.32}$$

The integral I_2 can be evaluated as

$$\begin{aligned}I_2 &= \frac{1}{WL}\int_{d_x/2-L/2}^{d_x/2+L/2} \mathrm{e}^{-\mathrm{i}(\omega_x x)}\left(\frac{\mathrm{e}^{-\mathrm{i}\omega_y y}}{-\mathrm{i}\omega_y}\right)\Bigg|_{-W/2}^{W/2} \mathrm{d}x \\ &= \left(\frac{\mathrm{e}^{-\mathrm{i}\omega_y W/2} - \mathrm{e}^{\mathrm{i}\omega_y W/2}}{-\mathrm{i}\omega_y W}\right)\left(\frac{\mathrm{e}^{-\mathrm{i}\omega_x L/2} - \mathrm{e}^{\mathrm{i}\omega_x L/2}}{-\mathrm{i}\omega_x L}\right)\mathrm{e}^{-\mathrm{i}\omega_x d_x/2}.\end{aligned} \tag{2.33}$$

Using *Euler's Theorem*, (2.33) simplifies to

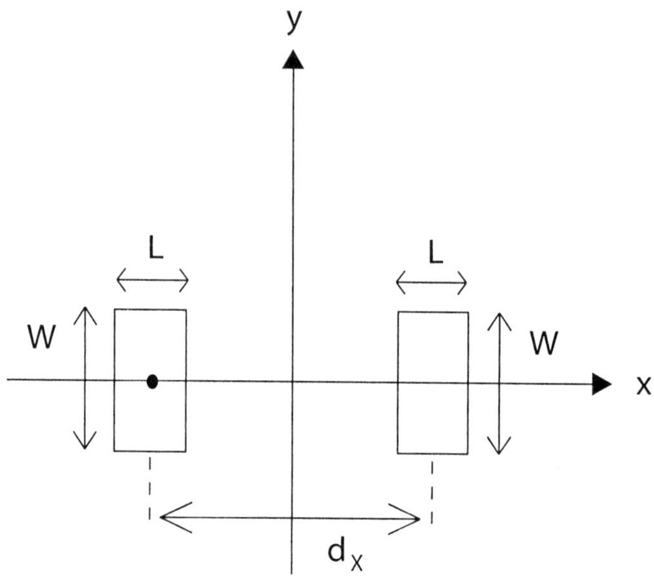

Fig. 2.5. Estimating the mismatch between transistors that lie on the x-axis separated by a distance d_x.

$$I_2 = \left(\frac{\sin(\omega_y W/2)}{\omega_y W/2}\right)\left(\frac{\sin(\omega_x L/2)}{\omega_x L/2}\right) \mathrm{e}^{-\mathrm{i}\omega_x d_x/2}. \tag{2.34}$$

Similarly, I_1 can be evaluated and differs only in the last term compared to (2.34) with the exponent being positive. Hence, we can write

$$\mathcal{F}_g(\omega_x, \omega_y) = 2\left(\frac{\sin(\omega_x L/2)}{\omega_x L/2}\right)\left(\frac{\sin(\omega_y W/2)}{\omega_y W/2}\right)\sin(\omega_x d_x/2). \tag{2.35}$$

Now let us consider the process-dependent term in (2.31). As discussed in Chap. 1, variations can be divided into two classes where either the variation is random across transistors or correlated. Since random and correlated variations behave differently, Pelgrom uses different modeling techniques to capture their impact.

The variations in the parameter P are assumed to be the result of many events of a random process. The random process is treated as independent across gates, and is assumed to behave as a source of white noise. Hence, random variations can be modeled as normally distributed noise sources [111]. If we assume that these variations have zero mean and are small enough such that the resulting variations in P can be assumed to be linear, then the variations in P can also be modeled as normally distributed zero mean RV. The

Fourier transform of a white noise random process is a constant in the frequency domain, and therefore, has equal contribution from all frequencies. This is intuitive, since a randomly varying signal should have equal components from all frequencies.

On the other hand, the correlated component of variation is a deterministic process. Again, assuming the impact of these variations on P can be captured using a linear relation, [49] expresses the correlated component of variation in P as

$$\Delta P_{corr} = \Delta P_{nom} + \alpha_1 x + \alpha_2 y \tag{2.36}$$

where the nominal value of the correlated variation ΔP_{nom} for a particular die can be estimated if the origin corresponding to these variations on the wafer is known (and is deterministic), and α_1 and α_2 are parameters which depend on the process and the choice of the coordinate system for the die. Hence, if the position of the dies on the wafer are known, then this component of variation can be precisely predicted from the knowledge of process gradients [49]. Unfortunately, the information regarding the placement of dies on a wafer is generally not available. Additionally, this information is not available at the design stage and cannot be used to *design for variability*. However, layout techniques such as *Quad Common-Centroid Configuration* [46] have been proposed to effectively cancel the impact of correlated variations. A reasonable approach is to model the correlated component as a stochastic process, with a low-frequency component whose frequency is inversely proportional to the correlation distance of the variation being considered. Again, due to the assumption of small variations, the variation in P can be assumed to be normally distributed [111].

From the characterization of a random process, we know that the variance in the samples of a random process is proportional to the power content of the process. Thus, the overall variation in parameter P can be expressed as

$$\sigma^2(\Delta P) = \frac{1}{4\pi^2}\int_{-\infty}^{\infty}\int_{-\infty}^{\infty} |P(\omega_x, \omega_y|^2 \; |\mathcal{F}_g(\omega_x, \omega_y|^2 \,\mathrm{d}\omega_x \,\mathrm{d}\omega_y. \tag{2.37}$$

The above integral can be evaluated using the definite integral properties of even and odd functions, and the following definite integral of a sinc function:

$$\int_{-\infty}^{\infty} \frac{\sin^2 x}{x^2}\,\mathrm{d}x = \pi \tag{2.38}$$

to finally obtain

$$\sigma^2(\Delta P) = \frac{A_p^2}{WL} + S_p^2 d_x^2 \tag{2.39}$$

where A_p and S_p are process-dependent parameters that capture the dependency of P on device area and spacing, respectively.

Using this model, simple expressions for the variance of a number of key parameters, such as threshold voltage (V_{th}) and gate oxide thickness (T_{ox}), can be expressed as

$$\sigma^2(V_{th}) = \frac{A_{V_{th}}^2}{WL} + S_{V_{th}}^2 D^2 \tag{2.40}$$

$$\sigma^2(T_{ox}) = \frac{A_{T_{ox}}^2}{WL} + S_{T_{ox}}^2 D^2 \tag{2.41}$$

where $A_{V_{th}}$, $S_{V_{th}}$, $A_{T_{ox}}$ and $S_{T_{ox}}$ are process dependent constants. Considering variations in W and L themselves, Pelgrom notes that the variations arise due to edge roughness and [111] uses a one dimensional variant of the analysis described above, concluding that the random variation in W and L can be expressed as

$$\sigma^2(L) = \frac{A_L^2}{W} \tag{2.42}$$

$$\sigma^2(W) = \frac{A_W^2}{L}. \tag{2.43}$$

Based on these expressions, variations in key performance metrics, such as device on-current, can be easily predicted. The coefficients used in the equations above are estimated using information from test structure measurements for the process.

Pelgrom's model is extensively used in analog design to analyze the mismatch between transistors that are required to match perfectly, as in analog designs. Based on (2.40), it can be inferred that large devices located close to each other will be well matched, and that increasing the device area is a possible approach to reduce the mismatch between devices [90].

2.2.2 Principal Components Based Modeling

This section will detail the variability modeling infrastructure based on Principal Component Analysis (PCA). This framework for simultaneously handling process random and correlated variations was first developed in [30] for statistical timing analysis and has since been used in a number of later works to model process variations. As discussed in Chap. 1, process parameters have an inter-die (which is fully correlated across a chip) and an intra-die component of variation. The intra-die component can again be categorized as being correlated and random. The overall intra-die variation is then expressed as a sum of correlated and random components and the sum of variances of both these components provides the overall variation in the process parameter.

To handle the correlated components of variations (inter-die and correlated intra-die) the overall chip area is divided into a grid as shown in Fig. 2.6. In the

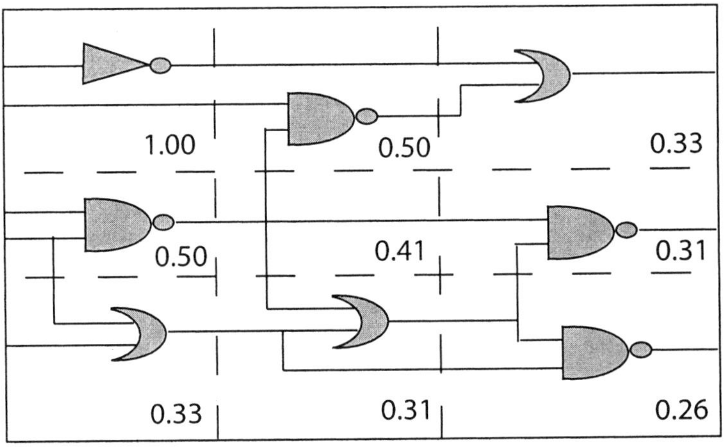

Fig. 2.6. Partitioning of a circuit using a 2D grid to model the correlated component of variation.

absence of inter-die process variations, the correlation coefficient varies from one (within the same square of the grid) and falls off to zero with increasing distance. Due to inter-die process variations, squares on the grid that lie at the opposite corners of a large design may have non-zero correlations and the correlation in this case falls off to a value higher than zero. This minimum value depends on the relative contribution of inter-die variations to the total correlated component of variation.

Let us now consider the RVs required to model variations in a given process parameter. Each square in the grid corresponds to a RV of the process parameter which has correlations with all other RVs corresponding to other squares on the grid. Squares that are much further away should demonstrate lower correlation compared to adjacent squares on the grid in this model. If we want to consider the impact of these RVs on the performance parameters of the design, we need to consider the correlations in these RVs at all points during the analysis. To simplify the problem, this set of correlated RVs is replaced by another set of mutually independent RVs with zero mean and unit variance using the *principal components* of the set of correlated RVs.

Principal Component Analysis

Principal Component Analysis (PCA) is a statistical technique that is used to identify patterns in data, and expresses the data in a much simpler and informative fashion. PCA maps a given set of correlated RVs to a new set of uncorrelated RVs, which are called the principal components, such that most of the variability in the original RVs is captured by the first few principal components as shown in Fig. 2.7.

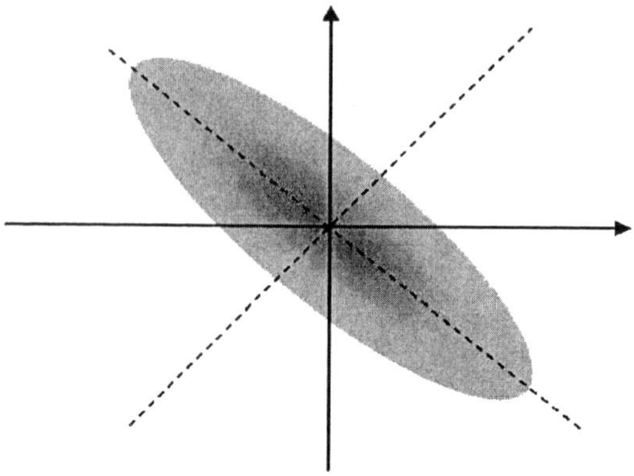

Fig. 2.7. Most of the variation in the shaded data is along one of the dotted axis which represents the first principal component.

Let us assume that we have a vector of n RVs $\mathbf{X'} = (X'_1, X'_2, \ldots, X'_n)^T$ which are distributed according to a given multivariate probability density function. Assume that this multivariate pdf has a mean vector $\mathbf{\Delta}$ and a covariance matrix $\mathbf{\Sigma}$. Let us generate a new vector of RVs $\mathbf{X} = (X_1, X_2, \ldots, X_n)^T$ such that $\mathbf{X} = \mathbf{X'} - \mathbf{\Delta}$, which implies that the new set of RVs are zero mean and have the same covariance matrix $\mathbf{\Sigma}$. The first principal component Y_1 of the components of $\mathbf{X}$ is a linear combination of the components and is expressed as

$$Y_1 = \alpha_{11}X_1 + \alpha_{12}X_2 + \cdots + \alpha_{1n}X_n = \boldsymbol{\alpha_1}^T\mathbf{X}. \tag{2.44}$$

This linear combination has the property that its sample variance is the greatest for all normalized $\alpha's$. Therefore, we can rewrite the problem of identifying the first principal component as

$$\begin{aligned} \max \quad & \sum_{i=1}^{n}\sum_{j=1}^{n} \alpha_{1i}\alpha_{1j}\Sigma_{ij} = \boldsymbol{\alpha_1}^T\mathbf{\Sigma}\boldsymbol{\alpha_1} \\ s.t. \quad & \sum_{i=1}^{n} \alpha_{1i}^2 = \boldsymbol{\alpha_1}^T\boldsymbol{\alpha_1} = 1. \end{aligned} \tag{2.45}$$

Introducing a Lagrange multiplier λ [15], we include the constraint into the objective function, and then take partial derivatives with respect to the components of $\boldsymbol{\alpha_1}$ to obtain

$$\frac{\partial}{\partial\boldsymbol{\alpha_1}}\left(\boldsymbol{\alpha_1}^T\mathbf{\Sigma}\boldsymbol{\alpha_1} - \lambda(\boldsymbol{\alpha_1}^T\boldsymbol{\alpha_1} - 1)\right) = 0$$

$$2(\mathbf{\Sigma} - \lambda\mathbf{I})\boldsymbol{\alpha_1} = 0. \tag{2.46}$$

Since $\boldsymbol{\alpha_1} = 0$ (which satisfies the above equation) corresponds to the minimization of the objective function, we know that $\mathbf{S} - \lambda\mathbf{I}$ has a non-empty null space. Hence the determinant of $\mathbf{S} - \lambda\mathbf{I}$ is zero, which implies that λ is an eigenvalue of the covariance matrix, with $\boldsymbol{\alpha_1}$ being the associated eigenvector. Let us look at the objective function again in light of this fact and note that

$$\boldsymbol{\alpha_1}^T\mathbf{\Sigma}\boldsymbol{\alpha_1} = \boldsymbol{\alpha_1}^T\lambda\boldsymbol{\alpha_1} = \lambda. \tag{2.47}$$

Thus we find that the objective function is maximized when λ is chosen to be the largest eigenvalue of $\mathbf{\Sigma}$. Also, we know that since $\mathbf{\Sigma}$ is a positive-definite symmetric matrix, all its eigenvalues are positive and real. Let us now estimate the second principal component, which is expressed as

$$Y_1 = \alpha_{21}X_1 + \alpha_{22}X_2 + \cdots + \alpha_{2n}X_n = \boldsymbol{\alpha_2}^T\mathbf{X} \tag{2.48}$$

and is a solution of the optimization problem

$$\begin{aligned} \max \quad & \sum_{i=1}^{n}\sum_{j=1}^{n} \alpha_{2i}\alpha_{2j}\Sigma_{ij} = \boldsymbol{\alpha_2}^T\mathbf{\Sigma}\boldsymbol{\alpha_2} \\ s.t. \quad & \sum_{i=1}^{n} \alpha_{2i}^2 = \boldsymbol{\alpha_2}^T\boldsymbol{\alpha_2} = 1 \\ & \sum_{i=1}^{n} \alpha_{1i}\alpha_{2i} = \boldsymbol{\alpha_1}^T\boldsymbol{\alpha_2} = 0 \end{aligned} \tag{2.49}$$

where the added constraint forces the new principal component Y_2 to be orthogonal to Y_1. Again, introducing the constraints into the objective function using Lagrange multipliers λ_1 and λ_2, and differentiating with respect to the components of $\boldsymbol{\alpha_2}$ we obtain

$$\begin{aligned} & \frac{\partial}{\partial\boldsymbol{\alpha_2}}\left(\boldsymbol{\alpha}^T\mathbf{\Sigma}\boldsymbol{\alpha_2} - \lambda_1(1 - \boldsymbol{\alpha_2}^T\boldsymbol{\alpha_2}) - \lambda_2\boldsymbol{\alpha_1}^T\boldsymbol{\alpha_2}\right) = 0 \\ & 2(\mathbf{\Sigma} - \lambda_1\mathbf{I})\boldsymbol{\alpha_2} - \lambda_2\boldsymbol{\alpha_1} = 0. \end{aligned} \tag{2.50}$$

If we multiply the above equation by $\boldsymbol{\alpha_1}^T$ on the right, we obtain $\lambda_2 = 0$, which implies that λ_1 is an eigenvalue of the matrix $\mathbf{\Sigma}$. Considering (2.47), we note that for the optimal solution to (2.49), λ_1 corresponds to the second largest eigenvalue of $\mathbf{\Sigma}$, and $\boldsymbol{\alpha_2}$ is the corresponding normalized eigenvector.

Extending this approach to estimate other principal components Y_j's (where we introduce additional constraints that Y_j is orthogonal to Y_i for $1 \leq i < j$), we find that the coefficients of the principal components correspond to the eigenvectors with decreasing magnitude. Hence we can write

$\mathbf{Y} = \mathbf{\Lambda X}$ where $\mathbf{\Lambda}$ is the $n \times n$ orthogonal matrix whose rows are the eigenvectors of $\mathbf{\Sigma}$.

Estimating the mean vector and covariance of the RVs $\mathbf{Y}$ we find

$$\begin{aligned} E[\mathbf{Y}] &= \mathbf{\Lambda} E[\mathbf{X}] = 0 \\ Cov[\mathbf{Y}] = E[\mathbf{Y}\mathbf{Y}^T] &= E[\mathbf{\Lambda X X}^T \mathbf{\Lambda}^T] = \mathbf{\Lambda} E[\mathbf{X}\mathbf{X}^T]\mathbf{\Lambda}^T = \mathbf{\Lambda \Sigma \Lambda}^T \\ &= \mathbf{\Lambda \Lambda}^T \mathrm{Diag}(\lambda_1, \ldots, \lambda_n) = Diag(\lambda_1, \ldots, \lambda_n) = \mathbf{D}. \end{aligned} \tag{2.51}$$

Using (2.51) we can generate a set of uncorrelated zero mean unit variance principal components $\mathbf{P}$ that are related to the original RVs ($\mathbf{X'}$) by the following relation

$$\mathbf{X'} = \mathbf{\Delta} + \mathbf{D}^{1/2}\mathbf{\Lambda}^{-1}\mathbf{P}. \tag{2.52}$$

The approach can also be used to compress data by using the first few principal components to express the data, since most of the variations in the data can be captured in the first few principal components. Hence PCA finds extensive use in areas such as image compression as well. For our purposes, we will use PCA-based techniques to simplify the correlation structure of variations in process parameters across a chip.

Another important fact regarding principal components of a set of random variables distributed according to a multi-normal distribution follows from the following properties of multi-normal distributions.

Property 2.1. Let the p-dimensional random vector $\mathbf{X}$ be distributed according to the multi-normal distribution with mean vector $\mathbf{\Delta}$ and covariance matrix $\mathbf{\Sigma}$ of rank p. If $\mathbf{A}$ is any $m \times p$ matrix of real numbers with rank $m \leq p$, the new m-component random vector $\mathbf{Y}$=$\mathbf{AX}$ is a multi-normal random vector with mean vector $\mathbf{A\Delta}$ and covariance matrix $\mathbf{A\Sigma A}^T$.

Property 2.2. Let X_1 and X_2 be random variables that are distributed according to a multivariate Gaussian distribution, then X_1 and X_2 are statistically independent if and only if their covariance is zero.

Based on Property 2.1, we note that principal components are a linear combination of the original RVs, and will be distributed according to a multi-normal distribution. From (2.51), we know that the principal components are uncorrelated, which, along with Property 2.2, implies that the principal components are independent RVs. Therefore any linear combination of the principal components will also be a Gaussian RV. This result will be very useful when we discuss statistical analysis techniques based on principal component analysis.

2.2.3 Quad-Tree Based Modeling

This approach to model process variations was first proposed in [4] and is also based on partitioning the overall die area into a number of regions. However, instead of adopting a covariance matrix based model to consider correlated components of variations, it uses an additive approach to consider the spatial dependence of process parameters. Let us consider the variation in a given process parameter; the value of the process parameter for a device i can be expressed as

$$X_{total,i} = X_{nom} + \Delta X_{inter} + \Delta X_{intra,i} \tag{2.53}$$

which is a sum of the nominal value of the process parameter (X_{nom}), the variation due to inter-die variation (X_{inter}), which is the same for all gates, and the intra-die variation corresponding to that particular gate ($X_{intra,i}$). The terms corresponding to the intra-die variation for different gates can be correlated. Similar to the PCA-based modeling approach, we seek to identify a set of uncorrelated RVs that can be used to model the overall variation in the process parameter.

This is achieved by recursively dividing the area of the die into four equal parts, which is known as *Quad-Tree partitioning*. As the regions of the die are recursively divided into parts, the number of parts increase by a factor of four for each additional level of partitioning as shown in Fig. 2.8. Each partition in this hierarchical scheme is then assigned to a RV, where the RVs are independent of each other. The intra-die RV associated with a gate i is now defined to be the sum of the RV associated with the lowest level partition that contains the gate i and RVs at each of the higher partitioning levels that intersect with the lowest level partition. This can be mathematically expressed as

$$\Delta X_{intra,i} = \sum_{0<l<k, r \text{ intersects } i} \Delta X_{l,r} + \Delta X_i^R \tag{2.54}$$

where $X_{l,r}$ are the RVs associated with the partitions in the multi-level Quad-Tree and X_i^R is the random component of variation of gate i, which is independent of the variation in any other gate. As an example, the term that is completely random between different gates could be used to model variations in threshold voltage arising due to random dopant fluctuations.

The pdfs of the individual partitions of the Quad-Tree are generated in the following fashion to ensure that the sum expressed in (2.54) always represents the correlated component of intra-die variation. All the RVs corresponding to a single level of partition are assumed to have the same distribution. The overall variation in the correlated component is distributed across different levels based on the degree of expected correlation. If the process parameters are known to be correlated over large distances then a larger fraction of the variation is assigned to higher partitioning levels. Based on this modeling scheme we note that gates that lie close to each other will be associated to the same RVs in the Quad-Tree for most levels and therefore have high correlation. On the other hand, gates that are far apart only have a few common RVs and hence their overall correlation is smaller.

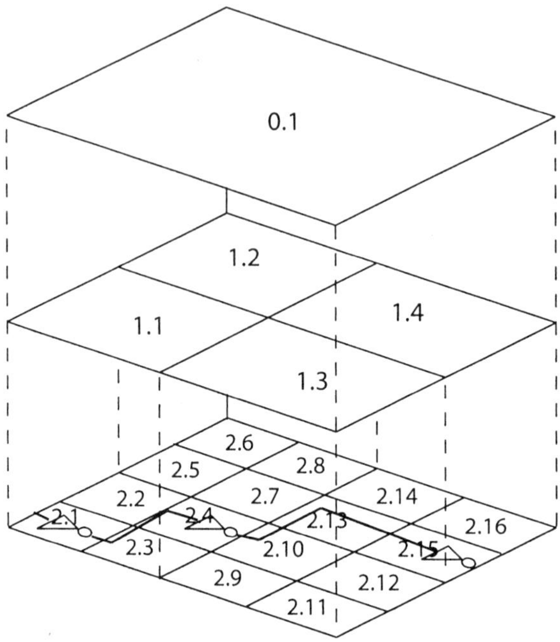

Fig. 2.8. Modeling spatial correlation using Quad-Tree partitioning. The numbering of regions in different levels is done as shown in the figure and a region (i, j) intersects the regions $(i+1, 4j-3) - (i+1, 4j)$.

2.2.4 Specialized Modeling Techniques

In this section we will discuss some of the modeling techniques that have been proposed to understand and quantify the impact of certain physical phenomena that results in variability. We will discuss models for random dopant variations that result in V_{th} variations, and are expected to have a strong influence in future technologies. We will also discuss modeling techniques for NBTI, which results in a change in V_{th} with time, and electromigration which results in opens and shorts in metal lines.

Random Dopant Variation

Since the number of doping impurities in the channel depletion layer has been reducing with technology scaling, both the number of dopants and their placement results in variations in the observed threshold voltage of the device.

If the event of different dopant atoms being introduced into the device is treated as independent *Bernoulli trials*, then the number of dopant atoms in a given volume v (N) across different devices can be shown to follow the *Poisson distribution* if the volume being considered is small compared to the total volume [91]. This follows from the fact that the probability of exactly obtaining a given number of successes, out of a large number of trials (tending to infinity) results in a Poisson distribution. Thus the probability that there are N_0 atoms within the volume v can be expressed as

$$\mathcal{P}(N = N_0) = \frac{(N_{av}v)^{N_0}}{N_0!} \exp(-N_{av}v) \tag{2.55}$$

where N_{av} is the mean value of the concentration of N_0. The mean and variance of this distribution is expressed as

$$\begin{aligned} \mu[N] &= N_{av}v \\ \sigma[N] &= \sqrt{N_{av}v}. \end{aligned} \tag{2.56}$$

Based on (2.56) we can infer, that although the absolute variance of the number of dopant atoms reduces as the concentration is reduced, the variation as a fraction of the mean, which is expressed as

$$\frac{\sigma[N]}{\mu[N]} = \frac{1}{\sqrt{N_{av}v}} \tag{2.57}$$

increases for lower doping concentrations. This implies that variations in dopant concentrations will result in larger variability in threshold voltage for low V_{th} devices. However, the variation in leakage power depends on the absolute variation in V_{th} and random dopant fluctuations will result in a smaller variability in leakage current. It is important to note that this considers variations arising due to random dopant variations. Variations in current technologies are dominated by gate length variations, and low V_{th} devices are considered to be more susceptible to variations due to worse V_{th} roll-off characteristics.

The volume in the above equations that we are interested in is the channel volume,

$$v = WLW_d \tag{2.58}$$

where W and L are the device width and length, respectively, and W_d is the depletion width, which is expressed as

$$W_d = \sqrt{\frac{4\epsilon_0 \epsilon_{Si} |\Phi_F|}{qN_{av}}} \tag{2.59}$$

where ϵ_0 is the permittivity of vacuum, ϵ_{Si} is the relative permittivity of Silicon (Si), q is the charge on an electron and Φ_F is the Fermi potential, which is expressed as

$$\Phi_F = \frac{kT}{q} \ln\left(\frac{N_{av}}{n_i}\right) \tag{2.60}$$

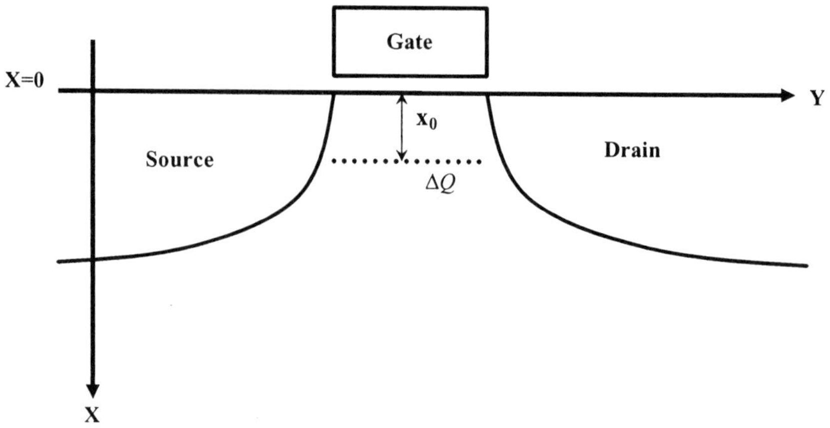

Fig. 2.9. Cross-section of a MOS device, showing a layer of dopants at a distance x_0 from the $Si - SiO_2$ interface.

where n_i is the intrinsic carrier concentration of Si ($1.1 \times 10^{10}\, cm^{-3}@300\, K$). Though this simple approach gives us insight into the impact of variation in the number of dopants, it does not consider the impact of the placement of these dopant atoms.

Now let us discuss one of the simple models proposed in [138] that considers the impact of the variation in placement of these dopant atoms along the depth of the device on threshold voltage. Assuming that the device is in inversion, the voltage drop from the surface of $Si - SiO_2$ interface to the edge of the depletion region (2.59) is constant. Let us now assume that we have an additional surface charge density ΔQ that is introduced at a distance x_0 from the surface, as shown in Fig. 2.9. This results in a change in the electric field, which increases in the region from $0 \leq x \leq x_0$. Since the voltage drop to the edge of the depletion region remains the same, the electric field in the region $x_0 \leq x \leq W_d$ reduces. This change in electric filed is illustrated in Fig. 2.10 for the simplified case, where the initial doping profile is uniform. This change in electric field will result in a change in threshold voltage, which can be simply expressed as (if the second order term due to region C shown in Fig. 2.10 is ignored)

$$\Delta V_{th} = \frac{\Delta Q}{C_{ox}} \left(1 - \frac{x_0}{W_d}\right) \tag{2.61}$$

which depends on x_0, the distance from the surface at which the extra charge sheet is present. Assuming that the doping process is a sequence of Bernoulli trials (the distribution is binomial), we can express the standard deviation of ΔQ as

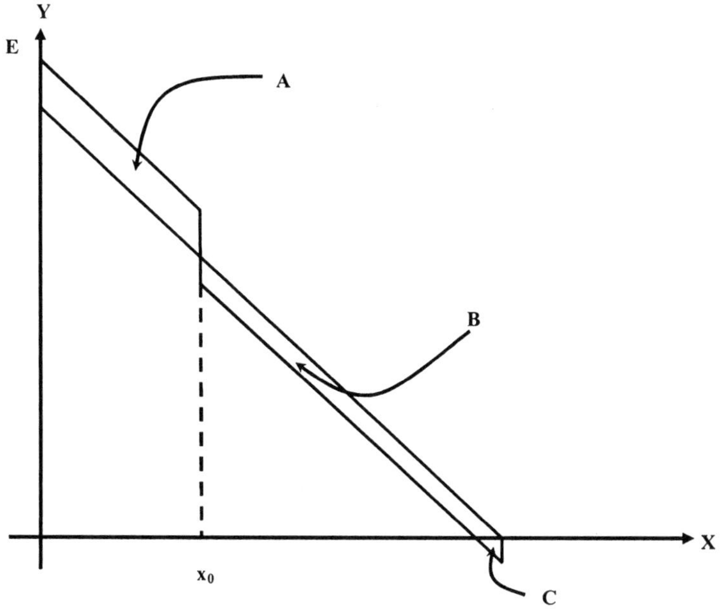

Fig. 2.10. Change in vertical electric field due to the additional sheet of charge in Fig. 2.9. *Poisson's* equation dictates that the area of Regions A and B be equal.

$$\sigma[\Delta Q] = q\sigma[N] = \frac{q\sqrt{N_{SUB}(x)LW\Delta x}}{LW} \tag{2.62}$$

where $N_{SUB}(x)$ is the doping concentration profile. The overall variance in threshold voltage can then be obtained by integrating the variances introduced by each of the infinitesimal charge sheets.

$$\begin{aligned}\sigma^2[V_{th}] &= \int_{x=0}^{x=W_d} (\sigma[\mathrm{d}V_{th}])^2 \\ &= \int_{x=0}^{x=W_d} \frac{q^2}{C_{ox}^2 LW} N_{SUB}(x) \left(1 - \frac{x}{W_d}\right)^2 \mathrm{d}x.\end{aligned} \tag{2.63}$$

Now let us define an effective doping concentration N_{EFF}, which is a weighted average of $N_{SUB}(x)$ as

$$N_{EFF} = 3\int_{x=0}^{x=W_d} \frac{N_{SUB}(x)}{W_d} \left(1 - \frac{x}{W_d}\right)^2 \mathrm{d}x \tag{2.64}$$

which yields a simple expression for the standard deviation of threshold voltage:

$$\sigma[V_{th}] = \frac{q}{C_{ox}} \sqrt{\frac{N_{EFF} W_d}{3LW}}. \tag{2.65}$$

From (2.64) we can infer that the strongest contribution to N_{EFF} is from the variations in doping concentrations that are closest to the surface, and the placement of the dopant atoms will have a strong impact on the variation in threshold voltage. Based on 3-D atomistic simulations, [13] showed that different models, including the one discussed here, provide good trends of threshold voltage variations with variations in key device characteristics. However, they fail to provide accurate quantitative estimates for the variation. Both [138] and [13] note that threshold voltage variation can also result from variations in doping profiles along the width of the transistor, which are not captured by the random dopant model (2.65). These variations can result in parts of the transistor associated with a section of the transistor width having a small number of dopants – this section therefore turns on earlier than the rest of the transistor. The probability of such events increases with the device width and results in a reduction in threshold voltage.

Negative Bias Temperature Instability (NBTI)

We saw in Chap. 1 that NBTI, which results from the generation of trap-sites at the Si/SiO_2 interface at elevated temperatures, yields a degradation (increase) in threshold voltage of PMOS devices with time. Experimental evidence has shown that trap-sites generated when $V_{gs} = -V_{dd}$ are partially annealed away when the PMOS device is in the off-state. Hence, to reduce the pessimism introduced through a constant negative-bias analysis, it is important that both modes of PMOS operation are considered.

For the inversion mode of PMOS operation, the rate of generation of interface traps (N_{IT}) is initially a function of the rate at which the Si – H bonds can be broken (k_f), and the rate at which hydrogen is annealed (k_r) at the interface [139]. During the latter stages, the rate is limited by the diffusion of hydrogen. The reaction-diffusion (RD) model used to capture this phenomena at the interface is expressed as

$$\frac{\mathrm{d}N_{IT}}{\mathrm{d}t} = k_f\left(N_0 - N_{IT}\right) - k_r N_H N_{IT} \qquad (x = 0) \tag{2.66}$$

$$\frac{\mathrm{d}N_{IT}}{\mathrm{d}t} = D_H \frac{\mathrm{d}N_H}{\mathrm{d}x} + \frac{\delta}{2}\frac{\mathrm{d}N_H}{\mathrm{d}t} \qquad (0 < x < \delta) \tag{2.67}$$

$$D_H \frac{\mathrm{d}^2 N_H}{\mathrm{d}x^2} = \frac{\mathrm{d}N_H}{\mathrm{d}t} \qquad (\delta < x < T_{ox}) \tag{2.68}$$

$$D_H \frac{\mathrm{d}N_H}{\mathrm{d}x} = k_p N_H \qquad (x > T_{ox}) \tag{2.69}$$

where x is the distance from the Si/SiO_2 interface (into the oxide), N_H and D_H are the concentration and diffusion coefficient of hydrogen, respectively,

N_0 is the number of $\mathrm{Si}-\mathrm{H}$ bonds that are unbroken at $t=0$, T_{ox} is the oxide thickness, δ is the interface thickness and k_p is the recombination velocity at the oxide-polysilicon interface. The most typical phase that is observed is when the rate of hydrogen diffusion controls trap generation. Under the additional assumption of slow trap generation, the number of interface traps as a function of time can be expressed as

$$N_{IT} \sim \sqrt{\frac{k_f N_0}{k_r}}(D_H t)^{1/4}. \tag{2.70}$$

In the annealing phase, $k_f = 0$ and the number of interface traps can be expressed in terms of the number of interface traps at the beginning of the annealing phase ($N_{IT(t=t_0)}$) and the time t spent in the annealing phase, as [139]

$$N_{IT} = N_{IT(t=t_0)}\left(1-\sqrt{\frac{\zeta t/t_0}{1+t/t_0}}\right). \tag{2.71}$$

Both the models ((2.70) and (2.71)) have been shown to be consistent with experimental data, and can be used to estimate the degradation in threshold voltage directly from the following relation

$$\Delta V_{th} = \frac{qN_{IT}}{C_{ox}}. \tag{2.72}$$

Considering the temperature sensitivity of N_{IT}, note that the term $k_f N_0/k_r$ in (2.70) is approximately temperature independent and the only dependence is through the diffusion coefficient of hydrogen D_H, which follows an Arrhenius relationship. Also NBTI is known to be electric-field dependent, and the dependence arises through the dependence of k_f on electric field in the oxide, which is expressed as

$$k_f = B\sigma_0 E_{ox}\exp\left(\frac{E_{ox}}{E_0}\right) \tag{2.73}$$

where B, the bond dissociation coefficient, and σ_0, the hole capture cross-section, are known to have weak electric field dependence. Effects such as NBTI are now seen to be reliability problems as well as performance issues. Since, traditional design margins are hard to suatain it is becoming increasingly important that additional pessimism is not introduced through considerations such as NBTI degradtion under DC conditions.

Electromigration Modeling

As discussed in Chap. 1, electromigration occurs when high energy electrons moving through the metal lines collide with the metal atoms. The transfer in momentum from the electrons to the metal atoms causes the metal atoms

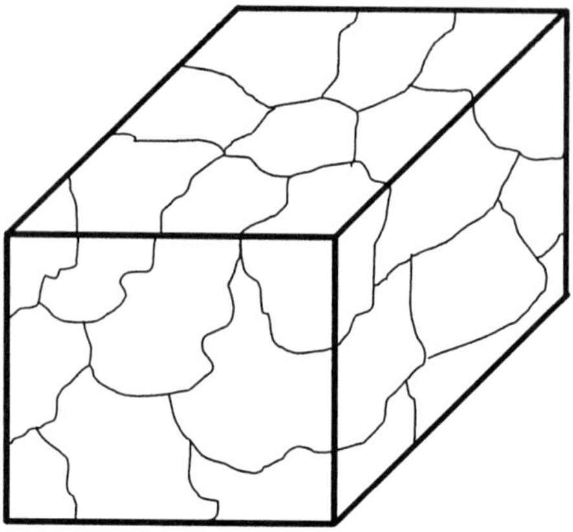

Fig. 2.11. Grain boundaries in polycrystalline materials.

to diffuse and creates an open in the wire, or alternatively, can cause shorts between adjacent wires due to metal atom pileup.

Electromigration occurs when there is a *flux divergence* of metal atoms, which generally occurs at points known as *triple points.* These triple points occur when three grain boundaries (bounding surface between crystals as shown in Fig. 2.11) meet within a wire. Failure time has generally been treated as a RV since the time to failure is dependent on the position of these grain boundaries. Additionally, the probability of failure of long wires is greater than the probability of a shorter wire, because there is a higher probability that a triple point lies on the longer wire than the shorter wire.

Traditional failure time distribution for electromigration based faults has been found to give a good fit to a lognormal, which provides the probability that the failure time of a wire t_f^{wire} is less than t as

$$\mathcal{P}(t_f^{wire} < t) = \Phi\left(\frac{\ln t - \mu}{\sigma}\right) \tag{2.74}$$

where Φ is the cdf of a standard Gaussian distribution function and can be expressed in terms of the *error function* as

$$\Phi(t) = \int_{-\infty}^{t} \left(\frac{1}{\sqrt{2\pi}} \exp\left(\frac{-t^2}{2}\right)\right) \mathrm{d}t = \frac{1}{2}\left(1 + \mathrm{erf}\left(\frac{x}{\sqrt{2}}\right)\right). \tag{2.75}$$

Different values of μ and σ in (2.74) can be used to capture the characteristics of a given wire, such as its length, width and current density. Note that since the RV is modeled as a lognormal, the probability for negative values of failure times is zero, which is physically consistent. However, a contradiction does arise when we consider a group of n wires. In this situation, the failure

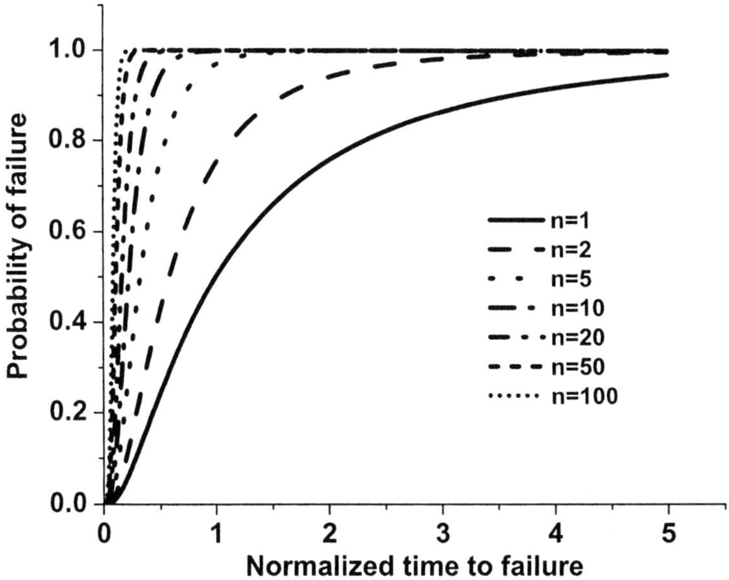

Fig. 2.12. The cumulative probability distribution of the failure time for a group of wires approaches a step function as the number of wires (n) increases.

time (t_f^{group}) is the time at which any of the wires in the group fails and can be expressed as

$$\mathcal{P}(t_f^{group} < t) = 1 - \prod_{i=1}^{n} \left(1 - \Phi\left(\frac{\ln t - \mu_i}{\sigma_i}\right)\right). \tag{2.76}$$

As we increase the number of wires in the group, the above expression tends to a unit step function, as shown in Fig. 2.12, and the probability density function becomes a delta function at $t = 0$. This implies, that as soon as the design is switched on, at least one of the wires will fail, which is inconsistent with observations.

Based on empirical evidence, namely that the mean time to failure of long wires approaches a non-zero value as the length of the wire increases, [153] proposes to use a shifted lognormal (SLN) to model electromigration failures. Based on the SLN model the failure time equation (2.74) can be rewritten as

$$\mathcal{P}(t_f^{wire} < t) = \Phi\left(\frac{\ln(t - \delta) - \mu}{\sigma}\right) \tag{2.77}$$

where δ is the amount of shift, and is process dependent. It has also be shown [153] that the SLN model is a simpler and more accurate approximation than the complicated physical eletromigration model when the activation energy for grain boundaries is assumed to have a truncated normal distribution, which directly results in a non-zero failure time for very long wires.

2.3 Performance Modeling

In this section, we will discuss modeling techniques that can be used to develop models for performance parameters (such as delay and power) in terms of process parameters. These models will allow us to analyze the impact of variations on the performance of a design and to perform optimization.

2.3.1 Response Surface Methodology

Response surface methodology (RSM) refers to a set of approaches that are used to approximate the relationship between a parameter of interest y and a set of underlying variables $\mathbf{X} = (x_1, x_2, \ldots, x_n)^T$ in a functional form. This unknown functional form is known as the *response surface model.* Though there is a vast amount of literature on RSM, with [92] serving as a good reference, we are mostly concerned with developing models for performance parameters that depend linearly on the underlying variables. This stems from the fact that variations in process parameters are generally small, therefore resulting performance variations can be assumed to be linearly related. In future generations, increasing levels of process variation may necessitate the use of higher order models. Therefore, we will also briefly look at the RSM techniques to handle quadratic response surfaces, which will be used in some sections later in this book.

Let us consider a parameter y that is approximated as $\widehat{y}$ and is assumed to vary linearly with the underlying variables $\mathbf{X}$. For the sake of discussion, assume that the zeroth component of this vector is 1, which gives $\mathbf{X} = (1, x_1, x_2, \ldots, x_n)^T$. Now we can write

$$\widehat{y} = \boldsymbol{\alpha}^T \mathbf{X} \tag{2.78}$$

where $\boldsymbol{\alpha} = (\alpha_0, \alpha_1, \ldots, \alpha_n)$ is a $n+1$-dimensional vector, whose components are known as *regression coefficients*, and the process of estimating these coefficients is known as regression analysis. Now assume that we make m observations of y for a set of m different combinations of values of X. Let

$$y_i = f(\mathbf{X}) \approx \alpha_1 x_{1i} + \cdots + \alpha_n x_{ni} = \boldsymbol{\alpha}^T \mathbf{X}_i \qquad 1 \leq i \leq m \tag{2.79}$$

where f is the unknown functional form that we seek to approximate using the linear expression (2.79). To establish a good approximation, a standard measure of the error in approximation is the sum of squared errors. Therefore, based on the information we have (i.e., the m observations) we seek a vector

$\boldsymbol{\alpha}$ that minimizes the sum of the squared errors for these m observations. The error term (ϵ), which we want to minimize, can be expressed as

$$\epsilon = \sum_{i=1}^{m} \left(y_i - \sum_{j=1}^{n} \alpha_j x_{ij} \right)^2 . \tag{2.80}$$

Taking the partial derivatives of (2.80), with respect to the α's and equating them to zero, we obtain the following set of equations

$$\frac{\partial \epsilon}{\partial \alpha_k} = \sum_{i=1}^{m} 2 \left(y_i - \sum_{j=1}^{n} \alpha_j x_{ij} \right) x_{ik} = 0 \qquad 0 \le k \le n \tag{2.81}$$

which gives

$$\sum_{i=1}^{m} \left(y_i x_{ik} - \sum_{j=1}^{n} \alpha_j x_{ij} x_{ik} \right) = 0 \qquad 0 \le k \le n. \tag{2.82}$$

These equations can be expressed in matrix form as

$$\begin{bmatrix} \sum_{i=1}^{m} y_i \\ \sum_{i=1}^{m} y_i x_{1i} \\ \sum_{i=1}^{m} y_i x_{2i} \\ \vdots \\ \sum_{i=1}^{m} y_i x_{ni} \end{bmatrix} = \begin{bmatrix} m & \sum_{i=1}^{m} x_{1i} & \sum_{i=1}^{m} x_{2i} & \cdots & \sum_{i=1}^{m} x_{ni} \\ \sum_{i=1}^{m} x_{1i} & \sum_{i=1}^{m} x_{1i}^2 & \sum_{i=1}^{m} x_{1i} x_{2i} & \cdots & \sum_{i=1}^{m} x_{1i} x_{ni} \\ \sum_{i=1}^{m} x_{2i} & \sum_{i=1}^{m} x_{2i} x_{1i} & \sum_{i=1}^{m} x_{2i}^2 & \cdots & \sum_{i=1}^{m} x_{2i} x_{ni} \\ \vdots & \vdots & \vdots & \vdots & \vdots \\ \sum_{i=1}^{m} x_{ni} & \sum_{i=1}^{m} x_{ni} x_{1i} & \sum_{i=1}^{m} x_{ni} x_{2i} & \cdots & \sum_{i=1}^{m} x_{ni}^2 \end{bmatrix} \begin{bmatrix} \alpha_0 \\ \alpha_1 \\ \alpha_2 \\ \vdots \\ \alpha_n \end{bmatrix} \tag{2.83}$$

which can also be succinctly expressed as

$$\mathbf{A}^T \mathbf{Z} = \mathbf{A}^T \mathbf{A} \boldsymbol{\alpha} \tag{2.84}$$

where $\mathbf{Z}$ is the column vector of the observed values of y for m observations, and $\mathbf{A}$ is a $m \times (n+1)$ matrix whose first column is all 1's while the rest of the columns correspond to the values of the variables ($x's$) used to make the observations of y. The above equation can also be obtained using ideas from vector projections. In a simplified scenario, we basically need to solve the equation $\mathbf{Z} = \mathbf{A}\boldsymbol{\alpha}$ for $\boldsymbol{\alpha}$. However, based on our data this equation might not have a solution. In this case, the best alternative is to find $\boldsymbol{\alpha}$ that reduces the mean squared error. This happens when the error vector $\mathbf{Z} - \mathbf{A}\boldsymbol{\alpha}$ is orthogonal to the space spanned by the columns of $\mathbf{A}$. This implies

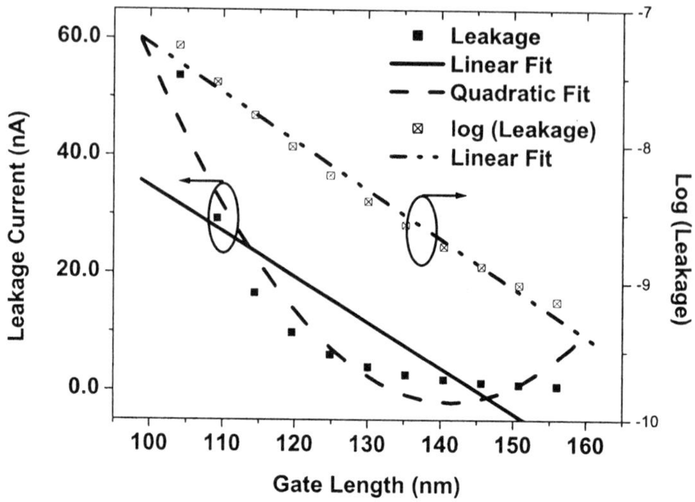

Fig. 2.13. The dependence of the log of leakage on gate length can be accurately approximated using a linear approximation whereas leakage itself is difficult to model, showing significant error even if a quadratic model is used.

$$\mathbf{A}^T(\mathbf{Z} - \mathbf{A}\boldsymbol{\alpha}) = 0, \tag{2.85}$$

which is identical to (2.84).

By solving the matrix equation in (2.84), we can estimate the components of $\boldsymbol{\alpha}$. Thus, using the above techniques we can approximate the impact of variations on performance parameters up to the first order. It is important to note that knowledge of the dependence of performance parameters on process parameters is extremely useful and provides information regarding what response surface models can be used as reasonable approximations. In particular, consider the case of subthreshold leakage power, which is known to have an exponential dependence on gate length variations. Subthreshold leakage is a strong function of gate length due to the dependency of V_{th} on L through DIBL, as discussed in Chap. 1. Variations in threshold voltage can be assumed to be linearly proportional to variations in gate length, for current level of variations. Therefore, while performing analysis of subthreshold leakage we can use linearized models to approximate $y = \ln I_{sub}$, where I_{sub} is the subthreshold leakage current. This is shown in Fig. 2.13 where both a linear and quadratic model to fit leakage are shown to result in huge inaccuracies, while a linear model is sufficient to model the dependence of the log of leakage on gate length.

If better accuracy is desired, then higher-order models can be used to approximate the response surface of the parameter y. Let us the consider the case of a second-order model, where we express the response surface as

$$\widehat{y} = \boldsymbol{\alpha}^T \mathbf{X} + \mathbf{X}^T \mathbf{B} \mathbf{X} \tag{2.86}$$

where $\mathbf{B}$ is a $n \times n$ matrix. The technique used to estimate $\boldsymbol{\alpha}$ in (2.79) can be easily extended to estimate the elements of $\boldsymbol{\alpha}$ and $\mathbf{B}$ in (2.86). This is achieved by treating the second-order terms in (2.86) as another component of $\mathbf{A}$ in (2.84). To simplify, let us assume that the original $\mathbf{X}$ consists of only two components x_1 and x_2, which gives

$$\widehat{y} = a_0 + a_1 x_1 + a_2 x_2 + a_3 x_1^2 + a_4 x_2^2 + a_5 x_1 x_2 \tag{2.87}$$

where the $a's$ are the parameters of the response surface we wish to generate. Treating the second-order terms as additional components of $\mathbf{X}$, we rewrite

$$\widehat{y} = a_0 + a_1 x_1 + a_2 x_2 + a_3 x_3 + a_4 x_4 + a_5 x_5 \tag{2.88}$$

where $x_3 = x_1^2$, $x_4 = x_2^2$ and $x_5 = x_1 x_2$. The $a's$ can then be approximated using m observations (2.83), which gives $\mathbf{P} = \mathbf{MN}$ where

$$\mathbf{P} = \begin{bmatrix} \sum_{i=1}^{m} y_i \\ \sum_{i=1}^{m} y_i x_{1i} \\ \sum_{i=1}^{m} y_i x_{2i} \\ \sum_{i=1}^{m} y_i x_{1i}^2 \\ \sum_{i=1}^{m} y_i x_{2i}^2 \\ \sum_{i=1}^{m} y_i x_{1i} x_{2i} \end{bmatrix} \qquad \mathbf{N} = \begin{bmatrix} a_0 \\ a_1 \\ a_2 \\ a_3 \\ a_4 \\ a_5 \end{bmatrix} \tag{2.89}$$

$$\mathbf{M} = \begin{bmatrix} m & \sum_{i=1}^{m} x_{1i} & \sum_{i=1}^{m} x_{2i} & \sum_{i=1}^{m} x_{1i}^2 & \sum_{i=1}^{m} x_{2i}^2 & \sum_{i=1}^{m} x_{1i}x_{2i} \\ \sum_{i=1}^{m} x_{1i} & \sum_{i=1}^{m} x_{1i}^2 & \sum_{i=1}^{m} x_{1i}x_{2i} & \sum_{i=1}^{m} x_{1i}^3 & \sum_{i=1}^{m} x_{1i}x_{2i}^2 & \sum_{i=1}^{m} x_{1i}^2 x_{2i} \\ \sum_{i=1}^{m} x_{2i} & \sum_{i=1}^{m} x_{1i}x_{2i} & \sum_{i=1}^{m} x_{2i}^2 & \sum_{i=1}^{m} x_{1i}^2 x_{2i} & \sum_{i=1}^{m} x_{2i}^3 & \sum_{i=1}^{m} x_{1i}x_{2i}^2 \\ \sum_{i=1}^{m} x_{1i}^2 & \sum_{i=1}^{m} x_{1i}^3 & \sum_{i=1}^{m} x_{1i}^2 x_{2i} & \sum_{i=1}^{m} x_{1i}^4 & \sum_{i=1}^{m} x_{1i}^2 x_{2i}^2 & \sum_{i=1}^{m} x_{1i}^3 x_{2i} \\ \sum_{i=1}^{m} x_{2i}^2 & \sum_{i=1}^{m} x_{1i}x_{2i}^2 & \sum_{i=1}^{m} x_{2i}^3 & \sum_{i=1}^{m} x_{1i}^2 x_{2i}^2 & \sum_{i=1}^{m} x_{2i}^4 & \sum_{i=1}^{m} x_{1i}x_{2i}^3 \\ \sum_{i=1}^{m} x_{1i}x_{2i} & \sum_{i=1}^{m} x_{1i}^2 x_{2i} & \sum_{i=1}^{m} x_{1i}x_{2i}^2 & \sum_{i=1}^{m} x_{1i}^3 x_{2i}^1 & \sum_{i=1}^{m} x_{1i}x_{2i}^3 & \sum_{i=1}^{m} x_{1i}^2 x_{2i}^2 \end{bmatrix}.$$

It is clear that the complexity of this approach increases exponentially with the order of the model used. However, delay and power in most cases can be accurately modeled using low-order models. Alternatively, techniques such as those used to model the dependence of I_{sub} on gate length as discussed above may be used. General techniques such as PCA can also be used to reduce the number of variables used to represent variations in process parameters. In this case, only the dominant eigenvectors of the correlation matrix are retained, simplifying the process of generating a response surface.

Additionally, using models of order higher than one results in non-normal distributions for the parameters of interest, and using non-normal probability distribution is computationally unattractive. In the next section, we will discuss the computational problems associated with second-order models of performance parameters.

2.3.2 Non-Normal Performance Modeling

As process technologies scale, the variations in some of the process parameters have continued to increase. The techniques we have discussed in the previous sections have modeled the impact of process variations on performance using a linearized model. However, in future technologies where variations can be as large 35%, the error introduced through the linear models may be unacceptable. Higher order models will have to be used to capture the impact of these variations on performance parameters. The first casualty of such a requirement is the Gaussian distrbution of performance metrics. Now, instead of dealing with linear combinations of Gaussian RVs, which can be mapped to another Gaussian RV, we will need to deal with complicated density functions.

Since the aim of modeling techniques is to enable efficient analysis and optimization, the probability distributions used to model performance parameters should, at the least, allow efficient evaluation of probabilities at different points. An approach to efficiently evaluate these probabilities, using a second order dependence of performance on process parameters called Asymptotic Probability EXtraction approach, **APEX**, was proposed in [79]. It uses ideas from interconnect simulation, such as *moment matching*, to approximate the complete distribution of the performance metric being considered.

Given a performance metric $p = f(\mathbf{X})$ that depends on n process parameters represented by the vector $\mathbf{X}$, and considering a second-order dependence of f on $\mathbf{X}$, we can express f as

$$f(\mathbf{X}) = f(\mathbf{X}_{nom}) + \mathbf{a}^T \Delta\mathbf{X} + \Delta\mathbf{X}^T \mathbf{B} \Delta\mathbf{X} \tag{2.90}$$

where $\mathbf{X}_{nom}$ is the nominal value of the process parameters and $\Delta\mathbf{X}$ represents the variation in process parameters, which can be approximated by Gaussian RVs with a correlation matrix $\mathbf{\Sigma}$. Using PCA (2.52) we write

$$\Delta\mathbf{X} = \mathbf{D}^{1/2} \mathbf{\Lambda}^T \mathbf{Y} \tag{2.91}$$

where $\mathbf{Y}$ is a set of uncorrelated Gaussian RVs with zero mean and unit variance, and the matrices $\mathbf{\Lambda}$ and $\mathbf{D}$ are as discussed in Sec. 2.2.2. Using (2.91), we rewrite (2.90) as

$$\begin{aligned} f(\mathbf{Y}) &= f(\mathbf{X}_{nom}) + \mathbf{a}^T \mathbf{D}^{1/2} \mathbf{\Lambda}^T \mathbf{Y} + \mathbf{Y}^T \mathbf{D}^{1/2} \mathbf{\Lambda} \mathbf{B} \mathbf{\Lambda}^T \mathbf{D}^{1/2} \mathbf{Y} \\ &= c_0 + \mathbf{c_1}^T \mathbf{Y} + \mathbf{Y}^T \mathbf{C} \mathbf{Y} \end{aligned} \tag{2.92}$$

where the matrix $\mathbf{C}$ will be symmetric.

Moment Matching Technique

Now we will discuss moment matching techniques that have been extensively used in interconnect simulation [27]. In these techniques, the impulse response of a RC tree is modeled as a pdf and the step response is modeled as the corresponding cdf. The pdf is then estimated by matching the moments of the pdf to the moments of the circuit expressed as a *Linear Time-Invariant* (LTI) system.

To estimate the pdf of the performance parameter p for a given circuit, the *Time-moments* of p are expressed as

$$m_k^p = \frac{(-1)^k}{k!} \int_{-\infty}^{\infty} f^k \, f_p(x) \, \mathrm{d}x \tag{2.93}$$

where $f_p(x)$ is the distribution function of p and m_k^p is the k-th order moment of p. Note that the time-moments defined in (2.93) differ from the standard definition of probabilistic moments only by a scaling factor of $(-1)^k/k!$. Also, if we consider the Fourier transform of $f_p(x)$, we obtain the *characteristic function* of p which can be expressed as

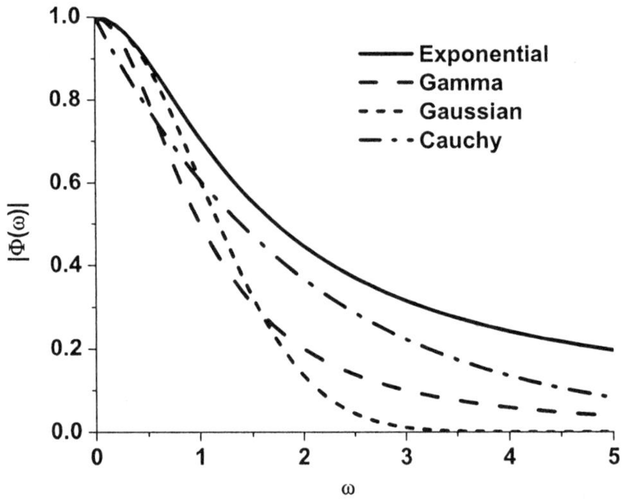

Fig. 2.14. Characteristic function for common distributions.

$$\mathcal{F}(\omega) = \int_{-\infty}^{\infty} f_p(x)\,\mathrm{e}^{-\mathrm{i}\omega x}\,\mathrm{d}x = \int_{-\infty}^{\infty} f_p(x) \sum_{k=0}^{\infty} \frac{(-\mathrm{i}\omega x)^k}{k!}\,\mathrm{d}x$$
$$= \sum_{k=0}^{\infty} m_k^p\,(-\mathrm{i}\omega)^k. \tag{2.94}$$

From (2.94), we have a power-series expansion for $\mathcal{F}(\omega)$ around the point $\omega = 0$. Also, we know that the characteristic function for typical distributions has a maximum magnitude at $w = 0$ and it tends to 0 as ω goes to infinity as shown in Fig. 2.14. Hence, the pdf of functions can be accurately estimated by estimating the first few moments [113].

Let us assume that we have an M-th order LTI system whose impulse response corresponds to f_p. We can express the transfer function of such a system in the Laplace domain as

$$H(s) = \sum_{i=1}^{M} \frac{a_i}{s - b_i} \tag{2.95}$$

which in the time domain gives

$$h(t) = \begin{cases} \sum_{i=1}^{M} a_i \mathrm{e}^{b_i t} & \text{for } t \geq 0 \\ 0 & \text{for } t < 0 \end{cases}. \tag{2.96}$$

The time-moments of H can be expressed as

$$m_k^H = \frac{(-1)^k}{k!}\int_{-\infty}^{\infty} t^k\, h(t)\,\mathrm{d}t = -\sum_{i=1}^{M}\frac{a_i}{b_i^{k+1}}. \tag{2.97}$$

We assume for now that we can estimate the moments of p. We can then match the estimated moments to the moments of H and solve for $b_i's$ and $a_i's$ using (2.97), and approximate f_p by h in (2.96).

If we consider only the first $2M$ moments of H and f_p, we get the following system of non-linear equations

$$\begin{aligned}
-\left(\frac{a_1}{b_1}+\frac{a_2}{b_2}+\cdots+\frac{a_M}{b_M}\right) &= m_0\\
-\left(\frac{a_1}{b_1^2}+\frac{a_2}{b_2^2}+\cdots+\frac{a_M}{b_M^2}\right) &= m_1\\
\vdots \quad \vdots &\\
-\left(\frac{a_1}{b_1^{2M}}+\frac{a_2}{b_2^{2M}}+\cdots+\frac{a_M}{b_M^{2M}}\right) &= m_{2M-1}.
\end{aligned} \tag{2.98}$$

The above set of equations can be solved using an iterative *Newton-Raphson* technique. However, iterative techniques can have convergence issues and tend to use heuristic techniques such as step-size control to achieve good performance. Reference [79] uses the technique proposed in [113] as part of the *Advanced Waveform Evaluation*, AWE, technique to solve the above set of equation, which is based on matching the first $2M$ moments to develop an mth-order approximation of the impulse response.

Thus the above technique can be used to approximate the pdf and cdf of the parameter p, which can be expressed as

$$cdf(p) \approx \int_0^t h(\tau)\,\mathrm{d}\tau = \begin{cases} \sum_{i=1}^{M}\frac{a_i}{b_i}(\mathrm{e}^{b_i t}-1) & \text{for } t \geq 0\\ 0 & \text{for } t < 0\end{cases}. \tag{2.99}$$

Now we will discuss an efficient technique proposed in [79] to estimate the moments of p.

Moment Evaluation

In the discussion above, we assumed that the moments of p were already provided to us. When we attempt to compute moments, we need to compute the expected values of powers of a quadratic function (2.92), which can be expressed as a higher order polynomial by

$$f^k(\mathbf{Y}) = (c_0 + \mathbf{c_1}^T\mathbf{Y} + \mathbf{Y}^T\mathbf{C}\mathbf{Y})^k = \sum_i \alpha_i\, y_1^{\beta_{1i}}\, y_2^{\beta_{2i}} \cdots y_n^{\beta_{ni}} \tag{2.100}$$

where y_i is the i-th component of $\mathbf{Y}$, and $\alpha_i' s$ and $\beta_{ij}' s$ are constants. The $y_i' s$ are uncorrelated since they are generated using PCA; therefore, we can express the k-th order moment as

$$E[f^k(Y)] = \sum_i \alpha_i \, E[y_1^{\beta_{1i}}] \, E[y_2^{\beta_{2i}}] \cdots E[y_n^{\beta_{ni}}]. \tag{2.101}$$

The expected values on the right hand side of (2.101) can be obtained by noting that, since the $y_i' s$ are Gaussian RVs with zero mean and unit variance, their moments can be expressed as

$$E[y_i^k] = \begin{cases} 0 & k \text{ odd} \\ 1 \cdot 3 \cdots (k-1) & k \text{ even} \end{cases}. \tag{2.102}$$

The odd moments in (2.102) are zero, since the Gaussian pdf is an even function. However, the computational complexity of this approach, called the *direct moment evaluation* approach, of determining the moments of p explodes for large values of k since the number of terms in (2.102) grows exponentially with increasing values of k.

We will now discuss a polynomial complexity algorithm to evaluate the moments of p. Consider (2.92) and note that C is a symmetric matrix. Due to the *Spectral Theorem*, [55] any symmetric matrix can be diagonalized and we can write

$$\mathbf{C} = \mathbf{S}\mathbf{L}\mathbf{S}^T \tag{2.103}$$

where $\mathbf{S}$ is an orthogonal matrix whose rows are the normalized orthogonal eigenvectors of $\mathbf{C}$, and $\mathbf{L}$ is a diagonal matrix with the eigenvalues of $\mathbf{C}$ on the diagonal. Now (2.91) can be written as

$$f(\mathbf{Y}) = c_0 + \mathbf{c_1}^T(\mathbf{S}\mathbf{S}^T)\mathbf{Y} + (\mathbf{Y}^T\mathbf{S})\mathbf{L}(\mathbf{S}^T\mathbf{Y}) \tag{2.104}$$

$$\begin{aligned} f(\mathbf{Z}) &= c_0 + \mathbf{q}^T\mathbf{Z} + \mathbf{Z}^T\mathbf{L}\mathbf{Z} \\ &= c_0 + \sum_{i=1}^{n}(q_i z_i + L_i z_i^2) \end{aligned} \tag{2.105}$$

where $\mathbf{q} = \mathbf{c_1}^T\mathbf{S}$ and $\mathbf{Z} = \mathbf{S}^T\mathbf{Y}$ is a vector of RVs that are linear combinations of $y_i' s$ and are therefore normally distributed. Since $\mathbf{L}$ is a diagonal matrix we do not have any cross-product terms in (2.105). In addition, if we consider the correlation matrix of $\mathbf{Z} = (z_1, z_2, \ldots, z_n)^T$ we get

$$\begin{aligned} E[\mathbf{Z}\mathbf{Z}^T] &= E[(\mathbf{S}^T\mathbf{Y})(\mathbf{S}^T\mathbf{Y})^T] = E[\mathbf{S}^T\mathbf{Y}\mathbf{Y}^T\mathbf{S}] \\ &= \mathbf{S}^T\mathbf{E}[\mathbf{Y}\mathbf{Y}^T]\mathbf{S} = \mathbf{S}^T\mathbf{I}\mathbf{S} = \mathbf{I}. \end{aligned} \tag{2.106}$$

Therefore, if the $y_i^{\prime}s$ are uncorrelated then the $z_i^{\prime}s$ are uncorrelated as well. Using this transformation we can simplify our problem and compute the moments of p recursively. First, let us define

$$\mu(u,v) = E\left[\left(c_0 + \sum_{i=1}^{v}(q_i z_i + L_i z_i^2)\right)^u\right]. \tag{2.107}$$

Using the above definition we can write the recursive relation

$$\begin{aligned}
\mu(u,v) &= E\left[\left(c_0 + \sum_{i=1}^{v}(q_i z_i + L_i z_i^2)\right)^u\right] \\
&= E\left[\left(c_0 + \sum_{i=1}^{v-1}(q_i z_i + L_i z_i^2) + q_v z_v + L_v z_v^2\right)^u\right] \\
&= E\left[\sum_{j=0}^{u}\binom{u}{j}\left(c_0 + \sum_{i=1}^{v-1}(q_i z_i + L_i z_i^2)\right)^j \left(q_v z_v + L_v z_v^2\right)^{v-j}\right] \\
&= \sum_{j=0}^{u}\left(\binom{u}{j} E\left[\left(c_0 + \sum_{i=1}^{v-1}(q_i z_i + L_i z_i^2)\right)^j\right] E\left[(q_v z_v + L_v z_v^2)^{v-j}\right]\right) \\
&= \sum_{j=0}^{u}\left(\binom{u}{j}\mu(j,v-1) E\left[\sum_{k=0}^{v-j}(q_v z_v)^k (L_v z_v^2)^{v-j-k}\right]\right) \\
&= \sum_{j=0}^{u}\left(\binom{u}{j}\mu(j,v-1)\sum_{k=0}^{v-j}\left(q_v^k L_v^{v-j-k} E\left[z_v^{2v-2j-k}\right]\right)\right).
\end{aligned} \tag{2.108}$$

Using the above recursive relation, we can estimate moments by first generating the lower order moments. Any lower order moment is generated by first generating moments corresponding to smaller v values. If we want to estimate the first M moments, we need to use the above recursive relation $O(Mn)$ times, where n is the dimension of the vector Z. In each of the recursive steps, we need to evaluate $O(M)$ terms. Thus, the overall complexity of the recursive *binomial moment evaluation* step in (2.108) is $O(M^2 n)$. If we consider the cubic complexity of the diagonalization in (2.103), the overall complexity for estimating moments is $O(M^2 n + n^3)$. Thus, binomial moment evaluation has polynomial complexity and is shown to provide a $10^6 X$ speedup over direct moment evaluation for the ISCAS'89 S27 circuit.

PDF shifting

Up to this point we have assumed that the pdf we are trying to estimate has the form of (2.96), and hence is positive only for non-negative values of t. This

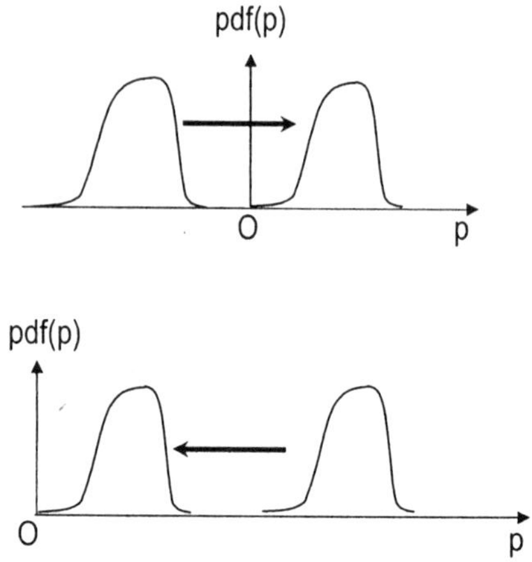

Fig. 2.15. The pdf of p in two different cases where a significant fraction of the pdf lies to the left of the origin or if it has a large positive mean value [79].

results from our assumption that the pdf can be approximated by the impulse response of a LTI system; the system would become non-causal if the impulse response has non-zero values for $t < 0$.

Therefore, we will need to shift the pdf's of parameters that have a significant probability of being negative. On the other hand, if we shift the pdf's to the right by a large amount (which could happen since many of the commonly used pdfs are non-zero over $(-\infty, \infty)$), we will have a large delay for the corresponding LTI system used to estimate the pdf. This degrades the accuracy of moment matching methods where we rely on matching the first few moments of the LTI system and the pdf of p. Additionally, we might also want to left shift a pdf that lies far away from the origin on the positive t-axis as shown in Fig. 2.15. Thus, we need to estimate the amount of shift s needed, such that the probability that $f_p < s$ (shaded region in Fig. 2.16) is smaller than a threshold ϵ, which can be mathematically expressed as

$$\mathcal{P}\left(p - s \leq 0\right) \leq \epsilon. \tag{2.109}$$

This probability can be estimated using C (2.9). A *Generalized Chebyshev Inequality* is also developed and used to obtain tighter bounds. Using the generalized inequality [79]

$$\mathcal{P}\left(|p - \mu| \geq \eta\right) \leq \frac{E[(p-\mu)^k]}{\eta^k} \tag{2.110}$$

where $\mu = E[p]$ and k is a positive even integer, we can upper bound the probability in (2.109) as

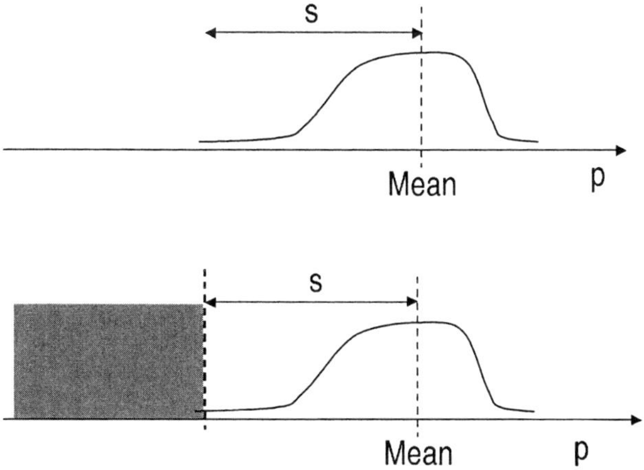

Fig. 2.16. Defining s such that the probability that p lies in the shaded region is below a given threshold [79].

$$\begin{aligned}\mathcal{P}(p - s \leq 0) &= \mathcal{P}(\mu - f_p \geq \mu - s) \\ &\leq \mathcal{P}(|p - \mu| \geq \mu - s) \leq \frac{E[(p-\mu)^k]}{(\mu - s)^k}.\end{aligned} \tag{2.111}$$

Using (2.109) and (2.111) we choose the shift amount $s = \mu - \delta$ such that

$$\epsilon \geq \frac{E[(p-\mu)^k]}{\delta^k} \tag{2.112}$$

which gives a total shift of

$$s = \mu - \min_{k=2,4,\cdots} \left(\left(\frac{\sum_{i=0}^{k} \binom{k}{i} E[p^i](-\mu)^{k-i}}{\epsilon} \right)^{1/k} \right) \tag{2.113}$$

and minimizes the required shift of the mean from zero such that the condition in (2.109) is satisfied. Note that when the pdf of p is far from the origin on the right, we obtain a left-shift ($s > 0$) that is smaller than the mean. Similarly, when the pdf is on the other extreme we obtain a right-shift ($s < 0$) with a magnitude larger than the mean value of the pdf. After an approximation to the altered pdf has been obtained, the original pdf can be simply recovered by shifting the t-axis in the opposite sense. Experimental results [79] show that the minimum in (2.113) is achieved for higher order moments, which points to the usefulness of the generalized inequality. Also, using moments higher than 10 results in values that are very close, and do not have a significant impact on reducing the shift.

If we consider the accuracy of the moment matching method we can infer that the approximated pdf will have better accuracy for higher confidence points. This follows from the observation that

$$F(\omega)|_{\omega=0} = \int_{\infty}^{\infty} f(t)\mathrm{d}t \tag{2.114}$$

where F is the Fourier transform of f, which implies that an approximation of the Fourier transform around $w = 0$ will provide a good approximation for integrals with a longer span on the t-axis. Hence [79] proposes that, when lower confidence points need to be estimated, better accuracy can be achieved by approximating $-f$ instead of f using the pdf shifting property to move the pdf of $-f$ into the desired region. The moments estimated for providing an approximation for higher confidence points can be reused, since $E[(-p)^k] = (-1)^k E[p^k]$.

Table 2.1 provides comparisons of the above approach while estimating the delay of ISCAS'89 benchmark circuits. The second column in the table (*Linear*) implies that the pdf has been approximated assuming a linear dependence of p on the process parameters (which incurs an error of 4.48% compared to 1.10% for a second-order response surface). The third column refers to Monte Carlo simulations using 10,000 runs. All the errors are with respect to the results obtained using Monte Carlo simulations with 1,000,000 runs. The table shows that APEX provides nearly a 100x and 10x improvement in estimation error compared to the linear model and Monte Carlo, respectively, while still providing a 200x speedup in run-time over Monte Carlo.

2.3.3 Delay Modeling

Until now, we have discussed general techniques that can be used to handle variations in any process and performance parameter. In this section, we specifically look at developing delay models for gates that can be used for statistical analysis.

Table 2.1. Estimation errors compared to Monte Carlo simulation with 1,000,000 runs [79].

Confidence point	Linear	Monte Carlo(10^4 runs)	APEX
1%	1.43%	0.34%	0.04%
10%	4.63%	0.64%	0.01%
25%	5.76%	0.47%	0.03%
50%	6.24%	0.32%	0.02%
75%	5.77%	0.25%	0.02%
90%	4.53%	0.66%	0.03%
99%	0.18%	0.78%	0.09%

Delay models for gates are generally based on table lookup or analytical equations. In a deterministic scenario, lookup table based gate delay models are generally used where transition time at the input and the output loading are used as indices to find the delay. The delay values for intermediate transition times and output loads are obtained using linear interpolation. The technique developed in [102] to develop statistical delay models uses both linear response surfaces and table-lookup based models to capture statistical variations in delay.

The approach proposed in [102] notes that it is difficult to develop response surfaces for modeling delay with varying transition times and output loads over a large range. A standard lookup table based approach is used to capture this variation in delay. However, each entry in the lookup table is modeled as a response surface of the form

$$d = d_0 + \boldsymbol{\alpha}^T \mathbf{X} \tag{2.115}$$

where $\mathbf{X} = (x_0, x_1, \ldots, x_n)^T$ is the vector of process parameters of interest and $(d_0, \boldsymbol{\alpha})^T = (d_0, \alpha_0, \alpha_1, \ldots, \alpha_n)^T$ is the vector obtained using response surface analysis, where the required data can be generated using SPICE simulations. This captures the dependence of delay on the variation in process parameters. If a straightforward approach is used, we will need $np+1$ parameters for each response surface in a p-transistor discharging/charging path while considering variations in n process parameters. The increase in the number of coefficients of the response surface results in an increase in characterization time needed to develop RSM models. Additionally, the time required for delay calculation also increases. The increase in number of variables results from intra-die variations that cause the individual transistors to vary independently. First, we will discuss the approach to model the delay of an inverter.

Let us write each process parameter as

$$x_i = \mu_i + \Delta(x_i)_{intra} + \Delta(x_i)_{inter} \tag{2.116}$$

where μ_i is the nominal value of the i-th process parameter x_i, $\Delta(x_i)_{intra}$ is the intra-die component of variation and $\Delta(x_i)_{inter}$ is the inter-die component of variation which is the same for all transistors on a die. Pelgrom's model for intra-die variations shows that this component of variation is independent of the location of the transistors and their separation, and therefore the RV associated with the random component of variation for each of the transistors can be treated as statistically independent. In addition, both components of variations can be modeled as Gaussian RVs.

Now let us consider the simplified case of the delay of a single inverter, where the delay of the gate is determined by a single NMOS or PMOS transistor depending on the direction of the transition. In this case, based on (2.115)-(2.116) we can write the delay of the inverter as

$$\begin{aligned} d &= d_0 + \sum_{i=1}^{n} \alpha_i(\mu_i + \Delta(x_i)_{intra} + \Delta(x_i)_{inter}) \\ &= d_0 + \boldsymbol{\alpha}^T \left(\boldsymbol{\mu} + \mathbf{X_{intra}} + \mathbf{X_{inter}}\right) \end{aligned} \tag{2.117}$$

where $\mathbf{X_{inter}}$ and $\mathbf{X_{intra}}$ are the vector of the inter- and intra-die variations of the transistors. Note that even though the random variables $\Delta(x_i)_{intra}$ for different transistors are independent for a particular i, the RV associated with a particular transistor for different i's may be correlated. Similarly the RVs that make up the vector $\mathbf{X_{intra}}$ may be correlated. Principal components prove useful here again, and the correlated set of random variables are transformed to a set of uncorrelated RVs using (2.52):

$$\begin{aligned} \mathbf{X_{inter}} &= \mathbf{D_g}^{1/2}\mathbf{\Lambda_g}^T\mathbf{P_g} \\ \mathbf{X_{intra}} &= \mathbf{D_l}^{1/2}\mathbf{\Lambda_l}^T\mathbf{P_l}. \end{aligned} \tag{2.118}$$

Note that the mean vector of the RVs in the above equations are zero and are already accounted for by the $\mu_i's$, and we can rewrite (2.117) as

$$\begin{aligned} d &= \left(d_0 + \sum_{i=1}^{n} \alpha_i \mu_i\right) + \boldsymbol{\alpha}^T\mathbf{D_g}^{1/2}\mathbf{\Lambda_g}^T\mathbf{P_g} + \boldsymbol{\alpha}^T\mathbf{D_l}^{1/2}\mathbf{\Lambda_l}^T\mathbf{P_l} \\ &= d_{nom} + \boldsymbol{\alpha}^T\mathbf{D_g}^{1/2}\mathbf{\Lambda_g}^T\mathbf{P_g} + \boldsymbol{\alpha}^T\mathbf{D_l}^{1/2}\mathbf{\Lambda_l}^T\mathbf{P_l} \end{aligned} \tag{2.119}$$

where d_{nom} is the nominal delay of the inverter when there are no variations in process parameters, and each component of $\mathbf{P_g}$ and $\mathbf{P_l}$ are independent Gaussian RVs with unit variance and zero mean. It is important to note that the random vector of $\mathbf{P_g}$ is common to all transistors in the design, and therefore the dependence of d_0 on this random vector needs to be maintained. This allows a statistical timing analyzer to consider the correlation in delays of different gates. On the other hand the RVs in the random vector $\mathbf{P_l}$ are independent across transistors and can be lumped together (being a sum of independent Gaussian RVs) into another Gaussian RV, which results in the following equation for the delay of the inverter

$$d = d_{nom} + \boldsymbol{\tau}_\mathbf{g}^T\mathbf{P_g} + \tau_l p_l \tag{2.120}$$

where

$$\begin{aligned} d_{nom} &= d_0 + \boldsymbol{\alpha}^T\boldsymbol{\mu} \\ \boldsymbol{\tau}_\mathbf{g}^T &= \mathbf{D_g}^{1/2}\boldsymbol{\alpha}^T\mathbf{\Lambda_g}^T \\ \tau_l &= \sqrt{\sum_{i=1}^{n} \left(\boldsymbol{\alpha}^T\mathbf{D_l}^{1/2}\mathbf{\Lambda_l}^T\right)_i^2} \\ p_l &\sim \mathcal{N}(0,1). \end{aligned} \tag{2.121}$$

The above expression captures the impact of both intra-die and inter-die variations on delay. The coefficients in the delay expression can be easily obtained using (2.121) from the simple RSM model (2.115) and the characteristics of process variations such as standard deviation and the correlation coefficients.

Complex Gates

Now let us the consider the case where we have several transistors involved simultaneously in charging or discharging the output capacitance. In this case, we require a term for the variation for each transistor in the path. The RSM for this gate is generated while assuming that all transistors are perfectly correlated, which again gives us an expression of the form (2.115).

Now let us rewrite (2.117) for a complex gate, while considering all components of variation as

$$d = d_0 + \sum_{k=1}^{m} \left(\boldsymbol{\beta_k}^T \left(\boldsymbol{\mu_k} + \mathbf{X_{inter,k}} + \mathbf{X_{intra,k}} \right) \right) \tag{2.122}$$

where we have m transistors on the charging/discharging path, and the vector $\boldsymbol{\beta_k}$'s are coefficients that express the dependence of delay on the process parameters related to different transistors. Since $\mathbf{X_{inter,k}}$ is the same for all transistors, we can express (2.122) as

$$d = \left(d_0 + \sum_{k=1}^{m} \boldsymbol{\beta_k}^T \boldsymbol{\mu_k} \right) + \left(\sum_{k=1}^{m} \boldsymbol{\beta_k}^T \right) \mathbf{X_{inter}} + \sum_{k=1}^{m} \boldsymbol{\beta_k}^T \mathbf{X_{intra,k}}. \tag{2.123}$$

Here we have assumed that the original RSM (2.115) for complex gates is generated while assuming that all transistors are perfectly correlated, which implies

$$\boldsymbol{\alpha}^T = \sum_{k=1}^{m} \boldsymbol{\beta_k}^T \tag{2.124}$$

using which, we can rewrite (2.123) as

$$d = d_{nom} + \boldsymbol{\tau_g}^T \mathbf{P_g} + \sum_{j=1}^{n} \sum_{k=1}^{m} \tau_{jk} p_{jk}. \tag{2.125}$$

In this equation, $\mathbf{P_g}$ is the vector of principal components of $\mathbf{X_{inter}}$ (2.118), d_{nom} and $\boldsymbol{\tau_g}$ are as expressed in (2.121), $p'_{jk}s$ are the principal components of the $\mathbf{X_{intra,k}}$'s that represent the intra-die variation in the jth principal component of intra-die variation for transistor k, and $\tau'_{jk}s$ are the coefficients for each of the principal components of intra-die variation for each transistor.

On the other hand, the coefficients for $p'_{jk}s$ cannot be obtained from the simple RSM that we have, since we do not have any information in our RSM regarding the dependence of delay on the parameters associated with a particular transistor. To simplify the problem, [102] assumes that given that variations are small, the variation in delay resulting from variations in a particular process parameter for different transistors can be assumed to be linearly related. This results in the assumption that

$$\tau_{jk} = \tau_{j0} \cdot s_k \tag{2.126}$$

where s_k determines the relative sensitivity of delay to variation in different transistors. Since the $p'_{jk}s$ are independent $N(0,1)$ variables, we can rewrite (2.125) using (2.126) as

$$\begin{aligned} d &= d_{nom} + \boldsymbol{\tau_g}^T \mathbf{P_g} + \sum_{j=1}^{n} \sum_{k=1}^{m} \tau_{j0} s_k p_{jk} \\ &= d_{nom} + \boldsymbol{\tau_g}^T \mathbf{P_g} + \sqrt{\sum_{k=1}^{m} s_k^2 \sum_{j=1}^{n} \tau_{j0} p_j} \end{aligned} \tag{2.127}$$

where $p'_j s$ are $N(0,1)$ RVs. If all transistors are assumed to be perfectly correlated we find

$$\tau_l p_l = \left(\sum_{k=1}^{m} s_k \right) \left(\sum_{j=1}^{n} \tau_{j0} p_j \right) \tag{2.128}$$

where τ_l and p_l are as defined in (2.121). Using the above equation we can rewrite (2.127) as

$$d = d_{nom} + \boldsymbol{\tau_g}^T \mathbf{P_g} + \frac{\sqrt{\sum_{k=1}^{m} s_k^2}}{\sum_{i=1}^{m} s_k} \tau_l p_l. \tag{2.129}$$

The above equation clearly shows the reduction in the impact of intra-die variation on gate delay compared to the impact of inter-die variations. This is due to the well known averaging effect that reduces the variance of a sum of uncorrelated RVs compared to correlated RVs. The model above requires the computation of the sensitivity constants $s'_k s$ beyond the computation of the RSM itself. The approaches used to estimate the sensitivity values present a direct trade-off in terms of accuracy and runtime. For example, estimation of the sensitivity values at typical values of transition time and load capacitance compared to the case where the sensitivity is evaluated for each transition time and load capacitance incurs an error of 0.76% in estimated delay for a 4-input NAND gate. Assuming that all sensitivity values are equal results in an error of 5.4%. However, if intra-die variability is neglected (all transistors are assumed to perfectly correlated), results in an overestimation in delay of 89%. Similar results for a multi-stage inverter chain are shown in Fig. 2.17.

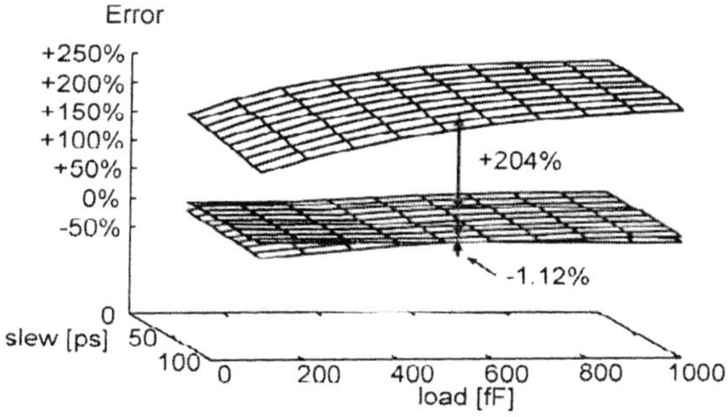

(a)

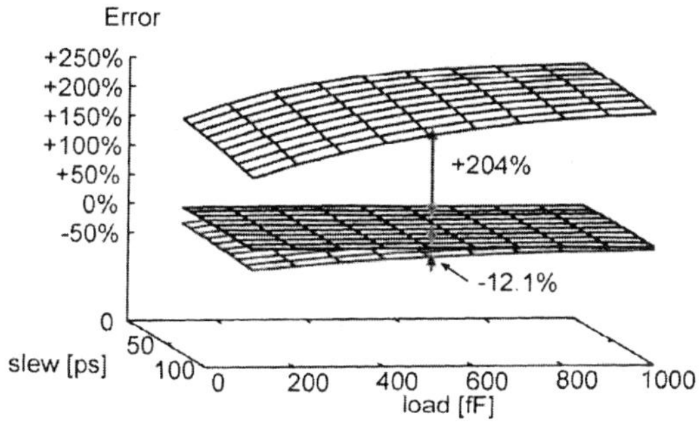

(b)

Fig. 2.17. Significant error in delay modeling occurs if intra-die variations are assumed to be perfectly correlated. Delay modeling when sensitivity calculation is performed (a) for all conditions (b) only for nominal conditions of load and input transition times. Results shown are for an inverter chain [102]. (©2005 IEEE)

2.3.4 Interconnect Delay Models

Due to technology scaling, the contribution of interconnect delay to the overall delay of a design has become significant. More interestingly, worst-case variations in interconnects often cannot be captured using worst-case corner models for interconnects [83] due to their context dependent nature. Significant work has therefore been done to capture the impact of interconnect delay considering statistical back-end variations. In this section, we discuss an analytical model to estimate the mean and variance of delay and an interval arithmetic based approach to estimate bounds on interconnect performance.

Statistical Delay Metrics

Delay metrics have been used in interconnect analysis to predict 50% delay and slew rates in a computationally efficient manner. Closed-form metrics are especially attractive since they reduce the computational overhead further, and are extensively used in incremental timing analysis and optimization engines. Most of the metrics, such as Elmore [45], D2M [10], S2M [8], Lognormal [11] (which are closed form), and h-gamma [80] and Weibull [82] (which are table lookup), are based on circuit moments. Circuit moments can be efficiently calculated using techniques such as path-tracing as employed in Rapid Interconnect Circuit Evaluator (RICE) [119]. Additionally, most of the closed-form metrics are based on the first few moments of the circuits. Delay estimation using a large number of moments are generally not closed-form and require nonlinear iterations, which can easily dominate the runtime required for delay estimation [112].

We now consider an RC interconnect, whose circuit moments are a function of the resistances and capacitances of its branches. These resistive and capacitive elements are a function of the interconnect geometry which is influenced by process variations. Based on the observation that process variations result in variations in interconnect delay which are normally distributed, [9] proposed to capture the effect of process variations on the resistive and capacitive elements using a linear model

$$\begin{aligned} R &= R_{nom} + \boldsymbol{\alpha}^T \mathbf{P} \\ C &= C_{nom} + \boldsymbol{\beta}^T \mathbf{P} \end{aligned} \tag{2.130}$$

where $\mathbf{P}$ is a p-dimensional vector of the variations in the process parameters of interest, and the vectors $\boldsymbol{\alpha}$ and $\boldsymbol{\beta}$ are weighting coefficients. In particular, considering variations in the width and thickness of the interconnect, which impacts both the resistive and capacitive components, and variations in inter-layer dielectric (ILD) thickness, which causes variations in the capacitive elements, we can write

$$\begin{aligned} R &= R_{nom} + \alpha_1 \Delta W + \alpha_2 \Delta T \\ C &= C_{nom} + \beta_1 \Delta W + \beta_2 \Delta T + \beta_3 \Delta H \end{aligned} \tag{2.131}$$

where R_{nom} and C_{nom} are the nominal values of the resistance R and capacitance C respectively. ΔW and ΔT represent the variation in interconnect width and thickness respectively, and ΔH represents the variation in ILD thickness. The coefficients $\alpha's$ and $\beta's$ can be estimated using SPICE simulation or by using empirical expressions such as those developed in [89] that relate the resistance (R), capacitance to ground (C_{gnd}) and coupling capacitance (C_{co}) to the geometrical parameters of an interconnect. For the simple case where the wire of interest has a wire on either side (as shown in Fig. 2.18)

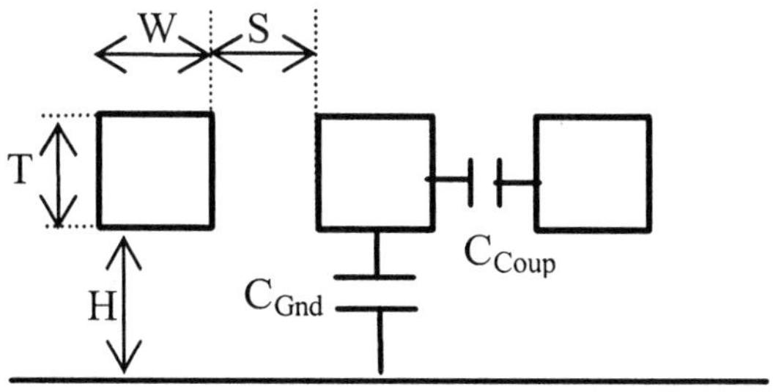

Fig. 2.18. Cross-section of a generic interconnect structure.

$$
\begin{aligned}
R &= \frac{\rho}{WT} \\
\frac{C_{gnd}}{\epsilon} &= \frac{W}{H} + 3.28\left(\frac{T}{T+2H}\right)^{0.023} + \left(\frac{S}{S+2H}\right)^{1.16} \\
\frac{C_{co}}{\epsilon} &= 1.064\frac{T}{S} + 3.28\left(\frac{T+2H}{T+2H+0.5S}\right)^{0.695} \\
&\quad + \left(\frac{W}{W+0.8S}\right)^{1.4148}\left(\frac{T+2H}{T+2H+0.5S}\right)^{0.804} \\
&\quad + 0.831\left(\frac{W}{W+0.8S}\right)^{0.055}\left(\frac{2H}{2H+0.5S}\right)^{3.542}
\end{aligned}
\tag{2.132}
$$

where W is the interconnect width, T is the interconnect thickness, H is the ILD thickness and S is the spacing between interconnects. Taking partial derivatives of the above equations with respect to the process parameter directly provides the values for the $\alpha's$ and $\beta's$ to be used in (2.131).

The next step is to estimate the impact of variation in R and C on the circuit moments. The path tracing technique, which is used for moment calculation of RC trees, relates the p-th order moment of a node i to the circuit elements and the lower order moments at node i as

$$m_p^i = \sum_{\text{all nodes k}} -R_{ik}\, C_k\, m_{p-1}^i \tag{2.133}$$

where C_k is the capacitance at node k and R_{ik} is the resistance of the intersection of paths from the source node to nodes i and k. Note that since we are concerned with tree-like structures when dealing with interconnects, these resistive paths are unique. Let us consider the specific circuit shown in Fig. 2.19. The first order moment at node 3 can be expressed as

$$m_1^3 = -R_1(C_1 + C_2 + C_3) - R_3C_3. \tag{2.134}$$

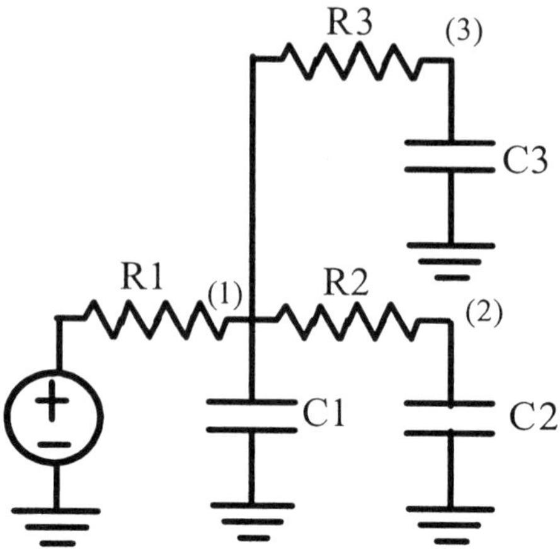

Fig. 2.19. Example RC circuit for variational analysis.

Introducing the equations for the resistive and capacitive terms in terms of variations in process parameters (2.130), we can rewrite (2.134) as

$$m_1^3 = m_{1(nom)}^3 + \mathbf{V_1}^T\mathbf{P} + \mathbf{P}^T\mathbf{V_2}\mathbf{P} \tag{2.135}$$

where $\mathbf{P}$ is the p-dimensional vector of the variation in process parameter, $\mathbf{V_1}$ is the p-dimensional vector of coefficients and $\mathbf{V_2}$ is a pxp matrix that captures the second order terms of the variations in the moment m_1^3. The entries of $\mathbf{V_1}$ and $\mathbf{V_2}$ can be expressed in terms of the parameters of the expressions of the form (2.130) for the resistive and capacitive elements in (2.134). Note that the number of higher-order terms increase with the order of the moment and increases the complexity in estimating the moments. Based on experimental measurements to be discussed later, [9] makes the assumption that neglecting the higher-order terms in (2.135) does not have a significant impact on the delay estimated using the delay metrics. This assumption simplifies the problem and allows the i-th moment of the j-th node to be written in the form

$$m_i^j = m_{i(nom)}^j + \mathbf{A_i}^T\mathbf{P}. \tag{2.136}$$

Having estimated the moments of the circuit in terms of the circuit and process variation parameters, the next step is to estimate the impact on delay. If the scaled Elmore delay model is used, which uses only the first moment, we can estimate the delay at the node j by

$$d_{Elmore} = (\ln 2)m_1^j = (\ln 2)(m_{1(nom)}^j + \mathbf{A_1}^T\mathbf{P}). \tag{2.137}$$

Note that (2.137) represents the delay as a linear combination of the variations in process parameters. If the original parameters are modeled as independent

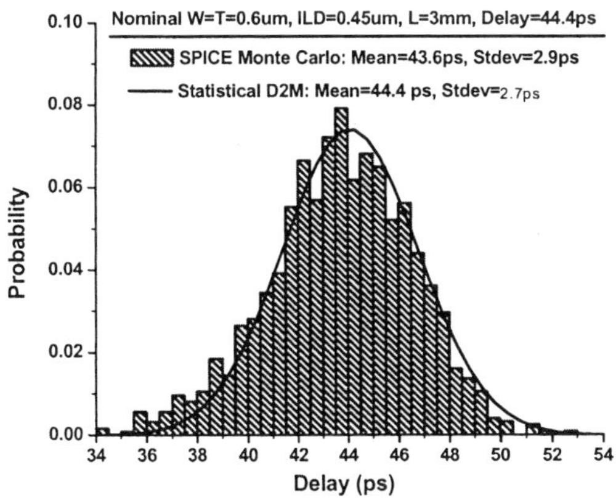

Fig. 2.20. Comparison of delay distribution obtained from Monte Carlo simulations and the statistical D2M metric [9].

Gaussian RVs, then the delay of the interconnect also becomes Gaussian. Even if the original process parameters are modeled as coming from a correlated multi-normal distribution, they can be easily mapped to a set of independent RVs using principal component analysis.

Now let us consider delay metrics, such as D2M that use higher order moments to provide better accuracy than the scaled Elmore delay model. The D2M delay metric is expressed as

$$\begin{aligned} d_{D2M} &= (\ln 2)\frac{(m_1^j)^2}{\sqrt{m_2^j}} \\ &= (\ln 2)\frac{(m_{1(nom)} + \mathbf{A_1}^T\mathbf{P})^2}{\sqrt{m_{2(nom)}^j + \mathbf{A_2}^T\mathbf{P}}}. \end{aligned} \tag{2.138}$$

Since the above expression is nonlinear in terms of the moments, it results in a non-normal distribution for delay. The above expression can be again linearized using Taylor's expansion while retaining the first order terms, and a Gaussian delay distribution is obtained.

The experimental results shown in [9] show that the linearity assumptions used in simplifying the variational delay metrics results in insignificant error. As shown in Fig. 2.20 the delay distribution of a simple line with nominal metal thickness of 600 nm and ILD thickness of 450 nm for a 30% 3σ variation in all dimensions obtained using Monte Carlo simulations. The figure clearly shows that the distribution remains Gaussian and corroborates the assumptions made in developing the variational delay metrics.

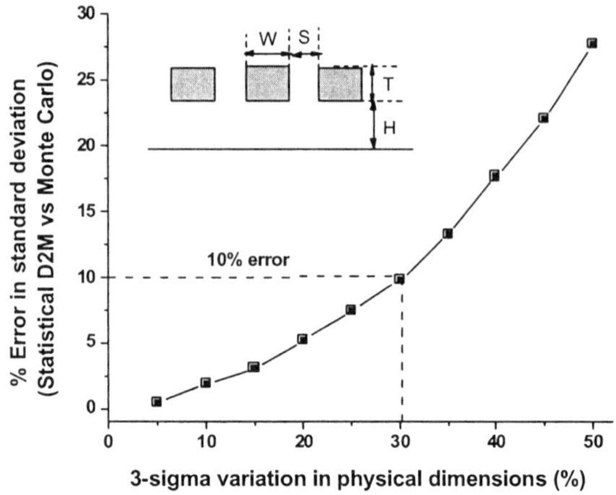

Fig. 2.21. Error in the statistical D2M model as a function of the magnitude of variation in physical dimensions [9].

The impact of increasing variability on the linearity assumption is considered in Fig. 2.21. The figure shows that increasing variability causes the error incurred by the analytical D2M delay metric to increase as expected. However, the errors are shown to be smaller than 10% even for significant process variations of as much as 30%.

Comparing the mean and standard deviations in delay obtained using D2M discussed above and those obtained using Monte Carlo simulations for a set of randomly generated test cases, the error in mean and standard deviation is found to be 1.2% and 3.8% respectively. The metal line widths were varied between 400 nm and 800 nm, and the ILD thickness was varied between 250 and to 550 nm to generate 2900 random test cases. These results show the validity of the assumptions made in developing the linear model. The approach is fairly simple and can also be easily extended to consider the distribution of slews using the S2M metric [8], and ramp inputs using the PERI metric [71].

Estimating Bounds on Interconnect Timing

The technique of estimating bounds on interconnect delay in the presence of variations was first proposed by [59]. The bounds, which were expressed as intervals or ranges, were estimated using the Rubenstein-Penfield model [122] which provided bounds on the Elmore delay.

The central idea of the approach in [59] is to use known bounds of the resistive and capacitive elements to estimate a bound on the delay of the interconnect. To achieve this, arithmetic operators $(+, -, \times, /)$ were replaced by interval arithmetic operators $(\oplus, \ominus, \otimes, \oslash)$ that act on ranges of real numbers instead of real numbers, and are expressed as

$$\begin{aligned}
[x_{11},\, x_{12}] \oplus [x_{21},\, x_{22}] &= [x_{11} + x_{21},\, x_{12} + x_{22}] \\
[x_{11},\, x_{12}] \ominus [x_{21},\, x_{22}] &= [x_{11} - x_{22},\, x_{12} - x_{21}] \\
[x_{11},\, x_{12}] \otimes [x_{21},\, x_{22}] &= [\min(x_{11}x_{21},\, x_{11}x_{22},\, x_{12}x_{21},\, x_{12}x_{22}), \qquad (2.139) \\
&\qquad \max(x_{11}x_{21},\, x_{11}x_{22},\, x_{12}x_{21},\, x_{12}x_{22})] \\
[x_{11},\, x_{12}] \oslash [x_{21},\, x_{22}] &= [x_{11},\, x_{12}] \otimes [1/x_{21},\, 1/x_{22}] \quad \text{if } 0 \notin [x_{21},\, x_{22}].
\end{aligned}$$

If the intervals are degenerate, i.e., the lower and upper bound of the interval are the same, then all interval arithmetic operators behave identically to their arithmetic counterpart. Hence, in the case where no variability is assumed, both forms of arithmetic provide the same result. Additionally it can be shown that the result obtained using the arithmetic expression and any of the real numbers from the ranges provided to the interval arithmetic expression is subsumed in the final range provided by the interval arithmetic expression [59]. Though the bounds provided by interval arithmetic are therefore correct, they are extremely loose. One of the main reasons for this stems from the fact that interval arithmetic does not have a multiplicative inverse and therefore instead of having

$$X/X = [x_{11},\, x_{12}] \oslash [x_{11},\, x_{12}] = [1,\, 1] \qquad (2.140)$$

where X represents some variable that is known to have a given range, we obtain

$$[x_{11},\, x_{12}] \oslash [x_{11},\, x_{12}] = [x_{11}/x_{12},\, x_{12}/x_{11}] \qquad (2.141)$$

for positive intervals. One of the improvements suggested in [59] was to minimize the use of the $\oslash$ operator by performing Taylor series expansion of the dividing terms. The problem can also be considered to arise from the lack of information regarding the correlation between operands of the interval arithmetic operators, and results in overestimation of the ranges whenever ranges of correlated variables are operated upon using interval arithmetic. As an example, consider the case where we subtract two perfectly correlated and identical ranges. If this interval arithmetic were cognizant of their correlation we will obtain a degenerate range as a result, which is $[0,\, 0]$. In reality we arrive at $[a - b,\, b - a]$ where $[a,\, b]$ is the range of the initial variable.

This problem was considered in [86], which used an improvement of interval arithmetic known as *affine arithmetic* [134]. The basic idea of modeling correlations in intervals has some similarity to the idea of principal components analysis used to analyze correlated RVs. Any given range of real numbers is expressed as an affine sum expressed as

$$X = [x_{11}.\, x_{12}] = \alpha_0 + \sum_{i=1}^{n} \alpha_i \epsilon_i \qquad (2.142)$$

where $\epsilon_i^{\prime}s$ are independent ranges from -1 to +1, and the $\alpha_i^{\prime}s$ are real numbers. We will call this form of expressing the range as the *canonical form.* Let us now redefine the interval arithmetic operators as affine arithmetic operators for addition and subtraction for two affine arithmetic variables which are expressed as

$$\begin{aligned} A_1 &= \alpha_0 + \sum_{i=1}^{n} \alpha_i \epsilon_i \\ A_2 &= \beta_0 + \sum_{i=1}^{n} \beta_i \epsilon_i \end{aligned} \tag{2.143}$$

and define the $\oplus$ and $\ominus$ as

$$\begin{aligned} A_1 + A_2 &= (\alpha_0 + \beta_0) + \sum_{i=1}^{n} (\alpha_i + \beta_i)\, \epsilon_i \\ A_1 - A_2 &= (\alpha_0 - \beta_0) + \sum_{i=1}^{n} (\alpha_i - \beta_i)\, \epsilon_i \end{aligned} \tag{2.144}$$

which are again in canonical form. The advantage of defining the result of operators in canonical form is that the operators can then be recursively applied, and any expression that can be written in terms of these operators can be easily evaluated.

Let us consider the effect of redefining the operators on the subtraction of perfectly correlated identical ranges as described above. Now the ranges are expressed as

$$X = Y = \alpha_0 + \sum_{i=1}^{n} \alpha_i \epsilon_i \tag{2.145}$$

and using (2.144) we obtain $X - Y = 0$ as desired. Let us now extend this idea for the operator $\otimes$ applied to the variables defined in (2.143)

$$\begin{aligned} A_1 \otimes A_2 &= \left(\alpha_0 + \sum_{i=1}^{n} \alpha_i \epsilon_i\right)\left(\beta_0 + \sum_{i=1}^{n} \beta_i \epsilon_i\right) \\ &= \alpha_0\beta_0 + \sum_{i=1}^{n} (\alpha_0\beta_i + \beta_0\alpha_i)\, \epsilon_i + \left(\sum_{i=1}^{n} \alpha_i \epsilon_i\right)\left(\sum_{i=1}^{n} \beta_i \epsilon_i\right) \end{aligned} \tag{2.146}$$

and we find a problem that the final result is not in canonical form. As noted above the results need to be in canonical form to be able to make any progress using this idea. As shown in [134] we can tradeoff the tightness of the bounds to achieve this goal. We rewrite the expression in (2.146) by approximating the problematic last term in terms of an additional independent range, which effectively adds a component to the $\epsilon_i' s$ which is independent of the other $\epsilon_i' s$. The coefficient of this additional term is simply approximated as the largest possible range that the last term can contribute, and is mathematically expressed as

$$\left(\sum_{i=1}^{n} \alpha_i \epsilon_i\right)\left(\sum_{i=1}^{n} \beta_i \epsilon_i\right) = \left(\sum_{i=1}^{n} |\alpha_i|\right)\left(\sum_{i=1}^{n} |\beta_i|\right) \epsilon_{n+1}. \tag{2.147}$$

Note that we have lost all correlation information of the last term in (2.146) when we use this approximation, which as shown above results in overestimation of the range of bounds. Details regarding the approximation of operations such as division, square-root and exponentials in canonical form can be found in [134]. To map any particular real number algorithm to a interval algorithm, we need to define one additional operation of comparison of real numbers for intervals. Comparison of intervals is not well defined when the intervals intersect and [86] notes that redefining this comparison to be the comparison of midpoints of the respective ranges results in an added advantage that using degenerate ranges provides the nominal case results. Using these basic operations any algorithm that uses only these basic operators can be mapped to a interval based algorithm. In particular, for interconnect analysis the moment generation techniques (2.133) and delay metrics (2.137)-(2.138) can be directly mapped to interval based expressions.

2.3.5 Reduced-Order Modeling Techniques

In deep-submicron (DSM) technologies a larger number of moments are required to provide a reasonable approximation to the actual response. Asymptotic waveform evaluation (AWE) [113] iteratively estimates the first $2M$ moments of the circuit in the Laplace domain, where M is much less than the actual order of the circuit. These moments are then used to generate an M-pole approximation of the transfer function, which can be directly mapped to the time domain response. A number of reduced-order modeling approaches have been proposed in the context of process variations. In [84] the congruence transformation-based PACT and PRIMA techniques were combined with matrix perturbation analysis, [60] used a balanced truncation realization based interconnect analysis, while [86] extended AWE and PRIMA to consider variations using interval arithmetic techniques. However, all these methods are unable to preserve passivity, and therefore time-domain simulations in combination with nonlinear devices can lead to numerical instability. A fast transistor level simulator (TETA) was proposed in [41] and was carefully coupled with variational interconnect models [84] in [2] to resolve passivity issues.

We will discuss the basic ideas used in reduced-order modeling techniques such as AWE, PRIMA and PACT and then discuss the proposed techniques to extend them to the case with process variations.

AWE

AWE is a *Pade approximation* based on using the first $2M$ moments of the circuit to develop an approximation for the transfer function of the network. These moments can be efficiently estimated using path-tracing (2.134) or using modified nodal analysis and provide a computationally efficient approach to approximate the characteristics of the network. Since any linear RLC circuit with a single input and output can be described using *Modified Nodal Analysis* (MNA) as

$$(\mathbf{G} + s\mathbf{C})\,\mathbf{X} = \mathbf{b} \tag{2.148}$$

$$y = \mathbf{c}^T\mathbf{X} \tag{2.149}$$

where $\mathbf{G}$ is the conductance matrix, $\mathbf{C}$ is the susceptance matrix, $\mathbf{b}$ is the excitation vector, $\mathbf{X}$ is the state-vector and $\mathbf{c}$ is the vector relating the state variables to the output variable y. The moments of the circuit are related to the state variables $\mathbf{X}$ as

$$\mathbf{X}(s) = \mathbf{x_0} + \mathbf{x_1}s + \mathbf{x_2}s^2 + \cdots \tag{2.150}$$

$$m_i = \mathbf{c}^T\mathbf{x_i} \tag{2.151}$$

where m_i is the i-th order moment of the circuit, and defines the transfer function as

$$H(s) = m_0 + m_1 s + m_2 s^2 + \cdots \tag{2.152}$$

Introducing (2.150) into (2.148) and comparing terms we note that the moments of the circuit can be recursively approximated as

$$\begin{aligned} \mathbf{G}\mathbf{x}_0 &= \mathbf{b} \\ \mathbf{G}\mathbf{x}_i &= -\mathbf{C}\mathbf{x}_{i-1} \end{aligned} \tag{2.153}$$

and (2.151). Note that these equations are not solved by explicitly inverting the matrix, but by LU factorization which can be very efficient if the involved matrices are sparse, which is generally the case when dealing with RC interconnects (however, this may not hold for RLC circuits). Having estimated the moments, the next step is to develop an q-th order approximation for the circuit that has the form

$$\hat{H}(s) = \frac{b_0 + b_1 s + \cdots + b_{q-1}s^{q-1}}{1 + a_1 s + \cdots + a_q s^q} \tag{2.154}$$

where $\hat{H}$ is an approximation of H. AWE computes the coefficients of this approximation by matching the estimated moments (using (2.153)) to the moments of the q-th order approximation. Equating the expressions in (2.152) and (2.154) and matching the coefficients of s results in a set of linear equations for the coefficients of the denominator and the numerator of the transfer function, which can be expressed as

$$
\begin{aligned}
b_0 &= m_0 \\
b_1 &= m_1 + m_0 a_1 \\
b_2 &= m_2 + m_1 a_1 + m_0 a_2 \\
&\vdots \\
b_{q-1} &= m_{q-1} + m_{q-2} a_1 + \cdots + m_0 a_{q-1}
\end{aligned}
\tag{2.155}
$$

for the first q powers of s. The remaining equations that result from matching coefficients of s^q to s^{2q-1} yield

$$
\begin{aligned}
m_q + m_{q-1} a_1 + \cdots + m_0 a_q &= 0 \\
m_{q+1} + m_q a_1 + \cdots + m_1 a_q &= 0 \\
&\vdots \\
m_{2q-1} + m_{2q-2} a_1 + \cdots + m_{q-1} a_q &= 0.
\end{aligned}
\tag{2.156}
$$

Equations (2.156) only involve the coefficients of the denominator of the approximate transfer function, which can now be solved as a set of simultaneous linear equations. The poles can then be found by finding the roots of the polynomial expression, that makes up the denominator of the transfer function. The coefficients of the numerator are found using an additional matrix-vector multiplication and the residues of the poles are found using a solution of a matrix equation.

Though AWE is computationally efficient, it is found to suffer from problems such as ill conditioning of higher moments and instability. Ill conditioning of AWE follows from (2.153) which shows that the moments of a circuit are approximated using a sequence of the form

$$\mathbf{R}, \mathbf{A}\mathbf{R}, \ldots, \mathbf{A}^{i-1}\mathbf{R}, \ldots \tag{2.157}$$

where $\mathbf{R} = \mathbf{G}^{-1}\mathbf{b}$ and $\mathbf{A} = -\mathbf{G}^{-1}\mathbf{C}$. Since the above sequence converges rapidly to an eigenvector of $\mathbf{A}$, a higher order AWE approximation does not add more information or accuracy to the reduced order model. Thus ill conditioning of higher moments occurs when moments beyond some high order are used to estimate the poles. These moments might not contain additional useful information and may affect AWE such that it starts producing poles in the right-half s-plane. This results in an unstable approximation for inherently stable systems. Although asymptotic stability can be guaranteed by dropping unstable poles and readjusting the residues to improve accuracy, the reduced order models provided by AWE cannot be guaranteed to be *passive*. Passivity ensures that when the system is connected to nonlinear devices, the simulation results in stable output responses. Projectional methods, such as Pole Analysis Via Congruence Transformation (PACT) [72] and Passive Reduced-order Interconnect Macromodeling Algorithm (PRIMA) [101], were later proposed that are numerically stable and guaranteed to produce passive reduced order models. We will now briefly describe PRIMA and PACT, which will also be extended to perform interconnect analysis while considering variations.

PRIMA

Consider the following set of MNA equations describing an N-port linear circuit in terms of $n \times n$ matrices $\mathbf{G}$ and $\mathbf{C}$, representing the conductance and susceptance matrix respectively, as

$$\mathbf{G}\mathbf{x} + \mathbf{C}\frac{\mathrm{d}\mathbf{x}}{\mathrm{d}t} = \mathbf{B}\mathbf{u}_{\mathbf{in}}(t) \qquad \mathbf{u}_{\mathbf{out}}(t) = \mathbf{L}^T\mathbf{x} \tag{2.158}$$

where $\mathbf{x}$ represents the variables of MNA, $\mathbf{u}_{\mathbf{in}}$ and $\mathbf{u}_{\mathbf{out}}$ represent the vector of input excitations and outputs, respectively, which are related to the MNA variables using the matrices $\mathbf{B}$ and $\mathbf{L}$. A set of sufficient conditions for passivity is that the matrices $\mathbf{G}$ and $\mathbf{C}$ are positive-definite and that $\mathbf{B} = \mathbf{L}$. If we are only interested in the voltage-current characteristics of the interconnect block, then using a Z-parameter formulation we have $\mathbf{B}$=$\mathbf{L}$. Also, the matrices of this system are generated using electromagnetic analysis and are typically positive-definite, implying that our system is passive. Defining $\mathbf{R} = \mathbf{G}^{-1}\mathbf{B}$ and $\mathbf{A} = -\mathbf{G}^{-1}\mathbf{C}$, we can rewrite (2.158) as

$$\mathbf{x} = \mathbf{A}\frac{\mathrm{d}\mathbf{x}}{\mathrm{d}t} + \mathbf{R}\mathbf{u}_{\mathbf{in}}(t) \tag{2.159}$$

which results in the following representation for the transfer function

$$H(s) = \mathbf{c}^T(\mathbf{I} - s\mathbf{A})^{-1}\mathbf{R} = \frac{\mathbf{c}^T\mathrm{adj}(\mathbf{I} - s\mathbf{A})\mathbf{R}}{\det(\mathbf{I} - s\mathbf{A})} \tag{2.160}$$

where adj and det refer to the adjoint and determinant of the matrix. Note that the denominator in the above equation is similar to the characteristic equation of matrix $\mathbf{A}$, and has roots that are the reciprocal of the eigenvalues of $\mathbf{A}$. Therefore the poles of the system can be approximated by finding the poles of matrix $\mathbf{A}$.

PRIMA is based on projecting the $n \times n$ matrices, used to define MNA, to a smaller subspace (V) of dimension q resulting in a set of matrices of dimension $q \times q$. This is achieved either through *orthogonal projection*, where the error vector is required to be orthogonal to subspace V, or *oblique projection*, where an additional q dimensional subspace (U) is defined (distinct from V) and the error vector is required to be orthogonal to the subspace U. PRIMA uses orthogonal projection, which results in a *congruence transformation* and is known to preserve positive-definiteness. This transformation results in

$$\mathbf{V_q}^T\mathbf{G}\mathbf{V_q}\mathbf{x_q} + \mathbf{V_q}^T\mathbf{C}\mathbf{V_q}\frac{\mathrm{d}\mathbf{x_q}}{\mathrm{d}t} = \mathbf{V_q}^T\mathbf{B}\mathbf{u_{in}}(t) \quad \mathbf{u_{out}}(t) = \mathbf{B}^T\mathbf{V_q}\mathbf{x_q} \tag{2.161}$$

where $\mathbf{V_q}$ is a $n \times q$ orthogonal matrix with the column vectors being the basis of subspace V on which the set of equations are projected. The subspace $\mathbf{V}$ is generally defined to be the *block Krylov subspace*, which is the span of the low order block moments

$$\mathrm{colsp}\,\mathbf{V_q} = \mathrm{Kr}\left(\mathbf{A}, \mathbf{R}, \lfloor\frac{q}{N}\rfloor\right) = \mathrm{span}(\mathbf{R}, \mathbf{AR}, \ldots, \mathbf{A}^{\lfloor q/N\rfloor - 1}\mathbf{R}) \tag{2.162}$$

where we have assumed that q is chosen such that $\lfloor q/N \rfloor$ is an integer. This matrix can be generated using the *block Arnoldi algorithm* which uses QR factorization to generate basis vectors for the subspace V. The poles can then be estimated by estimating the eigenvalues as in (2.160). Note that the congruence transformation is applied to (2.158) and not to (2.159). Therefore if the matrices in (2.159) define a passive system, then (2.161) also satisfies conditions of passivity.

PACT

PACT is based on two congruence transformations that are used to reduce the size of the admittance matrix of RC networks by dropping unwanted poles, while guaranteeing stability and passivity. We will discuss a slightly different form of PACT presented in [84]. PACT uses the admittance formulation for the state equations and partitions the conductance and susceptance matrices as

$$\mathbf{G} = \begin{bmatrix} \mathbf{G_P} & \mathbf{G_C}^T \\ \mathbf{G_C} & \mathbf{G_I} \end{bmatrix} \quad \mathbf{C} = \begin{bmatrix} \mathbf{C_P} & \mathbf{C_C}^T \\ \mathbf{C_C} & \mathbf{C_I} \end{bmatrix} \tag{2.163}$$

and rewrites (2.158) as

$$\left(\begin{bmatrix} \mathbf{G_P} & \mathbf{G_C}^T \\ \mathbf{G_C} & \mathbf{G_I} \end{bmatrix} + s\begin{bmatrix} \mathbf{C_P} & \mathbf{C_C}^T \\ \mathbf{C_C} & \mathbf{C_I} \end{bmatrix}\right)\begin{bmatrix} \mathbf{x_P} \\ \mathbf{x_I} \end{bmatrix} = \begin{bmatrix} \mathbf{b_P} \\ \mathbf{0} \end{bmatrix} \tag{2.164}$$

where $\mathbf{x_P}$ and $\mathbf{x_I}$ represent the m port node voltages and the n internal node voltages, respectively. The matrices $\mathbf{G_P}$ and $\mathbf{C_P}$ are referred to as port matrices, $\mathbf{G_I}$ and $\mathbf{C_I}$ are referred to as internal matrices and $\mathbf{G_I}$ and $\mathbf{C_I}$ are referred to as connection matrices. The right hand side of the above equation defines the currents injected into the system and is zero for the entries corresponding to internal nodes. The port and internal matrices can be shown to be symmetric, and if the system is known to have a unique DC solution then the connection conductance matrix can be shown to be positive semi-definite. Since $\mathbf{Y}(s)\mathbf{x_P} = \mathbf{b_P}$, using (2.158) we can write

$$\mathbf{Y}(s) = \mathbf{G_P} + s\mathbf{C_P} - (\mathbf{G_C} + s\mathbf{C_C})^T(\mathbf{G_I} + s\mathbf{C_I})^{-1}(\mathbf{G_C} + s\mathbf{C_C}). \quad (2.165)$$

The first congruence transformation is used to match the DC gain of the system, so that the DC behavior of the system does not change when unwanted poles are dropped from the system. This transformation is based on the matrix $\mathbf{X}$, which can be written as

$$\mathbf{X} = \begin{bmatrix} \mathbf{I} & \mathbf{0} \\ -\mathbf{G_I}^{-1}\mathbf{G_C} & \mathbf{I} \end{bmatrix} = \begin{bmatrix} \mathbf{I} & \mathbf{0} \\ \mathbf{V} & \mathbf{I} \end{bmatrix} \quad (2.166)$$

which results in a new conductance matrix $\mathbf{G}' = \mathbf{X}^T\mathbf{G}\mathbf{X}$ and a new susceptance matrix $\mathbf{C}' = \mathbf{X}^T\mathbf{C}\mathbf{X}$, of the form

$$\begin{aligned} \mathbf{G}' &= \begin{bmatrix} \mathbf{G_P} - \mathbf{G_C}^T\mathbf{V} & \mathbf{0} \\ \mathbf{0} & \mathbf{G_I} \end{bmatrix} = \begin{bmatrix} \mathbf{G'_P} & \mathbf{0} \\ \mathbf{0} & \mathbf{G_I} \end{bmatrix} \\ \mathbf{C}' &= \begin{bmatrix} \mathbf{C_P} + \mathbf{V}^T\mathbf{C_C} + \mathbf{C_C}^T\mathbf{V} + \mathbf{V}^T\mathbf{C_I}\mathbf{V} & (\mathbf{C_C} + \mathbf{C_I}\mathbf{V})^T \\ \mathbf{C_C} + \mathbf{C_I}\mathbf{V} & \mathbf{C_I} \end{bmatrix} \\ &= \begin{bmatrix} \mathbf{C'_P} & \mathbf{C'_C}^T \\ \mathbf{C'_C} & \mathbf{C_I} \end{bmatrix} \end{aligned} \quad (2.167)$$

which results in the following expression for $\mathbf{Y}(s)$

$$\mathbf{Y}(s) = \mathbf{G'_P} + s\mathbf{C'_P} - s^2\mathbf{C'_C}^T(\mathbf{G_I} + s\mathbf{C_I})^{-1}\mathbf{C'_C}. \quad (2.168)$$

The second congruent transformation in PACT is developed based on the following result from matrix theory.

Theorem 2.3. [84]. *Let* $\mathbf{A}$ *and* $\mathbf{B}$ *be two symmetric matrices and let* $\mathbf{B}$ *be positive-definite, then there exists a matrix* $\mathbf{U}$ *such that* $\mathbf{U}^T\mathbf{B}\mathbf{U} = \mathbf{I}$ *and* $\mathbf{U}^T\mathbf{A}\mathbf{U} = \mathbf{\Lambda}$*, where* $\mathbf{\Lambda}$ *is a real diagonal matrix.*

Based on the above theorem we can define $\mathbf{U}$ such that

$$\mathbf{U}^T\mathbf{G_I}\mathbf{U} = \mathbf{I} \quad \mathbf{U}^T\mathbf{C_I}\mathbf{U} = \mathbf{\Lambda} \quad (2.169)$$

and rewrite (2.168) as

$$\mathbf{Y}(s) = \mathbf{G'_P} + s\mathbf{C'_P} - s^2\mathbf{C'_C}^T\mathbf{U}^T(\mathbf{I} + s\mathbf{\Lambda})^{-1}\mathbf{U}\mathbf{C'_C}. \quad (2.170)$$

Note that matrix $\mathbf{U}$ has a column space that spans the generalized eigenspace of $(\mathbf{G_I}, \mathbf{C_I})$. Now PACT constructs a reduced order model

$$\mathbf{Y}(s) = \mathbf{G'_P} + s\mathbf{C'_P} - s^2\mathbf{C'_C}^T\mathbf{U_r}^T(\mathbf{I} + s\mathbf{\Lambda_r})^{-1}\mathbf{U}\mathbf{C'_C} \quad (2.171)$$

by considering only a few of the eigenvectors in $\mathbf{U}$ that correspond to dominant generalized eigenvalues. This desired matrices $\mathbf{U_r}$ and $\mathbf{\Lambda_r}$ can be efficiently generated using Cholesky decomposition and Lanczos method [55].

Details regarding AWE and other passive reduced order modeling techniques can be found in [27]. We will discuss two techniques that have been proposed to extend these reduced-order modeling techniques to consider variations.

Interval Analysis

In this section we will use interval arithmetic based techniques developed for interconnect delay analysis in the previous section to extend AWE and PRIMA. Additionally the $\epsilon_i' s$ as defined in (2.142) are now unit variance zero mean normal RVs, in keeping with canonical models used in statistical timing analysis. As we will see in the next chapter, this formulation to capture sources of variations is exactly the same as principal component analysis, which simplifies the steps involved in considering correlation among different RVs. We can use the resistive and capacitive elements expressions of (2.131). Note that though we have written these variables we will interpret them as intervals and use interval arithmetic based operations. Initial ranges for the $\epsilon' s$ can be chosen such that they capture most of the region of the Gaussian distribution.

To use interval arithmetic for AWE or PRIMA, we only need to define equivalent operations in interval arithmetic for the operations used in AWE or PRIMA. Let us first consider AWE and outline the basic steps. To develop an interval arithmetic based AWE, the LU decomposition steps are replaced by an interval LU solve and the roots of the polynomial can be estimated using *Laguerre's Method* [114]. This approach does not involve steps that require taking derivatives and can be implemented using simple interval operations, absolute value and square-root operations.

The key steps required to develop an interval arithmetic based PRIMA algorithm are basic matrix operations, orthogonalization, LU decomposition (used in block Arnoldi) and eigen-decomposition. Eigen-decomposition is performed in [86] using a QR decomposition followed by an inverse iteration, which can be performed using interval operations already defined. All of the remaining operations can also be performed using the interval operations previously defined.

The final poles and residues are then statistically sampled to generate the distribution of delay, by computing the 50% delay for each sample. Note that the final sampling is done in a much lower order space, and better accuracy can be obtained with fewer samples. An implementation of this approach [86] showed an error in mean and standard deviation of 4.1% and 5.6% for four designs compared to Monte Carlo based AWE simulation. Similar results for PRIMA showed errors of 4.9% and 5.9% for mean and standard deviation, respectively. Figure 2.22 demonstrates the pdf obtained using intervals and

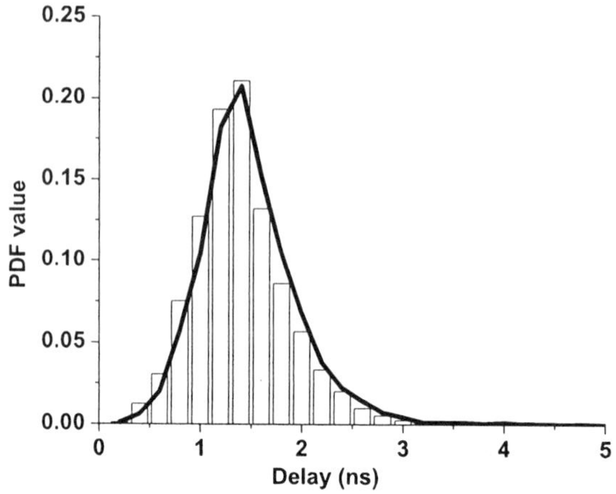

(a)

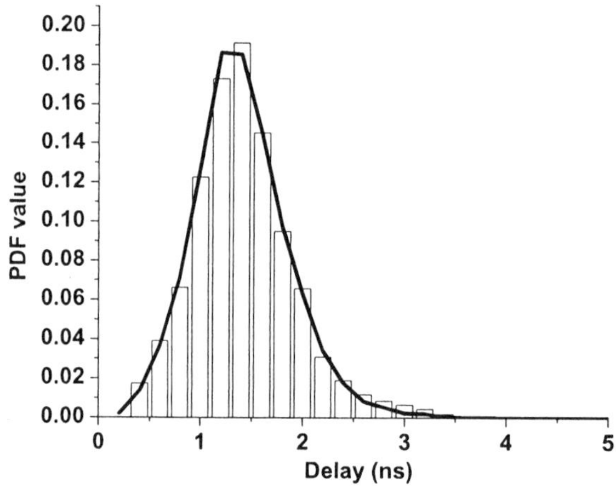

(b)

Fig. 2.22. Comparison of delay distribution obtained from Monte Carlo simulations and interval arithmetic based (a) AWE, (b) PRIMA [86].

that obtained using Monte Carlo simulation. The results are for 5% global variation, 30% local variation and 16 initial uncertainity symbols, and show good accuracy.

Perturbation Analysis

The key while considering variations in interconnects is to estimate the resulting fluctuations in the positions of the poles of the system, which impact critical circuit characteristics. To understand the impact of these [84] uses key ideas from matrix perturbation theory to assess the impact of these variations.

Let us consider PACT, which involves two congruence transformations, and define the same operations in a situation where we consider variations in process parameters. First, we assume that variations in process parameters are sufficiently small enough to allow us to express the susceptance and conductance matrix as

$$\begin{aligned} \mathbf{G} &= \mathbf{G_0} + \Delta\mathbf{G_1}w_1 + \Delta\mathbf{G_2}w_2 \\ \mathbf{C} &= \mathbf{C_0} + \Delta\mathbf{C_1}w_1 + \Delta\mathbf{C_2}w_2 \end{aligned} \tag{2.172}$$

where we have considered two sources of variations w_1 and w_2. Again, as in PACT, we partition these matrices such that

$$\begin{aligned} \mathbf{G_P} &= \mathbf{G_{P0}} + \Delta\mathbf{G_{P1}}w_1 + \Delta\mathbf{G_{P2}}w_2 \\ \mathbf{G_I} &= \mathbf{G_{I0}} + \Delta\mathbf{G_{I1}}w_1 + \Delta\mathbf{G_{I2}}w_2 \\ \mathbf{G_C} &= \mathbf{G_{C0}} + \Delta\mathbf{G_{C1}}w_1 + \Delta\mathbf{G_{C2}}w_2 \\ \mathbf{C_P} &= \mathbf{C_{P0}} + \Delta\mathbf{C_{P1}}w_1 + \Delta\mathbf{C_{P2}}w_2 \\ \mathbf{C_I} &= \mathbf{C_{I0}} + \Delta\mathbf{C_{I1}}w_1 + \Delta\mathbf{C_{I2}}w_2 \\ \mathbf{C_C} &= \mathbf{C_{C0}} + \Delta\mathbf{C_{C1}}w_1 + \Delta\mathbf{C_{C2}}w_2 \end{aligned} \tag{2.173}$$

The crucial step is determining $\mathbf{V} = \mathbf{G_I}^{-1}\mathbf{G_C}$, when $\mathbf{G_I}$ and $\mathbf{G_C}$ are influenced by variations. In [84], $\mathbf{V}$ is obtained using Taylor's expansions while retaining terms up to second order

$$\mathbf{V} \approx \mathbf{V_0} + \Delta\mathbf{V_{11}}w_1 + \Delta\mathbf{V_{21}}w_1^2 + \Delta\mathbf{V_{12}}w_2 + \Delta\mathbf{V_{22}}w_2^2 \tag{2.174}$$

where the parameters in the above equation are obtained by evaluating $\mathbf{V}$ for a set of sample points. Note that $\mathbf{V}$ can also be obtained at these points $\mathbf{V}$ from the DC solution of the circuit. Setting the susceptance matrix in (2.164) to zero, the port and internal voltages are related as:

$$\begin{aligned} \mathbf{G_C x_P} + \mathbf{G_I x_I} &= \mathbf{0} \\ \mathbf{x_I} = -\mathbf{G_I}^{-1}\mathbf{G_C x_P} &= \mathbf{V x_P}. \end{aligned} \tag{2.175}$$

Thus, the columns of $\mathbf{V}$ can be obtained by setting one of the port voltages to unity and grounding all the other nodes. Using (2.174) we can express the matrices corresponding to (2.167) as

$$\begin{aligned}\mathbf{G}'_{\mathbf{P}} &= \mathbf{G}_{\mathbf{P}} + \mathbf{G}_{\mathbf{C}}{}^T\mathbf{V} \\ &= (\mathbf{G}_{\mathbf{P0}} + \Delta\mathbf{G}_{\mathbf{P1}}w_1 + \Delta\mathbf{G}_{\mathbf{P2}}w_2) + (\mathbf{G}_{\mathbf{C0}} + \Delta\mathbf{G}_{\mathbf{C1}}w_1 + \Delta\mathbf{G}_{\mathbf{C2}}w_2)^T \\ &\quad \cdot(\mathbf{V}_0 + \Delta\mathbf{V}_{11}w_1 + \Delta\mathbf{V}_{21}w_1^2 + \Delta\mathbf{V}_{12}w_2 + \Delta\mathbf{V}_{22}w_2^2) \qquad (2.176)\end{aligned}$$

retaining only terms up to second order we obtain

$$\mathbf{G}'_{\mathbf{P}} \approx \mathbf{G}_{\mathbf{P0}} + \Delta\mathbf{G}_{\mathbf{P11}}w_1 + \Delta\mathbf{G}_{\mathbf{P21}}w_1^2 + \Delta\mathbf{G}_{\mathbf{P12}}w_2 + \Delta\mathbf{G}_{\mathbf{P22}}w_2^2 \qquad (2.177)$$

where

$$\begin{aligned}\mathbf{G}_{\mathbf{P0}} &= \mathbf{G}_{\mathbf{P0}} + \mathbf{G}_{\mathbf{C0}}{}^T\mathbf{V}_0 \\ \Delta\mathbf{G}_{\mathbf{P11}} &= \Delta\mathbf{G}_{\mathbf{P1}} + \Delta\mathbf{G}_{\mathbf{C1}}{}^T\mathbf{V}_0 + \mathbf{G}_{\mathbf{C0}}{}^T\Delta\mathbf{V}_{11} \\ \Delta\mathbf{G}_{\mathbf{P}}\mathbf{21} &= \Delta\mathbf{G}_{\mathbf{C0}}{}^T\Delta\mathbf{V}_{21} + \Delta\mathbf{G}_{\mathbf{C1}}{}^T\Delta\mathbf{V}_{11} \qquad (2.178) \\ \Delta\mathbf{G}_{\mathbf{P}}\mathbf{12} &= \Delta\mathbf{G}_{\mathbf{P2}} + \Delta\mathbf{G}_{\mathbf{C2}}{}^T\mathbf{V}_0 + \mathbf{G}_{\mathbf{C0}}{}^T\Delta\mathbf{V}_{12} \\ \Delta\mathbf{G}_{\mathbf{P}}\mathbf{22} &= \Delta\mathbf{G}_{\mathbf{C0}}{}^T\Delta\mathbf{V}_{22} + \Delta\mathbf{G}_{\mathbf{C2}}{}^T\Delta\mathbf{V}_{12}.\end{aligned}$$

Similar expressions can be derived for the partitions of $\mathbf{C}'_{\mathbf{P}}$ with many more terms than the expressions above. Now let us consider the second congruence transformation. However, before we begin the analysis we need a few results from matrix perturbation theory [55]

Theorem 2.4. [84]. *Let* $\mathbf{A}$ *be a symmetric matrix with eigenvalues* $\lambda_1 \geq \lambda_2 \geq \cdots \geq \lambda_n$, *and let* $\mathbf{P}$ *be a symmetric perturbation matrix with eigenvalues* $p_1 \geq p_2 \geq \cdots \geq p_n$. *Then* $\hat{\lambda}_i$, *the i–th largest eigenvalue of* $\mathbf{A} + \mathbf{P}$

$$\hat{\lambda}_i \in [\lambda_i + p_n,\ \lambda_i + p_1].$$

Theorem 2.5. [84]. *Let* $\mathbf{A}$ *and* $\mathbf{B}$ *be two symmetric matrices. Additionally, let* $\mathbf{B}$ *be positive-definite, and define matrices* $\mathbf{X_1}$ *and* $\mathbf{X_2}$ *such that*

$$\begin{bmatrix}\mathbf{X}_1 \\ \mathbf{X}_2\end{bmatrix}\mathbf{A}\,[\mathbf{X}_1\ \ \mathbf{X}_2] = \begin{bmatrix}\mathbf{A}_1 & 0 \\ 0 & \mathbf{A}_2\end{bmatrix}$$

$$\begin{bmatrix}\mathbf{X}_1 \\ \mathbf{X}_2\end{bmatrix}\mathbf{B}\,[\mathbf{X}_1\ \ \mathbf{X}_2] = \begin{bmatrix}\mathbf{B}_1 & 0 \\ 0 & \mathbf{B}_2\end{bmatrix}$$

where $\mathbf{A}_1$, $\mathbf{B}_1$, $\mathbf{A}_2$ *and* $\mathbf{B}_2$ *are diagonal matrices. Let* $\mathbf{P_1}$ *and* $\mathbf{P_2}$ *be symmetric matrices that represent the perturbations in* $\mathbf{A}$ *and* $\mathbf{B}$, *respectively. If the distance between eigenvalue clusters of* $(\mathbf{A}_1,\ \mathbf{B}_1)$ *and* $(\mathbf{A}_2,\ \mathbf{B}_2)$ *is large, and the matrix perturbations are small, then there exist matrices* $\mathbf{S_1}$ *and* $\mathbf{S_2}$ *such that the column space of* $[\mathbf{X}_1 + \mathbf{X}_2\mathbf{S}_1,\ \mathbf{X}_2 + \mathbf{X}_1\mathbf{S}_2]$ *spans the eigenspace of the perturbed matrices* $(\mathbf{A} + \mathbf{P}_1,\ \mathbf{B} + \mathbf{P}_2)$.

Theorem 2.4 tells us that for small variations the resulting eigenvalues are always close to the nominal eigenvalues, and Theorem 2.5 then applied to the second congruence transformation in PACT, tells us the behavior of the matrix $\mathbf{U_r}$ in (2.170) under process variation. However, direct application of Theorem 2.5 is not computationally feasible since matrices $\mathbf{S_1}$ and $\mathbf{S_2}$ are not known beforehand and require us to solve a set of generalized Sylvester equations. Since these matrices can be expected to depend on the perturbations themselves, [84] uses a Taylor's expansion for $\mathbf{\Lambda_r}$ and $\mathbf{U_r}$ (2.171) as

$$\begin{aligned} \mathbf{\Lambda_r} &\approx \mathbf{\Lambda_0} + \Delta\mathbf{\Lambda_{11}} w_1 + \Delta\mathbf{\Lambda_{21}} w_1^2 + \Delta\mathbf{\Lambda_{P12}} w_2 + \Delta\mathbf{\Lambda_{P22}} w_2^2 \\ \mathbf{U_r} &\approx \mathbf{U_0} + \Delta\mathbf{U_{11}} w_1 + \Delta\mathbf{U_{21}} w_1^2 + \Delta\mathbf{U_{P12}} w_2 + \Delta\mathbf{U_{P22}} w_2^2. \end{aligned} \tag{2.179}$$

The parameters in the above equation can again be obtained by estimating the dominant eigenvectors for a set of sample points. Finally, this allows us to express the variational reduced order model using (2.171), where the matrices are replaced with their variational counterpart as discussed in this section.

PRIMA involves calculation of a matrix $\mathbf{X}$ that can be used to perform a congruence transformation. As in PACT when variations are considered, a variational reduced order model in PRIMA can be constructed using a first-order Taylor series expansion of $\mathbf{X}$ as

$$\mathbf{X} = \mathbf{X_0} + \Delta\mathbf{X_1} w_1 + \Delta\mathbf{X_2} w_2 \tag{2.180}$$

where the parameters can be evaluated using a set of sample points.

Note that while considering variations we have performed a congruence transformation in which higher order terms associated with the varying parameters w_1 and w_2 are dropped. Thus the overall transformation does not remain a true congruence transformation and we lose passivity. However, [2] proposed that a transistor level timing simulator like TETA can be used in these situations. TETA obviates the need for passivity, since it uses a successive-chord method instead of Newton-Raphson's method to solve the system of nonlinear equations. This allows nonlinear devices to be mapped to a Norton equivalent model with constant impedance. Therefore, if the macromodeling approach is stable then the stability of the overall simulation can be guaranteed.

3

Statistical Timing Analysis

The focus of this chapter is on techniques to perform efficient timing analysis of circuit blocks while considering process variations. The result of such an analysis is invariably a probability distribution of delay. In the previous chapter, we looked at Monte Carlo techniques that can be used to estimate the distribution of circuit delay as well, and early approaches to perform statistical timing analysis were based on Monte Carlo techniques [61], [67]. However, even if only inter-die variations are considered, there already exist variations in three standard dimensions: process variations in devices, temperature and power supply variations. In addition, each metal layer contributes four other RVs corresponding to the metal line width, spacing, height and inter-layer dielectric (ILD) thickness variations. Thus, even in this highly simplified case where all intra-die variations are neglected, there are a significant number of RVs, making enumerative circuit simulations prohibitive, even as a golden model.

As process technologies have scaled, intra-die variations have grown to play a significant role in determining delay and power distributions. The RV that captures intra-die variations can be independent or spatially correlated (with correlation < 1) across gates, making Monte Carlo based techniques even more expensive. On the other hand, corner based models are rendered useless as well. Even if all corner cases are considered (which is a large number), we cannot guarantee that the worst-case is covered by these corner cases. Thus, it is imperative that the timing characteristics of a design are analyzed statistically.

The crucial difference between statistical and deterministic scenarios lies in the notion of critical paths. In a statistical sense, there is no single path in the circuit that can be identified as being *the* critical path of the circuit, or *the* path with the maximum delay. Any path in the circuit from the inputs to outputs can therefore become critical, depending on how variations manifest themselves on a particular sample of the chip. Generally, critical paths are redefined in a statistical setting as a set of paths that have a high probability (higher than a given threshold) of becoming the slowest path in the circuit.

Another frequently used definition is that a path is critical if the probability that the path delay exceeds a given deterministic critical delay for the circuit is higher than a certain threshold. Note that these two definitions are not equivalent, however both can be used to analyze and optimize circuits. The first definition is useful when the objective is to reduce the delay of a circuit, and the second definition becomes useful when we are analyzing/optimizing the circuit under delay constraints.

In this chapter, we will look at techniques to estimate the probability distribution of delay for circuit blocks. Most of the statistical timing analysis techniques fall into one of two categories - *block-based* timing analysis or *path-based timing* analysis. Block-based timing analysis is based on a topological traversal of the timing graph, whereas path-based techniques rely on extracting a set of paths from the circuit and performing timing analysis on all paths within this smaller circuit. We will also look at approaches that have been proposed to estimate timing yield using integration techniques in the parameter-space, and a Bayesian Networks based approach.

3.1 Introduction

We first discuss a simplified statistical model for the distribution of the maximum operating clock frequency (FMAX) of a chip [21], which corresponds to the distribution of maximum delay. This modeling technique provides insight on the influence of different components of variations on the distribution of FMAX.

Consider a single critical path within the circuit. If the delay of a single gate is modeled as a Gaussian RV, then the delay of a path can be modeled as a sum of Gaussian RVs. If these Gaussian RVs are part of a multinormal distribution, then the critical path delay can be expressed as a Gaussian RV as well.

Let us consider the delay of a single critical path within the circuit and also assume that the distribution of path delay under inter-die and intra-die random variations is a Gaussian RV with standard deviation σ_{inter} and σ_{intra}, respectively. However, the mean delay of a single path does not change under variations (assuming variations in delay can be captured using a linear function of process variations), and is assumed to be T_{nom}. It is important to note that the variance in path delay due to intra-die random variations depends on the number of gates in a path and reduces with increasing logic depth. This results from the fact that intra-die random variations are independent across gates, and therefore, if

$$\mu_{\text{path}} = \mu_{\text{gate}_1} + \mu_{\text{gate}_2} + \cdots + \mu_{\text{gate}_n} \tag{3.1}$$

then the variance due to intra-die variation can be expressed as

$$\sigma_{\text{intra,path}} = \sqrt{\sigma^2_{\text{intra,gate}_1} + \sigma^2_{\text{intra,gate}_2} + \cdots + \sigma^2_{\text{intra,gate}_n}} \tag{3.2}$$

whereas, the variance due to inter-die variations (which are identical across gates) can be expressed as

$$\sigma_{\text{inter,path}} = \sigma_{\text{inter,gate}_1} + \sigma_{\text{inter,gate}_2} + \cdots + \sigma_{\text{inter,gate}_n}. \tag{3.3}$$

This implies that the contribution of intra-die variations reduces with increasing depth. Assuming that all gate delays have the same variance, we can write

$$\frac{\sigma_{\text{intra,path}}}{\sigma_{\text{inter,path}}} = \frac{\sigma_{\text{intra}}}{\sqrt{n}\sigma_{\text{inter}}}. \tag{3.4}$$

Considering only intra-die random variations, the probability that the critical path meets timing can be expressed as

$$\mathcal{P}_{\text{intra,p}}(t \leq t_{\max}) = \int_0^{t_{\max}} f_{\text{intra}}(t)\text{d}t \tag{3.5}$$

where f_{intra} represents the pdf of a Gaussian RV with mean T_{nom} and variance σ^2_{intra}. Let us represent the cdf of f_{intra} as F_{intra}. Then, the probability in (3.5) can also be expressed as $F_{\text{intra}}(t_{\max})$. To estimate the timing yield of a circuit, we need to consider all critical paths in the circuit. Let us assume that the set of critical paths has N_p paths with identical mean and variance of delay. If all paths are perfectly correlated, then only a single path can be considered to capture the delay of the circuit as a whole. However, if paths are uncorrelated, we need to express the probability that the circuit meets timing as

$$\mathcal{P}_{\text{intra, circ}}(t \leq t_{\max}) = \mathcal{P}_{\text{intra,p}_1}(t \leq t_{\max}) \ldots \mathcal{P}_{\text{intra,p}_{N_p}}(t \leq t_{\max}) \tag{3.6}$$

$$= \left(\mathcal{P}_{\text{intra,p}}(t \leq t_{\max})\right)^{N_p}. \tag{3.7}$$

The pdf of circuit delay can then be computed by differentiating the cdf of delay in the above equation with respect to t, which gives

$$f_{\text{intra, circ}}(t) = \frac{\text{d}F_{\text{intra,circ}}}{\text{d}(t)} \tag{3.8}$$

$$= N_p\, f_{\text{intra}}(t)\left(F_{\text{intra}}(t)\right)^{N_p - 1}. \tag{3.9}$$

The pdfs obtained using the above expression are plotted in Fig. 3.1 for a varying number of critical paths. It can be observed that, as the number of critical paths increases, the pdfs shift to higher delay values. This implies that the probability that the design meets a timing constraint reduces. In addition, the pdfs becomes tighter as the number of critical paths increases and circuit delay becomes less sensitive to intra-die variations. Also, the circuit delay pdf sensitivity on the number of paths N_p reduces as the number of paths increase. It is important to note that though the pdfs seem to be Gaussian in Fig. 3.1,

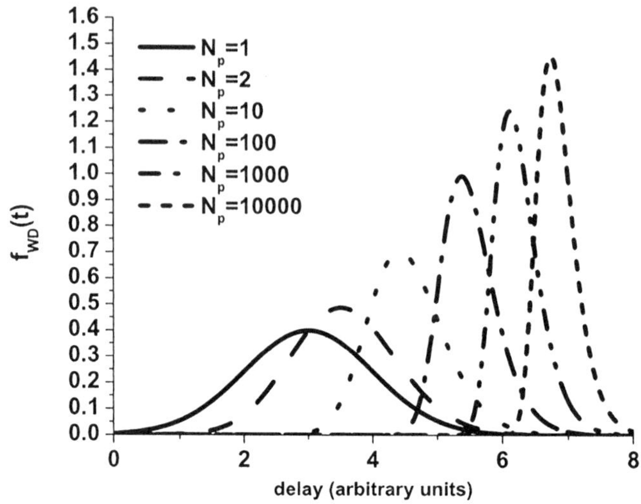

Fig. 3.1. Probability distribution function of circuit delay with varying number of uncorrelated critical paths considering intra-die variations. The mean delay increases while variance decreases with increasing number of critical paths.

the actual distribution is not Gaussian, which is obvious from (3.8). The tail towards the right of the pdfs is larger, than that to the left, for cases with many critical paths.

If we consider only inter-die variations, the delay pdfs of all paths are perfectly correlated since all paths are affected similarly. In this case, a single Gaussian RV can be used to capture the delay of a circuit. In the presence of both intra-die and inter-die variations, the circuit delay T_{circ} can be expressed as

$$T_{\text{circ}} = T_{\text{nom}} + \Delta T_{\text{intra}} + \Delta T_{\text{inter}} \tag{3.10}$$

where T_{nom} is the nominal delay of the circuit and ΔT_{intra} is the change in delay considering intra-die distribution and ΔT_{inter} is a Gaussian RV with zero mean and σ^2_{inter} variance. The distribution for total circuit delay can then be computed by convolving these three distributions:

$$f_{\text{path}}(t) = f_{\text{nom}} * f_{\text{intra}}(t) * f_{\text{inter}}(t) \tag{3.11}$$

where $*$ represents the convolution operator and f_{nom} is an impulse at $t = T_{\text{nom}}$.

Based on the analysis above and Fig. 3.1, we expect that for circuits with a large number of critical paths, intra-die variations will have a strong influence on the mean delay, while their impact on variance will be significantly reduced. Most of the variance in circuit delay will be contributed by inter-die variations.

The frequency of a circuit is inversely proportional to the delay of the circuit, therefore the RV FMAX can be expressed as $FMAX = 1/T_{\text{path}}$. The distribution function of a RV $Y = g(X)$ $(f_y(y))$ can be expressed as [109]

$$f_y(y) = \frac{f_x(x_1)}{|g'(x_1)|} + \cdots + \frac{f_x(x_n)}{|g'(x_n)|} \tag{3.12}$$

where $f_x(x)$ is the pdf of X, and $x_1, \ldots, x_n$ are the real roots of the equation $y = g(x)$ and g' represents the derivative of g. Using (3.12) we can write the distribution of FMAX as

$$f_{\text{FMAX}}(f) = \frac{1}{f^2} f_{\text{path}}\left(\frac{1}{f}\right). \tag{3.13}$$

Now let us consider circuit-specific techniques that can be used to perform statistical analysis of circuit delay.

3.2 Block-Based Timing Analysis

Block-based timing analysis techniques perform a topological traversal of the timing graph. The traversal of the timing graph is therefore exactly the same as in traditional static timing analysis (STA). The overall statistical timing analysis can then be expected to be computationally efficient if the computations required to perform timing analysis for each node in the timing graph are small. However, instead of deterministic delay values we propagate delay distributions, complicating the analysis. In addition, the delay distributions might be correlated due to spatial correlations in process parameters and reconvergent fanouts. Spatial correlations result from the fact that gates close together in the layout have similar variations in process parameters. On the other hand, correlations due to reconvergent fanouts can cause delay of far-away paths to be correlated, if they originate from a common node. Consider a gate that fans out to a set of paths and then some of these paths reconverge and fanin to a multiple input gate. These paths then have a component of delay that is identical and causes the delay at the input of the reconvergent node to become correlated.

To handle probability distributions, [81] introduced the idea of using discretized distribution functions. The approach was used to handle intra-die variations while assuming all distributions to be independent and is discussed in Sec. 3.2.1. Using the same general framework, [6] proposed an approach to consider correlations due to reconvergent fanouts and develops tight upper and lower bounds for the actual delay distribution, while [43] uses a heuristic enumerative approach to handle these correlations. These approaches are discussed in Sec. 3.2.2. However, all these approaches neglect correlations in

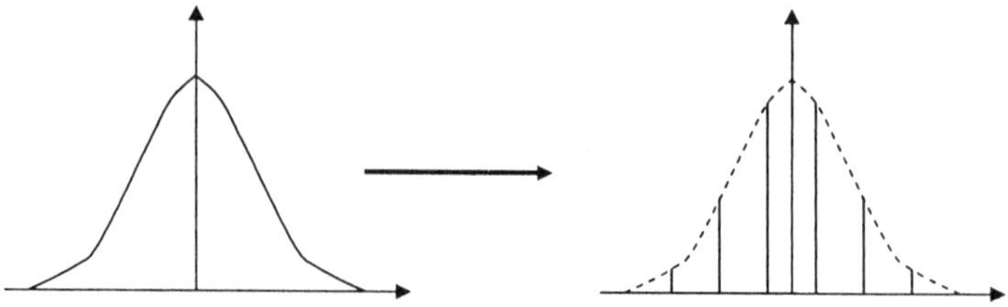

Fig. 3.2. Sampling a probability distribution of delay to generate a discrete probability distribution. The discrete probability distribution should be re-normalized so that it represents a valid pdf.

process parameters due to inter-die variations and the correlated component of intra-die variations. Canonical delay models pdfs were therefore proposed in [3], [30] and [146] to handle all components of process variation and build on the ideas of Quad-Tree and PCA-based analysis discussed in Chap. 2. These techniques are discussed in Sec. 3.2.3. Finally, in Sec. 3.2.4 we discuss an approach to handle multiple switching events while performing statistical static timing analysis (SSTA) [5].

3.2.1 Discretized Delay PDFs

The approach by [81] was one of the first techniques to perform SSTA using discretized pdfs to handle probability distributions. This technique performs SSTA in a computationally deterministic fashion, as opposed to Monte Carlo techniques which are inherently random. The gate delays are now defined as discrete delay distributions that are generated as shown in Fig. 3.2 given a sampling step (may be user specified). Note that the discrete pdfs should be renormalized so that the sum of the probabilities for the discrete events is equal to one. The sampling step provides a tradeoff in terms of runtime and accuracy. Using a small sampling step will result in good accuracy, since the discrete delay pdf will have a shape very close to the original continuous delay pdf. However, the larger number of samples in the discrete delay pdf increases the computational requirements of SSTA. Using a very large sampling step decreases accuracy, and in the case where the sampling window is larger than the width of the gate delay pdfs SSTA degenerates into a traditional STA approach. Thus, we have defined the delay of each node within the timing network using a discretized delay pdf. We will always assume that the delay pdf is non-zero only over a finite range of delay values.

The next step in SSTA is to propagate the distribution of circuit delay from the primary inputs to the primary outputs. Hence, as in STA we need to define operations that *sum* the delay distribution at the input of the gate with the

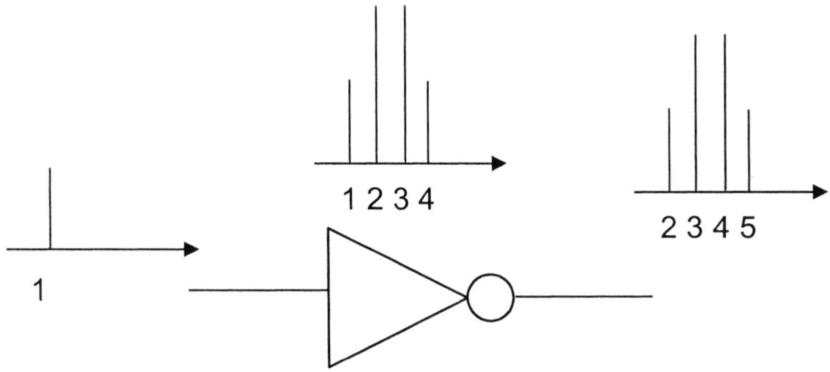

Fig. 3.3. Computing the delay pdf at the output of a gate for a degenerate input delay pdf. The numbers on the x-axis represent the delay value associated with the particular discrete probability distribution sample.

gate delay distribution or perform the *max* of delay distributions in the case of multiple input gates. In the case of a degenerate or deterministic input delay distribution, the sum operation is fairly simple and involves the computation of the output delay pdf by simply *shifting* the gate delay distribution as shown in Fig. 3.3. However, in the case when the input delay pdf is non-degenerate, we can generate a set of these shifted output delay distributions. Each of these shifted pdfs occur with a probability corresponding to the probability of the discrete event in the input delay pdf that resulted in this output delay pdf. This set of shifted pdfs can then be combined using *Bayes' Theorem* which states that

$$\mathcal{P}(B) = \sum_i \mathcal{P}(B|A = i) \cdot \mathcal{P}(A = i) \tag{3.14}$$

where $\mathcal{P}(B|A = i)$ refers to the probability of Event B, given that $A = i$. Thus, we need to generate the shifted pdfs with *scaling*, where the scaling factor is the probability of the discrete input event. These events can then be *grouped* according to the above equation by summing the probability at each of the discrete points, as shown in Fig. 3.4, where the probability at the top of each discrete event corresponds to a non-normalized probability. The actual probability of an event in the figure can be obtained by dividing the number by the sum of the numbers corresponding to all the discrete events in each discrete pdf.

The same idea can be used to analyze this case more formally using the definition of a *timing graph*.

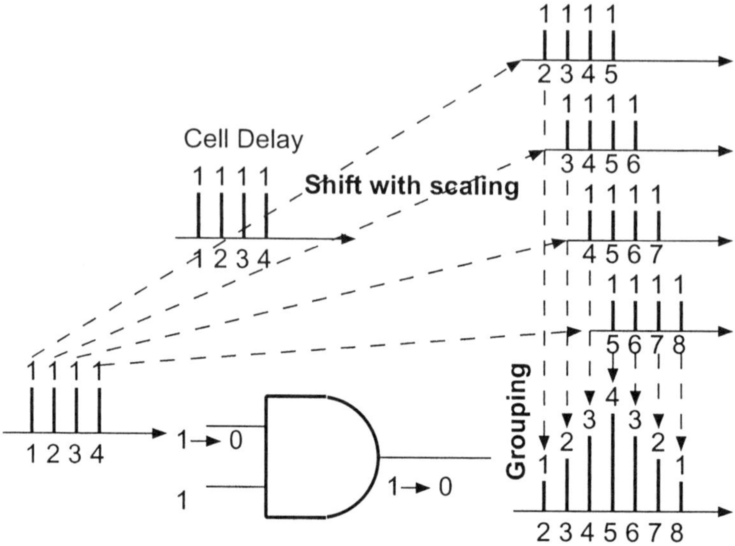

Fig. 3.4. Shift with scaling and grouping techniques to perform convolution of input and gate delay pdfs to compute the output delay pdf. (©2005 IEEE)

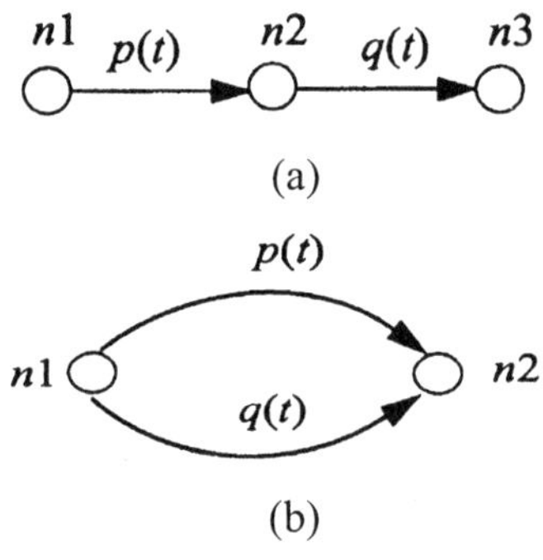

Fig. 3.5. Various combination scenarios of edges in a DAG (a) Series (b) Parallel. (©2005 IEEE)

Definition 3.1. *A timing graph is a directed acyclic graph having exactly one source and one sink:* $G = \{N, E, n_s, n_f\}$, *where* $N = \{n_1, n_2, \ldots, n_k\}$ *is a set of nodes,* $E = \{e_1, e_2, \ldots, el\}$ *is a set of edges,* $n_s \in N$ *is the source node and* $n_f \in N$ *is the sink node and each edge* $e \in E$ *is an ordered pair of nodes* $e = (n_i, n_j)$.

The edges in the timing graph correspond to connections from gate inputs to outputs and the nodes correspond to nets in a circuit. A *probabilistic timing graph* is defined as a timing graph where each edge is associated with a distribution corresponding to the delay of that edge. The source and sink nodes are imaginary nodes and are connected to the nodes representing the primary inputs and outputs of a circuit, respectively. Each of the edges connecting the source node has a delay corresponding to the arrival time at the input node, to which this edge is connected at the other end.

Consider a timing graph with a set of series arcs. These arcs can be reduced to a single arc, which has a delay pdf equal to the *sum* of the individual path delay distributions of the two arcs. Consider Fig. 3.5(a) which shows two timing arcs with node delay pdfs $p(t)$ and $q(t)$ (we will refer to the corresponding cdfs with their capital letters). These two arcs can be replaced by a single arc with a delay distribution $r(t)$ that satisfies

$$\begin{aligned} R(t) = \int_{-\infty}^{t} r(t)\mathrm{d}t &= \int_{t_1+t_2\leq t} p(t_1)q(t_2)\mathrm{d}t_1\,\mathrm{d}t_2 \\ &= \int_{-\infty}^{\infty}\int_{-\infty}^{t-t_1} p(t_1)q(t_2)\mathrm{d}t_1\,\mathrm{d}t_2 \\ &= \int_{-\infty}^{\infty} p(t_1)Q(t-t_1)\mathrm{d}t_1. \end{aligned} \tag{3.15}$$

Differentiating both sides in the above equation we get the standard result that

$$r(t) = \int_{-\infty}^{\infty} p(t_1)q(t-t_1)\mathrm{d}t_1, \tag{3.16}$$

which implies that the pdf of a sum of two RVs is expressed as a convolution of the two pdfs.

In the case of multiple fanin gates, we can generate an output delay pdf for each input node using the scaling and grouping / convolution technique described above. The next step is to calculate the *max* of these individual pdfs to estimate the final delay at the output of the last multi-input gate. In the case of Fig. 3.5(b), we can replace the two arcs with a single arc having an edge delay pdf $r(t)$. Assuming independence of $p(t)$ and $q(t)$, the probability that $r(t) < t_0$ is the probability that both $q(t)$ and $p(t)$ are less than t_0. This is mathematically expressed as

$$R(t) = P(t)Q(t) \tag{3.17}$$

differentiating with respect to t we obtain the pdf $r(t)$ as

$$r(t) = P(t)q(t) + p(t)Q(t). \tag{3.18}$$

The SSTA approach proposed in [6] propagates discretized pdfs through the circuit as explained above. However, [43] propagates piecewise linear (PWL) cdfs through the circuit, while gate delays are maintained as delay pdfs. The sum and max operations are performed using (3.15) and (3.17), respectively. Each multiplication in the convolution and max computation results in a quadratic function, generating a total of $O(n^2)$ quadratic functions, where n is the number of linear elements in the piecewise linear pdf. These quadratic functions are then re-sampled at preset probability values to generate a PWL cdf that is propagated through the circuit.

Both these approaches require $O(|E| + |V|)$ sum or max operations where E and V are the sets of arcs and nodes in the timing graph. Each sum and max operation requires $O(n^2)$ computations, where n is the number of discretization of the delay pdf. Thus the overall complexity is $O(n^2(|E| + |V|))$. Note that as delay pdfs are propagated through the circuit they become wider and the number of discretizations increase for a given sampling step. The long tails of the discretized pdfs are typically characterized by very small probabilities and significant improvement in runtime can be achieved by pruning them. The tradeoff involved in pruning and the loss in accuracy is investigated in [81]. It is shown that the runtime improves by more than 10X when the minimum probability threshold for pruning is increased from 1e-10 to 1e-5. The inaccuracy for the case where the minimum probability of pruning is 1e-5 is 0.11% and 3.24% for mean and variance, respectively (compared to Monte Carlo).

Having discussed the basics of SSTA, let us now attempt to consider correlations in the propagated delay pdfs at different points in the timing graph arising due to gates with multiple fanouts. These correlations need to be handled properly at points of reconvergence to maintain good accuracy in SSTA.

3.2.2 Reconvergent Fanouts

The approach discussed above assumes that the distribution of delay at each input of a node is independent. Even if spatial correlations are ignored, reconvergent fanouts cause delay pdfs to be correlated which must be considered to maintain accuracy of statistical timing analysis. Figure 3.6 illustrates a simple circuit with a reconvergent fanout node. If we backtrace paths from the inputs of Gate 4 to the primary inputs (PIs), we notice that Gate 1 lies in both these paths and causes the input delays of Gate 4 to be correlated. Therefore (3.17)-(3.18) can no longer be used. Also, note that if the arrival time at the node of Gate 3 connected directly to the PI is much greater than the arrival time at the other node, then the output of Gate 3 is determined by

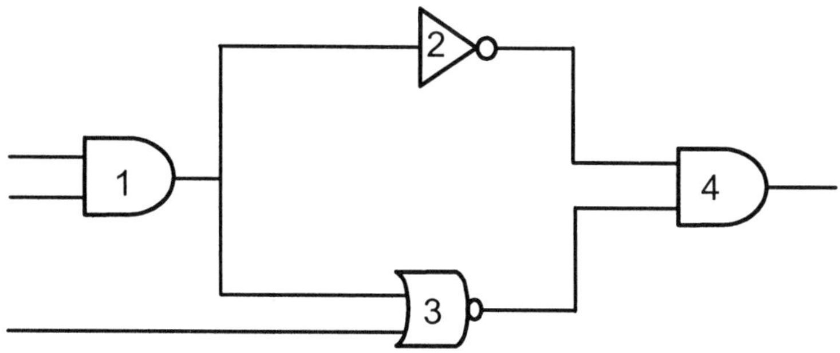

Fig. 3.6. Multiple fanouts originating from Gate 1 reconverge at Gate 4 and result in correlation between the inputs of Gate 4.

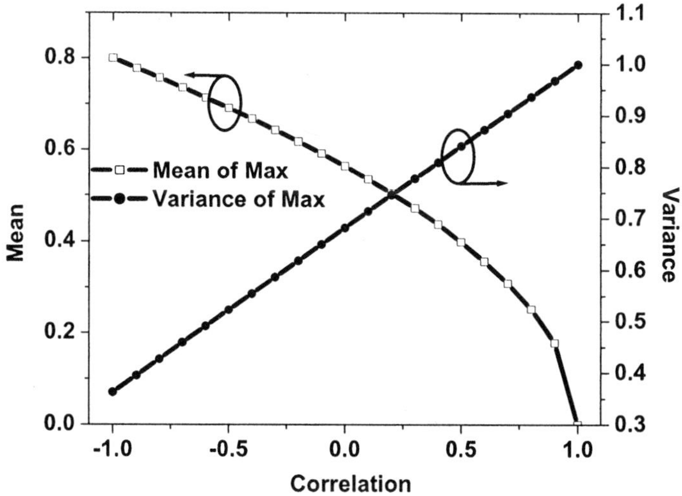

Fig. 3.7. Mean and variance of the max of two identical Gaussian RVs with zero mean and unit variance for varying correlation coefficients.

the PI and the inputs of Gate 4 become independent (assuming arrival times at PIs are independent). This masking of correlation at reconvergence further complicates the analysis.

The correlations in input delay pdf have a strong impact on the mean and variance of delay at the output of a gate. Figure 3.7 shows the mean and variance of the max of two identical Gaussian RVs with zero mean and unit variance, as their correlation is varied. If the RVs are perfectly correlated then the max of the two RVs is essentially equal to one of the RVs and the max has zero mean and unit variance. As the correlation reduces, the distribution of the max tightens towards the right of the original distribution of the RVs, and the mean of the max progressively increases while variance decreases.

An exact approach to handle reconvergent fanouts was proposed in [6], which is based on generating multiple copies of the timing graph for each dependence node (to be defined later) and for each discretization of the delay pdf, and has a worst-case runtime complexity that increases exponentially with circuit size. Hence, we will also consider two computationally feasible approaches that consider reconvergent fanouts. The approach presented in [6] develops exact lower and upper bounds on the pdf of delay while [43] proposes a heuristic approach to handle reconvergent fanout nodes.

Exact Approach

To consider the impact of correlation at the inputs of a reconvergent fanout node [6] defines the notion of a *dependence set*, which is based on the intersection of *fanin subgraphs*. A fanin subgraph is defined as:

Definition 3.2. *A* fanin subgraph $G_{s,n}$ *of a timing graph* G_p *at node* n *is a timing graph consisting of all edges and nodes of* G_p *that lie on a path from the source node* n_s *of* G_p *to node* n*, and where node* n *is set as the sink node* n_f *of* $G_{s,n}$*.*

Definition 3.3. *Consider a pair of fanin nodes* $n_{p,1}$ *and* $n_{p,2}$ *of node* n*, with fanin subgraphs* $G_{s\,1}$ *and* $G_{s,2}$*. The intersection graph* G_I *consists of edges and nodes shared by* $G_{s\,1}$ *and* $G_{s,2}$*, excluding the source node* n_s*. The set of dependence nodes for the fanin node pair* $n_{p,1}$ *and* $n_{p,2}$ *is the set of nodes* $\{n_1, n_2, \ldots, n_d, \ldots\}$*, such that* n_d *lies on the intersection graph* G_I*, and such that* n_d *has one or more fanout edges that lie on either* $G_{s\,1}$ *or* $G_{s,2}$*, but not both. The set of* dependence nodes *for node* n *is the union of the dependence sets over all possible pairs of its fanin nodes.*

Definition 3.4. *A node in a timing graph* G_p *with a non-empty dependence set is defined to be a* reconvergent node*. The union of the dependence set of all reconvergent nodes in a timing graph is the* dependence set *of the timing graph.*

Consider Fig. 3.8(a) and note that the timing graph has reconvergence at Nodes d and f. Nodes a, e, and h are not reconvergent nodes since the only node in the intersection of their fanin subgraphs is the source node, which does not contribute to delay. Consider Node f and note that the set of dependence nodes f is $\{b, d\}$, since these nodes lie in the fanin subgraphs of both the fanins and have an edge that does not lie in both the fanin subgraph (Node a does not lie in the dependence set since all its fanout edges lie in both the

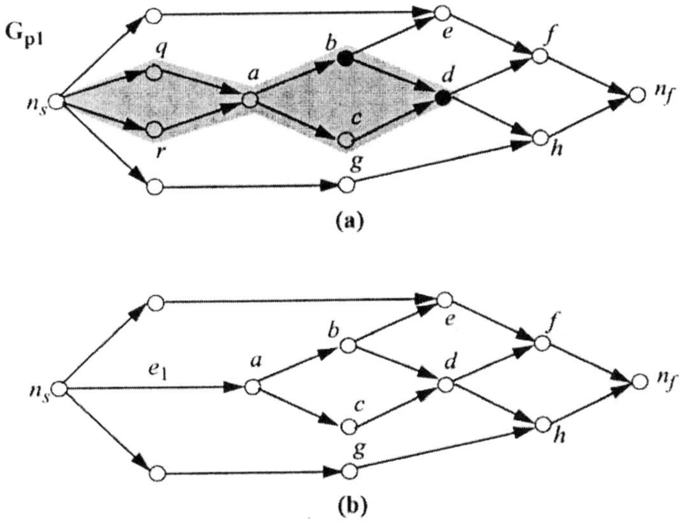

Fig. 3.8. DAG for a circuit with nodes {a,b,d} forming the set of dependence nodes (a) Shaded intersection graph for the fanin subgraphs of Node n_f (b) subgraph from n_s to a can be replaced by a single edge e_1. (©2005 IEEE)

fanin subgraphs). Similarly, the dependence set of d can be found to be $\{a\}$. Thus, the dependence set for this timing graph is $\{a, b, d\}$, which represent the nodes that have multiple fanouts and lie on the edge of the intersection of fanin subgraphs and result in correlation at some later reconvergent node in the timing graph.

To perform timing analysis on this timing graph, the set of nodes in the dependence set is first topologically sorted. The subgraph from the source node to the first node in the sorted dependence set can be replaced by a single edge using the series and parallel reduction techniques described above. This step can be performed since none of the nodes in this subgraph have fanouts that result in correlated inputs at reconvergence. At each node in the dependence set, a set of timing graphs is generated corresponding to each discrete sample of the pdf at this node. This single discrete probability event is then propagated through the timing graph, with more timing graphs being generated at each dependence node encountered downstream in the DAG. Each of the timing graphs generated is associated with a probability of occurrence that corresponds to the product of the probabilities of the discrete events from which the timing graph originated. The final arrival time pdf at a node can then be obtained by performing a weighted sum of the pdfs at that node from all the timing graphs generated. The weighting factor is the probability associated with the particular timing graph. Note that this procedure follows from Bayes' Theorem, which states that

$$p_x(t) = \sum_{i=0}^{k} p_i \, p_{x,\,i}(t) \tag{3.19}$$

where $p_{x,\,i}(t)$ is the timing pdf at node x for a timing graph generated with probability p_i. Since each timing graph is generated from a series of discrete events, the product of these discrete events corresponds to p_i in the above equation. It has also been shown in [6] that the set of dependence nodes at which the timing graphs are generated, through enumeration of the discrete probabilistic events, is both sufficient and necessary for the computation of the exact pdf of delay.

Statistical Bounds

Let us now discuss a computationally efficient approach to compute lower and upper bounds on the cdf of delay. We first define the stochastic upper bound of a cdf.

Definition 3.5. *Consider a cdf P(t). A cdf Q(t) is said to be a stochastic upper bound of P(t) if*

$$Q(t) \leq P(t), \quad \forall t. \tag{3.20}$$

The upper bound of a cdf is illustrated in Fig. 3.9. It can be observed that for a given probability (which corresponds to a timing yield) the upper bound always predicts a larger delay, and is therefore a conservative bound. Similarly, we can define a lower bounding cdf that gives a smaller delay for a given probability or a higher probability for a fixed delay. The upper bound on the latest arrival time is important for critical path analysis since we are interested in the worst possible delay for the circuit and an overestimate is preferable. Similarly, lower bounds for earliest arrival times are preferable for fast path analysis to identify potential hold time violations. We will discuss the approach to generate upper and lower bounds for late arrival times, and the approach can be easily extended to the case of early arrival times.

The authors in [6] prove that if all correlations arising due to reconvergent fanouts are neglected, then the resulting delay cdf is an upper bound on the exact delay cdf. This simplifies the analysis as well, since all timing pdfs can be propagated through the circuit while assuming independence, as discussed above. Now, let us prove the theorem upon which this result is based.

Theorem 3.6. *Let x, y and z be independent RVs and assume that their pdfs are non-zero for a finite range of delay values. Let x_1 and x_2 be independent RVs that are distributed identically to x, then the cdf of* $\max(x_1 + y,\, x_2 + z)$ *is an upper bound for the cdf of* $\max(x + y,\, x + z)$.

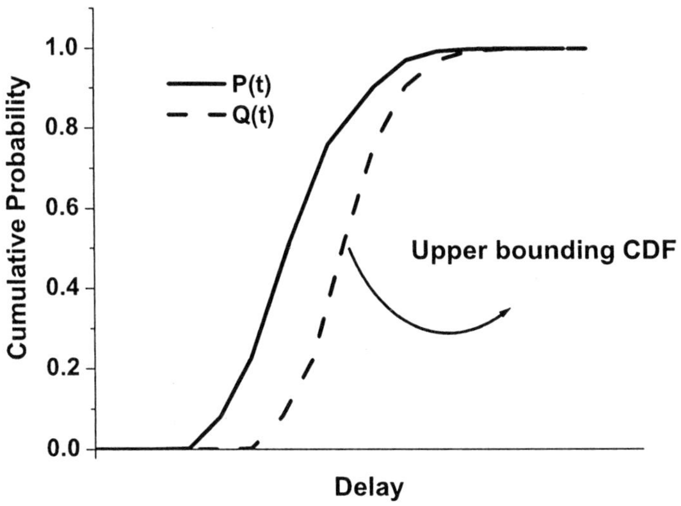

Fig. 3.9. The upper bound of a delay cdf provides a conservative estimate of circuit delay for a given timing yield.

Proof. The cdf of P can be expressed as

$$\begin{aligned} P(t) &= \int_{x+\max(y,z)\le t} p(x)q(y)r(z)\,\mathrm{d}x\,\mathrm{d}y\,\mathrm{d}z \\ &= \int_0^\infty \int_0^\infty q(y)r(z) \int_{x\le\min(t-y,\,t-z} p(x)\,\mathrm{d}x\,\mathrm{d}y\,\mathrm{d}z. \end{aligned} \tag{3.21}$$

Similarly, the cdf of Q can be expressed as

$$\begin{aligned} Q(t) &= \int_{x_1+y\le t,\,x_2+z\le t} p_1(x_1)p_2(x_2)q(y)r(z)\,\mathrm{d}x_1\,\mathrm{d}x_2\,\mathrm{d}y\,\mathrm{d}z \\ &= \int_0^\infty \int_0^\infty q(y)r(z) \int_{x_1\le t-y} p(x_1)\mathrm{d}x_1 \int_{x\le(t-z)/a} p(x)\,\mathrm{d}x\,\mathrm{d}y\,\mathrm{d}z. \end{aligned} \tag{3.22}$$

Let us consider the case when $t-y < t-z$, for which (3.21) can be simplified as

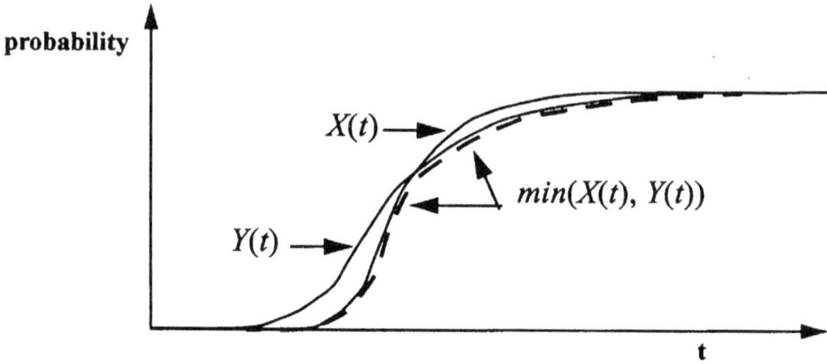

Fig. 3.10. Lower bound computation for two dependent arrival times. (©2005 IEEE)

$$P(t) = \int_0^\infty \int_0^\infty q(y) r(z) \left(\int_{x \le t-y} p(x) \mathrm{d}x \right) \mathrm{d}y \, \mathrm{d}z. \tag{3.23}$$

Comparing (3.23) and (3.22), we note that the integrand for Q has an additional term compared to the integral for P. The additional term represents the probability of a RV being less than a given value and is always less than 1, therefore the integrand for Q is always less than the integrand for P, which implies that $Q(t) \le P(t)$.

The other case when $t - y < (t - z)/a$ can also be similarly analyzed to obtain the same condition. This implies that $Q(t)$ defines an upper bound for $P(t)$. A detailed version of the proof can be found in [6].

This proves the result that neglecting correlations results in an upper bound of the exact delay pdf. Now let us develop a lower bound on the delay pdf. The lower bound on the cdf for $z = \max(x, y)$ can be obtained from the relation that

$$\begin{aligned} P(Z(t) \le t) = P(\max(x(t), y(t)) \le t) &\le P(x(t) \le t) \\ P(Z(t) \le t) = P(\max(x(t), y(t)) \le t) &\le P(y(t) \le t), \end{aligned} \tag{3.24}$$

which implies that $Z'(t) = \min(X(t), Y(t))$, as illustrated in Fig. 3.10, is a lower bound for the cdf of z. The computation of the lower bound can now be performed by selecting the minimum of the values of the cdf of x and y for all discrete time points. The number of computational steps required for the computation of both the lower and upper bounds increases linearly with circuit size and is computationally feasible. If the lower and upper bounds are close, then these bounds provide a good approximation to the actual delay pdf. However, if these bounds are very different, then the technique of *selective*

Table 3.1. Comparison of Monte Carlo and upper and lower bounds of mean delay for the ISCAS'85 benchmark circuits. The difference in the bounds is observed to be small [6].

Circuit	Monte-Carlo	Lower Bound	Upper Bound	Difference (%)
c17	1.399	1.369	1.428	4.2
c432	7.740	7.448	8.060	7.6
c499	5.168	4.730	5.282	10.5
c880	9.253	9.057	9.448	4.1
c1355	10.232	9.444	10.444	9.6
c1908	14.540	14.250	14.782	3.6
c2670	12.829	12.469	13.112	4.9
c3540	16.995	16.651	17.351	4.0
c5315	17.381	17.251	17.649	2.3
c6288	46.911	45.242	48.591	6.9
c7552	15.851	15.558	16.081	3.3

enumeration proposed in [6] can be used. Selective enumeration is based on the exact approach discussed above, and selects a small subset of dependence nodes at which the timing graphs are enumerated.

Table 3.1 compares the mean delay obtained using the upper and lower bounds and those obtained using Monte Carlo simulations for the ISCAS'85 benchmark circuits [23]. The table shows that the Monte Carlo results always lie between the lower and upper bounds, verifying the concept of computing lower and upper bounds. Moreover, the difference between the lower and upper bounds is typically small and the bounds can be used as a close approximation to the exact delay cdf.

Dependency Lists

The approach proposed in [43] is based on maintaining *dependency lists* for all nodes in the timing graph. The dependency list for a node n corresponds to the nodes in the DAG on which the arrival time at the inputs of node n depends. In the worst-case, this list can grow to be as large as the size of the circuit itself. Hence, [43] proposes a heuristic approach to limit the size of these lists, and performs timing analysis while considering reconvergence arising only from nodes within the dependency lists.

Consider Fig. 3.11 and note that the delay at the inputs of the gate can be written as

$$D_i = A_r + D_1$$

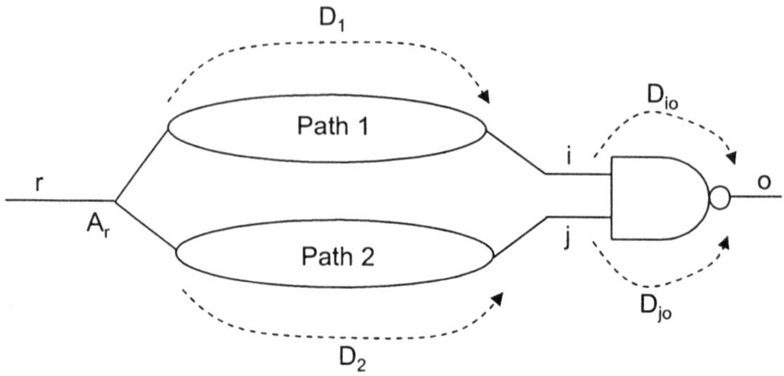

Fig. 3.11. Reconvergence can be handled by estimating the maximum of the sum of path delay and adding the delay up to the node that acts as the dependence node.

$$D_j = A_r + D_2 \tag{3.25}$$

where A_r is the cdf of the arrival time at node r and D_1 and D_2 correspond to the delay pdf of paths 1 and 2, respectively. Note that the sum operation in the above equation actually represents a convolution. Now, the delay at the output of the gate can be expressed as

$$A_o = \max(A_r + D_1 + D_{io}, A_r + D_2 + D_{jo}). \tag{3.26}$$

The presence of A_r causes the two delays, whose max needs to be computed in the above expression, to be correlated. Therefore a simple multiplication of their corresponding cdfs will not provide the correct result. However, we can rewrite (3.26) as

$$A_o = A_r + \max(D_1 + D_{io}, D_2 + D_{jo}). \tag{3.27}$$

Now, the max operation can be performed as previously discussed and the resultant delay pdf can be convolved with the arrival time cdf at node r to estimate the delay cdf at node o. To perform the above computation we need to establish the multiple-fanout node in the circuit that causes the input delay pdfs to be correlated, and moreover we need to calculate the delay pdf of the paths from node r to the inputs of the gate. These path delay pdfs can be computed by traversing the path from one end to the other. Another approach that is computationally efficient is to note that

$$\begin{aligned} D_1 &= A_i - A_r \\ D_2 &= A_j - A_r. \end{aligned} \tag{3.28}$$

The statistical subtraction in the above equation is performed using moment matching in [43]. The computation of A_i is performed as a convolution of A_r and D_1 (which are independent), therefore we can write

$$\mu[A_i] = \mu[A_r] + \mu[D_1]$$
$$\sigma^2[A_i] = \sigma^2[A_r] + \sigma^2[D_1]. \quad (3.29)$$

Using (3.29) we can write

$$\mu[D_1] = \mu[A_i] - \mu[A_r]$$
$$\sigma^2[D_1] = \sigma^2[A_i] - \sigma^2[A_r]. \quad (3.30)$$

Note the negative sign in the above expression for the variance of D_1. If these expressions are developed directly from (3.28), then we need to consider the correlation between A_j and A_r. The approach in [43] uses the first two moments of path delays to fit a Gaussian pdf to the distribution of path delay. However, the approach is general and can be extended to perform higher order moment matching using Pade approximation techniques (discussed in Chap. 2) to determine the distribution of path delay. Although path delay computation can be performed efficiently as discussed above, identifying the multiple-fanout nodes for all inputs in a multiple-input gate is not straightforward. This is the case because an input of a gate typically depends on more than one previous node that may have correlations to other inputs due to the sharing of sub-paths.

In [43], the authors propose to tackle this problem by maintaining dependency lists for each node in the timing graph. The dependency list is ideally a list of all nodes on which the arrival time at that node depends. However, to limit the size of these lists the size of the list at a node is maintained below a user-specified limit. To ensure that nodes that are important to capture the correlation (due to reconvergence) are not removed while truncating the list, the lists are stored in a levelized fashion with nodes having the highest level appearing first. Thus, when lists are truncated nodes with the lowest levels, or that are far away from the current node, are removed. While performing a levelized traversal of the timing graph, at each node we look at the dependency list of all the inputs of the node, which are then inserted into the dependency list of the current node using insertion sort. In addition, while generating the dependency list of a multiple-input gate, the dependency list associated with nodes that have a much smaller arrival time compared to other inputs are not included.

The pseudo-code for arrival time computation at a node o based on the arrival time at its inputs is shown below as *depMax*. To propagate the arrival time in a multi input gate to the output, the first step is to identify the set of nodes that occur in two or more dependency lists of the inputs. If there is no common node then there are no dependency nodes for the current node, and the analysis proceeds as discussed previously. However, if there are reconvergent nodes then the analysis is performed by computing the max of the path delays (A_{ov}) from the multiple fanout node (v) to the node under consideration (o). This max path delay pdf is then convolved with the arrival time delay pdf at node v to compute the arrival time pdf at output node v. This procedure is then repeated over all such nodes v to compute the arrival time delay pdf at node o.

```
depMax(o)
  A_o = -∞
  L = NULL
  for each input i of o
    for each vertex v in DL_i
      if (v occurs in DL of other inputs)
        insert (v, L, level(v))
        make a list of inputs in whose DL's v occurs
  if (L is NULL)
    inputs are independent: Proceed as described previously
  else
    for each v in L
      A_ov = -∞
      for each input i such that v ∈ DL_i
        A_ov = max(A_ov, A_i - A_v + D_io)
      A_ov = A_v + A_ov
      A_o = max(A_o, A_ov)
  return A_o
```

3.2.3 Canonical Delay PDFs

The previous section showed that discretized pdfs can be used to handle intra-die variations and correlations due to reconvergent fanouts. However, the approach becomes cumbersome when it is used to handle correlated intra-die and inter-die variations. A number of SSTA approaches based on canonical delay models have been proposed that allow efficient handling of the correlated component of variation. The approaches in [30][146][77][3] are based on the assumption that the delay at all nodes in the circuit can be expressed in a canonical form. In addition, [30][146][77] make the assumption that the canonical delay model has a Gaussian form.

The Gaussian approximation for delay is based on the assumption that variations in process parameters are typically small and their impact on gate/circuit delay is linear. The Gaussian approximation introduces inaccuracies due to two reasons: 1) In addition to the statistical *sum*, we also need to perform the statistical *max* of node delay pdfs and the max of two Gaussian RVs is not an exact Gaussian RV. 2) Process variations are expected to grow in future technologies, making the assumption of linearity between gate delay variations and process variations less accurate. However, in practice this assumption does not lead to large errors in current technologies and we

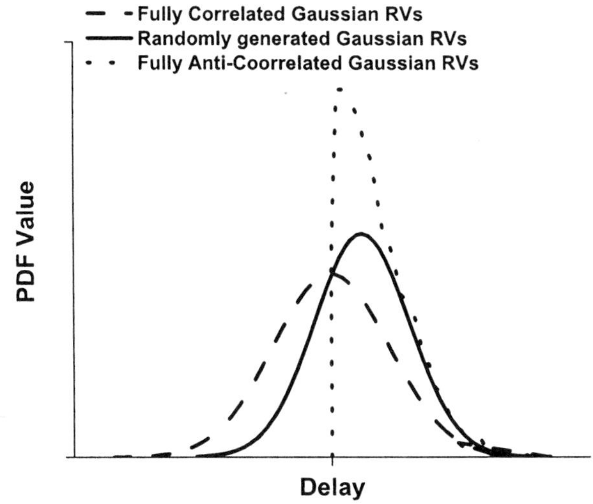

Fig. 3.12. The pdf of the max of two identical Gaussian RVs for varying correlation coefficients. The error introduced through the Gaussian delay assumption increases as the correlation in gate delay reduces from +1 to -1.

make this assumption for most of our analysis while discussing SSTA using canonical delay models.

The error introduced due to the Gaussian approximation depends very strongly on various characteristics of the pdfs whose max is being computed. If one of the pdfs is defined only for values that are much lower than the values for which the other pdf is defined, then its values are dominated by the larger pdf and the maximum has a Gaussian shape, leading to small errors. In the case where the pdfs are defined for comparable values, then the error depends on their variance and correlation. Figure 3.12 plots the maximum of two Gaussian RVs with identical mean and variance as their correlation coefficient is varied. As can be clearly observed, with decreasing correlation coefficient the distribution progressively tightens and loses its Gaussian nature. In the extreme case where the two pdfs are negatively correlated the pdf has a shape that resembles a tight Gaussian distribution with its left half removed.

Before we discuss canonical delay models, we examine some of the work done in SSTA using continuous delay pdfs which will help us in performing SSTA using canonical delay expressions. One of the first works to perform SSTA using the Gaussian delay assumption for each node [17], was based on expressions for the mean and variance of a max of two Gaussian RVs. These parameters, along with the assumption of normality for node delay pdfs, were used to propagate gate delay pdfs through the circuit. Note that the sum of two Gaussian RVs is Gaussian and can be easily handled within such an analysis. This approach was extended in [141] to consider arbitrary correlations in gate delay pdfs, arising due to reconvergent fanouts and correlation between the delay of different gates within timing graphs. However, when correlations are considered, the complexity of the algorithm increases from $O(|V| + |E|)$ to $O(|V||E|)$, where $|V|$ and $|E|$ are the number of nodes and edges in the graph.

The approach maps a combinational logic block to a DAG, which has a vertex for each pin in the circuit and an edge for each net in the circuit or timing arc of a gate. The edge delays represent the delay of the timing arcs and the node delay pdf represent the distribution of the delay from the primary inputs to that node. Each edge is associated with a rising and falling delay, which are normally distributed based on the distribution of gate length. In addition, the delay of any two edges that correspond to the same gate are also assumed to have a known correlation coefficient. Since, the rising and falling delay of a gate are associated with different types of transistors (NMOS and PMOS), they are assumed to be independent. This assumption does not hold true for interconnect delay where the rising and falling delays will be correlated depending on the variation in physical dimensions of the wire and its environment. However, we will assume that the rising and falling delays are independent for all edges in the graph. The delays are then propagated through the graph to estimate the delay distribution from the primary inputs to any node within the graph. Depending on the type of timing arc (inverting or non-inverting), the rising delay at the input defines a rising or falling delay at the output of the node.

Consider Fig. 3.5(a) and assume that the edge delays $p(t)$ and $q(t)$ are Gaussian RVs. As in Sec. 3.2.2 the two series edges p and q in the graph can be replaced by a single edge r such that,

$$E[r] = E[p] + E[q] \tag{3.31}$$

$$Var[r] = Var[p] + Var[q]. \tag{3.32}$$

Note that in the above equations we used the assumption that the delays of two edges across gates are not correlated. Using this technique we reduce the initial graph to a graph such that all series edges are replaced by a single edge. However, when performing such a reduction we need to maintain the correlation between the delay of some edge x and r, based on the correlation between x and p (ρ_{xp}) and between x and y (ρ_{xy}). The correlation coefficient of the delay distributions u and v for a pair of edges is defined as

$$\rho_{uv} = \frac{E[uv] - E[u]E[v]}{\sigma_u \sigma_v} \tag{3.33}$$

where σ_i^2 represents the variance of distribution i. Now, let us consider the correlation of the reduced edge r with an arbitrary edge x,

$$\rho_{xu} = \frac{E[x(p+q)] - E[x]E[p+q]}{\sigma_x \sigma_r}. \tag{3.34}$$

Using (3.31) we can simplify (3.34) as

$$\begin{aligned}
\rho_{xu} &= \frac{(E[xp] - E[x]E[p]) + (E[xq] - E[x]E[q])}{\sigma_x \sigma_r} \\
&= \frac{\rho_{xp}\sigma_x\sigma_p + \rho_{xq}\sigma_x\sigma_q}{\sigma_x\sqrt{\sigma_p^2 + \sigma_q^2}} \\
&= \frac{\rho_{xp}\sigma_p + \rho_{xq}\sigma_q}{\sqrt{\sigma_p^2 + \sigma_q^2}}
\end{aligned} \tag{3.35}$$

which expresses the correlation of edge delay x in terms of its correlation with the edges removed (p and q) and their variances. Thus, using this approach we can reduce a series of edges to a single edge while maintaining correlation with other edges in the graph.

In the case of parallel edges the situation becomes much more complex since we need to estimate the maximum of delay distributions, which is known to have a non-Gaussian distribution. In addition, we need to handle the correlation in node delays to account for reconvergent fanouts. The standard assumption is that the max of two Gaussians has a Gaussian shape. Then, based on the estimated mean and variance of the max of Gaussian RVs, the complete distribution of the output node is defined. However, note that as in the case of series edges we cannot simplify the graph by merging parallel edges since the correlation of the delay distribution for the fanin nodes is unknown. Consider two edges (p, r) and (q, r) and assume that the delay from the primary inputs to node p and q are known and have a correlation of ρ_{pq}. The edge delays to node r from p and q can be combined with the node delay (exactly as the case for series devices) to obtain two delay distribution x and y with a correlation ρ. Now, if $z = \max(x, y)$, then the mean and variance of z can be expressed using expressions developed by Clark [35]:

$$E[z] = \mu_x \Phi(\beta) + \mu_y \Phi(-\beta) + \alpha\varphi(\beta) \tag{3.36}$$

$$\begin{aligned}
Var[z] &= \left(\mu_x^2 + \sigma_x^2\right)\Phi(\beta) + \left(\mu_y^2 + \sigma_y^2\right)\Phi(-\beta) \\
&\quad + \left(\mu_x + \mu_y\right)\alpha\varphi(\beta) - E^2[z]
\end{aligned} \tag{3.37}$$

where

$$\alpha = \sqrt{\sigma_x^2 + \sigma_y^2 - 2\rho\sigma_x\sigma_y}$$

$$\beta = \frac{\mu_x - \mu_y}{\alpha}$$

$$\begin{aligned}\varphi(x) &= \frac{1}{\sqrt{2\pi}} \exp\left(\frac{-x^2}{2}\right) \\ \Phi(x) &= \frac{1}{\sqrt{2\pi}} \int_{-\infty}^{x} \exp\left(\frac{-y^2}{2}\right) \mathrm{d}y.\end{aligned} \tag{3.38}$$

If z is assumed to be Gaussian then the above equations completely define the distribution of delay at the output node. In addition, the correlation between the delay at an arbitrary node t and z (ρ_{tz}) can also be estimated given the correlation between node delays at t and x (ρ_{tx}) and between t and y (ρ_{ty}) as

$$\rho_{tz} = \frac{\sigma_x \rho_{tx} \Phi(\beta) + \sigma_y \rho_{ty} \Phi(-\beta)}{\sqrt{Var[z]}}. \tag{3.39}$$

If a node has more than two inputs, then the approach can be used recursively

$$z = \max(in_1, \max(in_2, \cdots)) \tag{3.40}$$

to calculate the final delay distribution at the output.

Let us now outline the timing algorithm using the ideas discussed. To perform SSTA, [141] first simplifies the graph using the series edge reduction technique. Next, let us define a *front*, which is the set of nodes in the circuit for which the distribution of delay from the primary inputs is known. In addition, the correlation in delay between any pair of nodes on the front is also known. This set is initialized to be the set of primary inputs of a combinational logic block at the start of the algorithm. In each step of the algorithm a node is selected such that all the nodes which are its immediate predecessors lie in the front. Now, timing analysis is performed for this node using the max operation described above. The correlation of the output node with any node that lies in the front and has a fanout node that goes to any other node not in the front is also calculated. The node is then added to the front, and the procedure is continued as long as the set of nodes in the front is not the same as the set of primary outputs.

Consider Fig. 3.13, which represents a graph that has been obtained by reducing all series edges to a single edge. At the start of the algorithm nodes 1, 2 and 3 define the front. At this point, only node 4 satisfies the condition that all its immediate predecessors lie in the front set. The delay at node 4 is then computed by adding the node delay of 2 to the edge delay (2,4), and the correlation of the delay at node 4 with nodes 1 and 3 is calculated. Note that we do not need to calculate the correlation of node 4 with node 2 since node 2 does not fanout to any other node in the graph that does not lie in the front. This implies that at any later stage of the algorithm node 2 is not going to act as a fanin edge and we will not need its correlation with any other node. The next step is to add node 4 to the front and calculate the delay pdf of nodes 5 and 6 using the max operation as defined above. Similarly the delay at node 7 is calculated, which defines the delay distribution of the circuit.

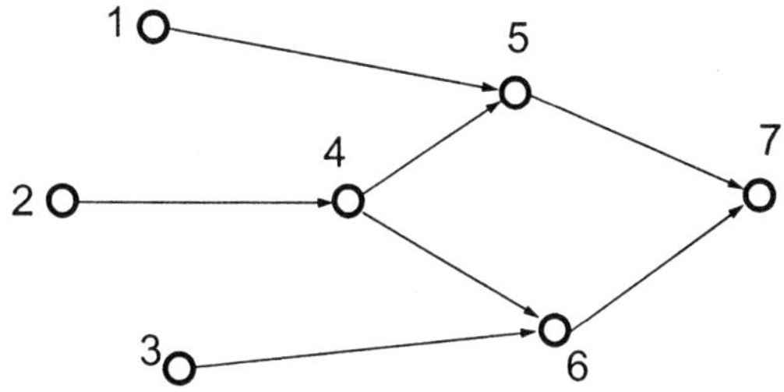

Fig. 3.13. A front-based technique to consider correlations in path delays due to reconvergent fanouts.

At the end of the algorithm we have a set of delay distributions with their respective correlations that can be used to calculate the timing yield (the probability that a sample of the design meets the timing constraint) for a given cycle time T as

$$\mathcal{P}(Delay < D) = \int_{-\infty}^{D} \int_{-\infty}^{D} \cdots \int_{-\infty}^{D} f(x_1, x_2, \ldots, x_n) \mathrm{d}x_1 \, \mathrm{d}x_2 \cdots \, \mathrm{d}x_n \quad (3.41)$$

where f represents the joint multinormal distribution of delay at the output nodes. The above expression also defines a sample of the cdf of delay and can be numerically computed at a set of points to define the complete cdf, which can then be differentiated to find the pdf of the maximum delay of the circuit.

Hence, we see that considering correlation within a gate alone results in an increase in complexity. However, as discussed in Chap. 1 the gate length of transistors within a gate are generally very strongly correlated and the correlation drops off rapidly as the distance between gates increase. Therefore, a better approach to capture the influence of correlated variations is to assume that transistors within a gate are perfectly correlated and use a distance-based map to define the correlation structure across gates. This kind of delay modeling is extremely cumbersome using the approach we have at hand. Now, let us discuss canonical delay model based timing analysis techniques that allow for efficient delay computation when considering spatially correlated process variations.

Tightness Probability

The concept of *tightness probability* was proposed in [146] and models delay as a function of n global variations and a random component as

$$d = d_{\text{nom}} + \sum_{i=1}^{n} \alpha_i \Delta X_i + \alpha_{n+1} \Delta R \tag{3.42}$$

where d_{nom} is the nominal delay, ΔX_i represents the fluctuation in the *ith* global parameter ($\rho = 1$ across gates) X_i, R represents the random variation ($\rho = 0$ across gates) and the coefficients $\alpha's$ represent the sensitivity of delay to the corresponding parameter. The above equation represents a canonical model of delay and provides Gaussian distributions for edge delays if the global parameter variations X_i and random variation R, are Gaussian. Without loss of generality we can assume these RVs to have zero mean and unit variance.

These Gaussian delay pdfs in canonical form are then propagated through the circuit to estimate the node delay pdf at each of the nodes of the graph while maintaining the node delay in the same form as (3.42). Let us consider an edge (u, v) with an edge delay pdf of

$$d_e = d_{\text{nom, e}} + \sum_{i=1}^{n} \alpha_{i,\,e} \Delta X_i + \alpha_{n+1,\,e} \Delta R \tag{3.43}$$

and let the node delay pdf at node u be

$$d_u = d_{\text{nom, u}} + \sum_{i=1}^{n} \alpha_{i,\,u} \Delta X_i + \alpha_{n+1,\,u} \Delta R \tag{3.44}$$

then the delay pdf at node v can be simply obtained by summing the two pdfs, which consists of arithmetically adding the coefficients that correspond to the same process parameters. Thus, the delay pdf at v is

$$d_v = d_{\text{nom, u}} + d_{\text{nom, e}} + \sum_{i=1}^{n} (\alpha_{i,\,u} + \alpha_{i,\,e}) \Delta X_i + \sqrt{\alpha_{n+1,\,u}^2 + \alpha_{n+1,\,e}^2}\, \Delta R \tag{3.45}$$

where we have assumed that the random component of delay is independent across gates. Consider two edges $e1$ and $e2$ with canonical pdfs and assume that the delay from the primary inputs to their source node are p and q, respectively, and are known in canonical form. The edge delays to the output node from p and q can be combined with the node delay (exactly as in the case for series nodes) to obtain two delay distributions x and y. Let x and y have the form

$$d_x = d_{\text{nom, x}} + \sum_{i=1}^{n} \alpha_{i,\,x} \Delta X_i + \alpha_{n+1,\,x} \Delta R$$

$$d_y = d_{\text{nom, y}} + \sum_{i=1}^{n} \alpha_{i,\,y} \Delta X_i + \alpha_{n+1,\,y} \Delta R. \tag{3.46}$$

The correlation of x and y and their variance can be expressed as

$$Var[x] = \sum_{i=1}^{n+1} \alpha_{i,\,x}^2 \qquad Var[y] = \sum_{i=1}^{n+1} \alpha_{i,\,y}^2$$
$$\rho_{xy} = \sum_{i=1}^{n+1} \alpha_{i,\,x}\alpha_{i,\,y}. \tag{3.47}$$

Based on these expressions we can use Clark's expression to estimate the mean of $z = \max(x, y)$, which defines the first term of the canonical delay expression. To estimate the remaining components [146] uses the concept of tightness probability, which is defined as the probability that a sample of a RV x has a value greater than a sample of RV y, and is mathematically expressed as

$$\mathcal{P}_{x>y} = \int_{-\infty}^{\infty} \varphi\left(\frac{x - d_{\text{nom, x}}}{\sqrt{Var[x]}}\right) \Phi\left(\frac{\left(\frac{x - d_{\text{nom, y}}}{\sqrt{Var[y]}}\right) - \rho_{xy}\left(\frac{x - d_{\text{nom, x}}}{\sqrt{Var[x]}}\right)}{\sqrt{1-\rho_{xy}^2}}\right) dx$$
$$= \Phi\left(\frac{d_{\text{nom, x}} - d_{\text{nom, y}}}{\alpha}\right) \tag{3.48}$$

where φ, Φ and α are as defined in (3.38). Using the concept of tightness probability and the fact that in traditional timing analysis the delay at the output is completely defined by either of the inputs, [146] proposes to use a weighted sum of the coefficients of the input delays to define the coefficients of output delay. The weighting parameter is chosen to be the tightness probability, and thus we can write

$$\alpha_{i,\,z} = \mathcal{P}_{x>y}\alpha_{i,\,x} + (1 - \mathcal{P}_{x>y})\alpha_{i,y} \quad 1 \le i \le n. \tag{3.49}$$

The coefficient of the random component of the max is computed such that the variance of $z = \max(x, y)$ estimated using Clark's expressions and that estimated using the canonical expression are identical. This defines the canonical expression completely.

The delay pdfs can now be propagated using the above approach to handle series and parallel nodes. This approach neglects the correlations arising due to reconvergent fanouts and thus results in a conservative estimate as shown in the previous section. However, the canonical model can be extended to handle these correlations at the cost of additional computational complexity. This is achieved by maintaining the list of source nodes for the random component of delay while delay pdfs are propagated through the circuit [155]. In general the number of terms in these expressions will be equal to the number of nodes in the timing graph with a random delay component and result in substantial overhead.

Principal Components

In Chap. 1 process parameters were expressed as a sum of a nominal value and an intra-die and inter-die variation coefficient. In addition, intra-die variation has correlated and random components, with the contribution of each being defined by the maturity of the process and the particular process parameter. In Chap. 2 we discussed principal component analysis and found that the correlated variations can be handled by dividing the chip area using an $n \times n$ grid and associating a RV to each square in the grid to represent the variations in that grid. The correlation among these RVs is captured by defining an $n^2 \times n^2$ correlation matrix. Using principal component analysis we transform this set of correlated RVs to a set of uncorrelated RVs. This step also involves the eigen-decomposition of the correlation matrix and is computationally expensive. However, this step needs to be performed only once for each correlation structure and does not add to the computational complexity of SSTA itself.

If the delay of an edge i is initially defined to be linearly dependent on the process parameters as in (3.42), then after performing PCA we can write delay as a function of the principal components Y, which are Gaussian RVs with zero mean and unit variance, as

$$d_i = a_{i,0} + a_{i,1}\Delta y_1 + \cdots + a_{i\,n}\Delta y_n. \tag{3.50}$$

This expression now has the same form as (3.42) but captures variations due to spatially correlated variations as well. The additional cost is paid in terms of principal component analysis which has a complexity of $O(pn^6)$, where p is the number of process parameters and n^2 is the number of squares in the grid. Once the edge delays are defined as in (3.50), the sum and max operations can be defined in a fairly straightforward manner. The approach proposed in [30] uses Clark's expressions to estimate the mean, variance and the correlation with the principal components (y) and equates it to the respective quantity for the max expression.

Let us consider the case where $d_z = \max(d_i, d_j)$ and outline the steps required for this computation. We assume that the delay pdfs are defined in the form (3.50). The mean is expressed by the first term and the variance and correlation of d_i and d_j can be calculated as

$$\begin{aligned} Var[d_i] = \textstyle\sum_{k=1}^{n+1} a_{k,\,i}^2 \qquad Var[d_j] = \textstyle\sum_{k=1}^{n+1} a_{k,\,j}^2 \\ \rho_{d_i d_j} = \textstyle\sum_{k=1}^{n+1} a_{k,\,i}a_{k,\,j}. \end{aligned} \tag{3.51}$$

In addition, the correlation of d_i with the principal component y_k is simply $a_{i,\,k}$. The first term in the expression for d_z is defined to be the mean, calculated using Clark's expressions. The remaining coefficients are the correlations coefficients of d_z with the principal components and are defined to be the correlation coefficient obtained using Clark's expression. This results in an expression that maintains the mean and the first-order correlation with the principal components. However, this approach may result in a mismatch in variance. To handle this error [30] proposes to scale the coefficients of the principal components in the expression for d_z by a factor calculated as the ratio of the variance of d_z obtained after the above steps and that predicted by Clark's expression. This results in a small mismatch in the correlation coefficient with the principal components. Note, that we have neglected random variations in the principal component based approach – if random variations are considered, then timing analysis can be performed while maintaining the mean, variance and correlation coefficient [130]. We will discuss this approach in more detail in Chap. 5 where a PCA-based analytical approach is used to determine the parametric yield of a design under delay and power constraints.

The overall complexity of the approach can be estimated by observing that we need to map the delay expression to the canonical delay model for $O(|V|+|E|)$ delay elements, which are the timing arcs and interconnects. Since each mapping requires $O(n)$ computation, the overall complexity of generating the delay models is $O(n(|V|+|E|))$. The complexity of both the sum and max computations is $O(n)$, and they are performed $O(|V|+|E|)$ and $O(|E|)$ times, respectively. Thus the overall complexity of the approach is $O(n(|V|+|E|))$. There is an additional cost in terms of PCA itself, as mentioned before this is a one time investment for all future analysis and is not considered to be a part of the overall SSTA complexity. In the case where we consider p process parameters, the overall complexity becomes $O(np(|V|+|E|))$.

Quad-Tree Modeling

The Quad-Tree modeling scheme, which was introduced in Chap. 2, modeled the intra- and inter-die components of variation by generating a tree-like structure that successively divides each region of the chip into four smaller pieces. Each piece was assigned to a RV from a set of independent RVs and correlation at the gate level was captured by the squares that were common to the delay expression for a pair of gates. The canonical delay model used in the Quad-Tree based SSTA technique [3] is similar to the ones in (3.42) and (3.50) used in the previous two analysis methodologies. However, the analysis technique is different from the canonical delay model based techniques discussed above. It bears more similarity to the discrete pdf propagation technique discussed in Sec. 3.2.2 and does not make the Gaussian assumption for delay pdfs.

In Chap. 2 the variation in the process parameter for a particular gate i was expressed based on Fig. 2.8 as

$$\Delta X_{intra,i} = \sum_{0<l<k, r \text{ intersects } i} \Delta X_{l,r} + \Delta X_i^R \tag{3.52}$$

using which we can write the canonical form for delay as

$$d_i = d_{\mathrm{nom,\,i}} + \sum_k \alpha_{i,\,k} \Delta P_k + \Delta D_{\mathrm{random,\,i}}. \qquad (3.53)$$

To propagate the node delays through the graph we need to define the sum and max operation such that the final expression is also in canonical form. The sum expression can be defined as $d_z = d_i + d_j$:

$$d_z = d_{\mathrm{nom,\,i}} + d_{\mathrm{nom,\,j}} + \sum_k (\alpha_{i,\,k} + \alpha_{j,\,k}) \Delta P_k + \Delta D_{\mathrm{random,\,i}} + \Delta D_{\mathrm{random,\,j}} \quad (3.54)$$

where the sum involving the nominal delay and the coefficient of the spatially correlated component is the standard arithmetic operation. However, the sum involving the random component is a sum of two RVs and a simple convolution is performed to calculate the distribution of this sum. Since the random component of an edge is independent of the random component of the input node delay, the above computation is exact.

Since there is no straightforward way to compute the max of two RVs expressed in canonical form as expressed above, [3] proposes to generate a bound for the max operation using the following theorem.

Theorem 3.7. *For any given numbers* $a_1, a_2, \ldots, a_n$ *and* $x_1, x_2, \ldots, x_n$

$$\max\left(\sum_{i=1}^{n} a_i, \sum_{i=1}^{n} x_i\right) \leq \sum_{i=1}^{n} \max(a_i, x_i).$$

Using Theorem 3.7 max $d_z = \max(d_i, d_j)$, can be conservatively approximated as

$$d_z = \max(d_{\mathrm{nom,\,i}}, d_{\mathrm{nom,\,j}}) + \sum_k (\mathrm{absmax})(\alpha_{\mathrm{i,\,k}}, \alpha_{\mathrm{j,\,k}}) \Delta \mathrm{P_k} + \max(\Delta D_{\mathrm{random,\,i}} + \Delta D_{\mathrm{random,\,j}}) \qquad (3.55)$$

where absmax selects the value with the largest arithmetic absolute value while retaining the sign. The max of the nominal value is also an arithmetic max operation, however the max of the random components is a max of RVs. These RVs are correlated due to reconvergent fanouts and computing their max while neglecting the correlation results in an upper bound (as shown in Sec. 3.2.2).

To reduce the pessimism introduced due to the above conservative bounds, [3] notes that the above bound is exact if one of the delays (say d_i) *completely dominates* delay d_j. Complete domination is said to occur if all of the following conditions hold:

(1) $d_{\mathrm{nom,\,i}} > d_{\mathrm{nom,\,j}}$

(2) $\alpha_{i,\,k} > \alpha_{j,\,k}$

(3) The minimum value of $\Delta D_{\mathrm{random,\,i}}$ with non-zero probability is greater than the maximum value of $\Delta D_{\mathrm{random,\,j}}$ with non-zero probability

However, if these conditions are not met then the max operation as defined above defines a conservative upper bound. In this case multiple arrival time delay pdfs are propagated through the circuit. At each node a subset of the pdfs are propagated through the node, while the other pdfs are merged using the max operation as defined in (3.55). The subset pdfs to be merged are selected to minimize the conservatism introduced, which is achieved by selecting the pdfs with the smallest mean delays. This strategy propagates the pdfs with large mean delay through the circuit, which can be expected to have a strong influence on the overall pdf of delay and thus effectively reduces the error due to the conservative upper bound.

At the primary outputs (POs), the delay of each PO can be defined as a pdf by convolving each of the terms in the canonical expression and then calculating their maximum numerically. However, these delay pdfs are correlated and we require the following theorem from [3] to show that the max arrival time through the circuit can be bounded by ignoring the correlation in the delay pdfs at the POs.

Theorem 3.8. *Let x, x_1, x_2, y and z be positive, independent RVs with pdfs $p(x)$, $p(x_1)$, $p(x_2)$, $q(y)$, $r(z)$ noting that x_1 and x_2 have the same pdf as RV x. For any positive constant value a, the cdf of RV $\max(x+y,\, ax+z)$ is upper bounded by the cdf of RV $\max(x_1+y,\, ax_2+z)$.*

Proof. The cdf of P can be expressed as

$$
\begin{aligned}
P(t) &= \int_{x+y\le t,\, ax+z\le t} p(x)q(y)r(z)\mathrm{d}x\,\mathrm{d}y\,\mathrm{d}z \\
&= \int_0^\infty \int_0^\infty q(y)r(z) \int_{x\le \min(t-y,\,(t-z)/a)} p(x)\mathrm{d}x\,\mathrm{d}y\,\mathrm{d}z. \qquad (3.56)
\end{aligned}
$$

Similarly, the cdf of P' can be expressed as

$$
\begin{aligned}
P'(t) &= \int_{x_1+y\le t,\, ax_2+z\le t} p_1(x_1)p_2(x_2)q(y)r(z)\mathrm{d}x_1\,\mathrm{d}x_2\,\mathrm{d}y\,\mathrm{d}z \\
&= \int_0^\infty \int_0^\infty q(y)r(z) \int_{x_1\le t-y} p(x_1)\mathrm{d}x_1 \int_{x\le (t-z)/a} p(x)\mathrm{d}x\,\mathrm{d}y\,\mathrm{d}z. \qquad (3.57)
\end{aligned}
$$

Let us consider the case when $t-y < (t-z)/a$, for which (3.56) can be simplified to

$$
P(t) = \int_0^\infty \int_0^\infty q(y)r(z) \left(\int_{x\le t-y} p(x)\mathrm{d}x \right) \mathrm{d}y\mathrm{d}z \qquad (3.58)
$$

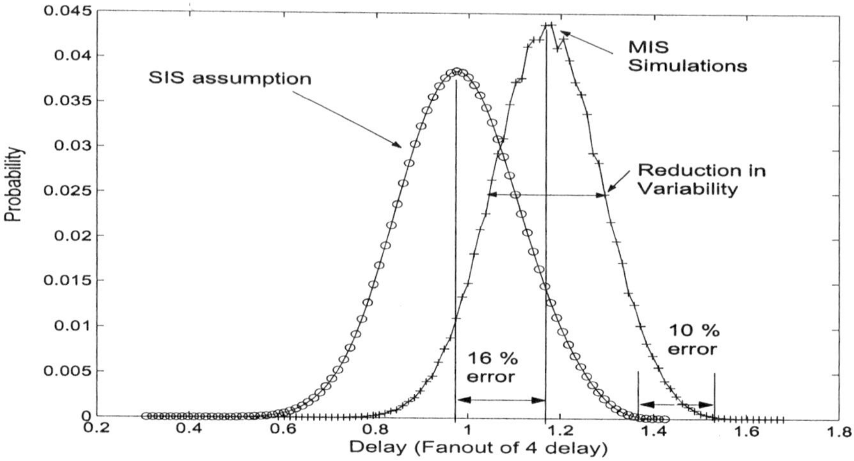

Fig. 3.14. Neglecting multiple input switching cases results in a 16% error in mean delay estimation and a 10% error in the delay at higher percentile points. Timing simulations considering MIS also show a smaller variance in delay [5]. (©2005 IEEE)

Comparing (3.58) and (3.57), we note that the integrand for P' has an additional term compared to the integral for P. The additional term represents the probability of a RV being less than some value and is always less than 1, therefore the integrand for P' is always less than the integrand for P, which implies that $P'(t) \leq P(t)$.

The other case when $t - y < (t - z)/a$ can also be similarly analyzed to obtain the same condition. Therefore P' defines an upper bound for P.

3.2.4 Multiple Input Switching

A number of issues complicate traditional timing analysis such as false paths, multiple input switching and input slope effects. In this section we discuss an approach to handle multiple input switching (MIS) in statistical timing analysis that was proposed in [5]. Multiple input switching refers to the scenario when multiple inputs of a gate switch in close temporal proximity of each other, resulting in an increase in propagation delay of the gate. The probability of such an event becomes higher in SSTA since we deal with delay pdfs instead of a deterministic arrival time, and thus the chances that two switching events overlap is much higher.

For all analysis discussed previously, we assumed that only one input is switching, known as the single input switching (SIS) assumption. In the case of SIS, we handle multi-input gates by propagating the input delay pdfs to the output by convolving it with the delay pdf of the appropriate timing arc, and then computing the maximum of all such delay pdfs over all inputs.

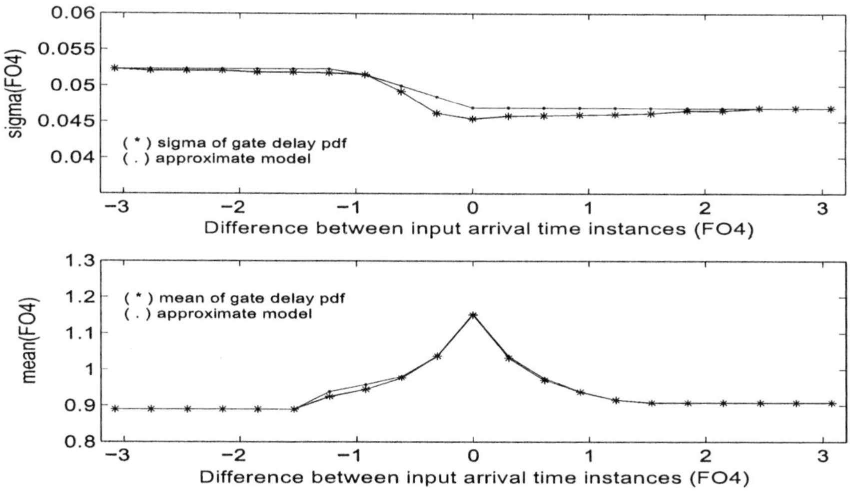

Fig. 3.15. Change in the mean and variance of gate delay as a function of the difference between the switching times of the two inputs of the gate [5]. (©2005 IEEE)

Figure 3.14 shows the inaccuracy introduced by this assumption. The figure is generated by performing normal SSTA. However, for each multi-input gate the actual output delay pdf (considering proximity switching) is computed through Monte Carlo simulations.

The modeling approach proposed in [5] is based on a deterministic MIS model proposed in [29]. This deterministic model proposed to use a delay push-out (D.PO.) factor for different combinations of input arrival times at a multi-input gate, which increases the delay by a constant factor when nodes with MIS are encountered. Figure 3.15 shows the change in mean gate delay and standard deviation as a function of the difference between the delay of the switching times of the two inputs, obtained by performing Monte Carlo simulations for variations in gate length for transistors in the logic gate. As can be observed, the mean gate delay attains a maximum when both inputs switch simultaneously and reduces as the separation between the switching times increases, finally saturating to the mean delay of the SIS case. However, the standard deviation shows a minimum when both inputs are perfectly aligned. This results from the fact that the devices corresponding to these inputs within the logic gate are not perfectly correlated and thus the overall variance in delay reduces from the case of SIS. As the difference between the switching times of the two inputs increases, the standard deviation saturates to the standard deviation for the SIS case.

To model these effects analytically [5] notes that the increase in mean gate delay is a weak function of the standard deviations of the process variations themselves. Neglecting this dependence, the increase in mean delay can be approximated using a deterministic MIS model. The standard deviation is modeled as a piecewise linear function that is exactly equal to the SIS case for large separation in switching times and approximates the MIS region using a simple linear interpolation of the variance for the two SIS cases, as illustrated in Fig. 3.15. Mathematically, the standard deviation considering MIS σ_m is expressed as

$$\sigma_m = \begin{cases} \sigma_{s1} & \text{if} \quad \Delta DA \leq -X \\ \sigma_{s2} & \text{if} \quad \Delta DA \geq 0 \\ \alpha \min(\sigma_{s1}, \sigma_{s2}) + (1-\alpha)\max(\sigma_{s1}, \sigma_{s2}) & \text{o.w.} \end{cases} \tag{3.59}$$

where ΔDA is the separation in the switching time of the two inputs, X is the difference between the arrival times beyond which the effect of MIS become negligible and $\alpha = |\Delta DA|/X$. We can now use this model to perform MIS-aware SSTA.

Consider a two input gate with a discretized delay pdf for each of the input pins. For each combination of the discrete events in the two input pdf we get a different value for ΔDA, and therefore a different gate delay pdf. Assuming that the gate delay pdf is Gaussian we can generate the gate delay pdf based on the model for mean and variance developed above. Using this gate delay an output delay pdf is generated by scaling the gate delay pdf by the maximum of the input delay events. Thus, for $O(n^2)$ combinations of discrete events on the input we generate $O(n^2)$ output delay pdfs. Each of these output pdfs is then scaled by the product of the probability of the two discrete events on the input to which this output delay pdf corresponds. These scaled pdfs are then grouped by summing the probabilities of all events occurring at a given time point. Since we need to combine $O(n^2)$ delay pdfs, each with n discretizations, we get an overall complexity of $O(n^3)$. Since the number of discretizations are typically small, ranging from 5-10, this increase in complexity is reasonable.

To extend the above analysis for more than two input gates, note that a straightforward extension would result in computational complexity that increases exponentially with the number of inputs. The approach proposed in [5] iteratively considers a pair of input pins to generate the final output delay pdf. Let us consider the steps involved in performing MIS-aware SSTA for a three input gate with input pins A, B and C.

1) The first step is to order the nodes based on the mean delay at the input pins. We refer to the ordered set of input pins as 1, 2 and 3, with input 1 having the smallest mean delay.

2) Considering the two earliest switching inputs (1 and 2) we can generate the output delay pdf using the technique described above.

3) Next, compute the output delay pdf assuming a SIS occurring on input pin 2 and compare it to the output delay pdf calculated assuming MIS on inputs 1 and 2. Compute the increase in mean delay as μ_{12} and the decrease in variance as σ_{12}.

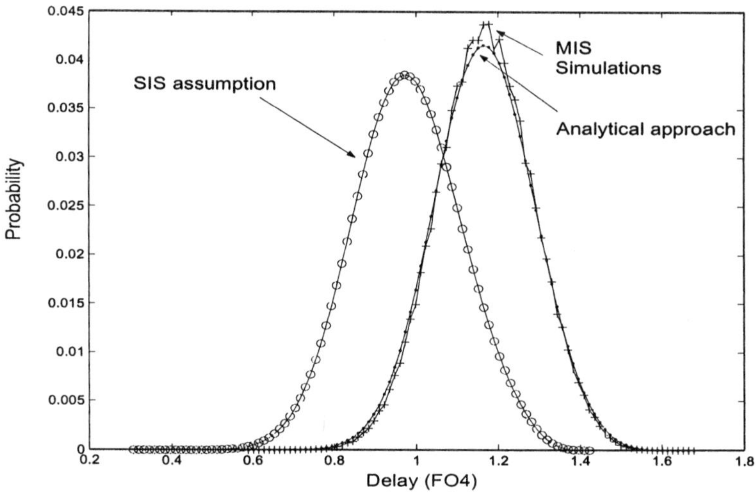

Fig. 3.16. The approach proposed in [5] to handle MIS shows good accuracy compared to results obtained using Monte Carlo SPICE simulations. (©2005 IEEE)

4) Now, reduce the variance of the input delay pdf of node 2 by σ_{12}. Then, for all pairs of discrete events on input pins 2 and 3 (with delay d_2 and d_3) compute the gate delay pdf. Instead of shifting the delay pdf by $\max(d_2, d_3)$ as in the case of two input gates, it is shifted by $\max(d_2 + \mu_{12}, d_3)$.

5) The next two steps of scaling the output delay pdfs by the product of the probabilities of the discrete events on the inputs and the grouping step remain the same as in the case of two input gates.

Using this heuristic approach, gates with more than two inputs can be handled efficiently and the number of computations required grows linearly with the number of inputs.

Figure 3.16 shows the delay pdf of a circuit obtained using the SIS assumption and considering MIS through Monte Carlo and compares these with results from the MIS-aware SSTA approach discussed in this section. Over a set of benchmarks [5] found that the SIS assumption results in an average error of 13.2% and -10.2% in the mean delay and standard deviation, respectively, with the maximum error being 26% and -20%. MIS-aware SSTA was found to provide good accuracy with an average error of 0.01% and 2.07% in mean delay and standard deviation, respectively. The maximum error in this case was found to be 0.2% and 7.0% for the mean and standard deviation, respectively.

3.3 Path-Based Timing Analysis

Path-based statistical timing analysis techniques are based on performing timing analysis on a selected set of paths in a given circuit. These paths are expected to have a significant probability of becoming critical and therefore have the strongest influence on the circuit delay pdf. The goal of path-based SSTA is to estimate the distribution of the maximum delay of the selected set of paths. If the delay of each gate is assumed to be Gaussian, then the delay of a single path is Gaussian since it is a sum of Gaussian RVs. The crucial step is to estimate the maximum of these Gaussian RVs in order to compute the circuit delay pdf. The simplified timing analysis approach discussed in Sec. 3.1 is an example of path-based statistical timing analysis.

Let us consider a circuit where we select a set of N paths to perform path-based SSTA. Using these paths we can define the cdf of circuit delay as

$$F(t) = \mathcal{P}(\max D_1, \ldots, D_N \leq t) = \mathcal{P}_{\mathbf{\Sigma}}(\bigcap\{D_i \leq t\}) \tag{3.60}$$

where $\mathbf{\Sigma}$ is the correlation matrix for the vector of path delays. The above equation can be rewritten by normalizing the path delays to standard Gaussian RVs Z_i as

$$F(t) = \mathcal{P}_{\mathbf{\Sigma}}\left(\bigcap\{Z_i \leq \frac{t - E[D_i]}{Var[D_i]}\}\right). \tag{3.61}$$

Note that even if gate delays are not assumed to be Gaussian, path delays can be assumed to be Gaussian since a sum of independent RVs rapidly converges (for most practical correlation structures involved in circuit delay computation) to a Gaussian RV due to the *Central Limit Theorem* [109].

A path-based statistical delay computation was proposed in [54]. The approach was based on the delay computation of each path and was able to account for signal transition times and output loading. However, the analysis is performed on one path at a time and the number of critical or near-critical paths in an optimized circuit can be large. In general, path-based techniques suffer from the fact that it is unclear how to select the initial set of paths, since a path with a significantly smaller delay may become critical for a particular combination of process parameters. In addition, performing timing analysis on one path a time is computationally very expensive.

In this section we will discuss the approach proposed in [103] to compute bounds for the delay cdf, which is based on the theory of *stochastic majorization*. The first step is to extract a subgraph G' from the complete timing graph G that contains the k longest paths of the circuit in terms of their deterministic delay, and then perform a topological traversal of the subgraph to estimate the bounds. The approach has a complexity of $O(|V| + |E|)$, where V and E are the sets of nodes and edges in a DAG, respectively. The improvement in computational complexity of this path-based approach compared to other approaches rests on the use of a topological traversal to establish bounds on the cdf of delay of all paths in the network. The path extraction step is performed using the approach proposed in [152], which can be used to list the k most critical paths in the circuit.

Theorem 3.9. *Let* $\mathbf{X}$ *be an* n*-dimensional centered multinormal Gaussian distribution. Let* $\mathbf{\Sigma_1}$ *and* $\mathbf{\Sigma_2}$ *be two* $n \times n$ *correlation matrices such that*

$$(\mathbf{\Sigma_1})_{ij} \geq (\mathbf{\Sigma_2})_{ij}, \quad \forall i, j \in \{1, \ldots, n\} \tag{3.62}$$

then

$$\mathcal{P}_{\mathbf{\Sigma_1}}\left(\bigcap_{i=1}^{n}\{X_i \leq a_i\}\right) \geq \mathcal{P}_{\mathbf{\Sigma_2}}\left(\bigcap_{i=1}^{n}\{X_i \leq a_i\}\right) \tag{3.63}$$

is true for all vectors $\mathbf{a} = (a_1, \ldots, a_n)^T$.

Using the above theorem we can bound the probability in (3.61) as (assuming a correlation matrix for path delays to be $\mathbf{\Sigma}$)

$$\begin{aligned} \mathcal{P}_{\mathbf{\Sigma}}\left(\bigcap\{Z_i \leq \frac{t - E[D_i]}{Var[D_i]}\}\right) &\geq \mathcal{P}_{\mathbf{\Sigma_{min}}}\left(\bigcap\{Z_i \leq \frac{t - E[D_i]}{Var[D_i]}\}\right) \\ \mathcal{P}_{\mathbf{\Sigma}}\left(\bigcap\{Z_i \leq \frac{t - E[D_i]}{Var[D_i]}\}\right) &\leq \mathcal{P}_{\mathbf{\Sigma_{max}}}\left(\bigcap\{Z_i \leq \frac{t - E[D_i]}{Var[D_i]}\}\right) \end{aligned} \tag{3.64}$$

where $\mathbf{\Sigma_{min}}$ is the correlation matrix generated by setting all the off-diagonal terms to $\min_{i,j}(\Sigma)_{ij}$, and $\mathbf{\Sigma_{max}}$ is the correlation matrix generated by setting all the off-diagonal terms to $\max_{i,j}(\Sigma)_{ij}$. The computation of the bounds is simpler because all off-diagonal terms are equal, which implies that all RVs have the same correlation coefficient. However, since the probability computation requires an integral over a non-equi-coordinate (length of each axis in the region is different) region, the above step is still computationally expensive.

The authors in [103] use the ideas of strong and weak stochastic majorization to compute the above developed probability bounds.

Definition 3.10. *Let* $\mathbf{X}$ *and* $\mathbf{Y}$ *be two* n*-dimensional RVs.* $\mathbf{X}$ *is said to* strongly stochastically majorize $\mathbf{Y}$ *or* $\mathbf{X} \succ \mathbf{Y}$*, if*

$$\mathcal{P}[X \in A] \geq (\leq) \mathcal{P}[Y \in A] \tag{3.65}$$

for every Borel-measurable Schur-convex (Schur-concave) set A.

Definition 3.11. *Let* $\mathbf{X}$ *and* $\mathbf{Y}$ *be two* n*-dimensional RVs.* $\mathbf{X}$ *is said to* weakly stochastically majorize $\mathbf{Y}$ *or* $\mathbf{X} \succ\succ \mathbf{Y}$*, if*

$$\mathcal{P}[X \in A] \geq (\leq) \mathcal{P}[Y \in A] \tag{3.66}$$

for every Borel-measurable increasing Schur-convex (decreasing Schur-concave) set A.

Using the above definition and properties of multinormal distributions it can be shown that if $\mathbf{t} = (t_1, t_2, \ldots, t_n)$ and $\check{\mathbf{t}} = (\bar{t}, \bar{t}, \ldots, \bar{t})$, where

$$\bar{t} = \frac{1}{n} \sum_{i=1}^{n} t_i \tag{3.67}$$

then $\mathbf{t} \succ \check{\mathbf{t}}$ [103]. In addition, if $\hat{\mathbf{t}} = (\tilde{t}, \tilde{t}, \ldots, \tilde{t})$, where

$$\tilde{t} = \min_{i=1}^{n} t_i \tag{3.68}$$

then $\hat{\mathbf{t}} \succ\succ \mathbf{t}$. This implies,

$$\mathcal{P}\left(\bigcap_{i=1}^{n} \{Z_i \le \tilde{t}\}\right) \le \mathcal{P}\left(\bigcap_{i=1}^{n} \{Z_i \le t\}\right) \le \mathcal{P}\left(\bigcap_{i=1}^{n} \{Z_i \le \bar{t}\}\right). \tag{3.69}$$

Note that the above bounds are defined using the probability of a multinormal RV (with same off-diagonal terms in the correlation matrix) over an equi-coordinate region that can be efficiently computed. To compute the above bounds, the only required information is the maximum and minimum value of the correlation between any two paths.

Now let us discuss a technique that can be used to compute the maximum and minimum of the correlation between two paths in a DAG. Assume that the delay of a node i can be expressed as

$$d_i = d_{\text{nom,i}} + a_i P_i + b_i P \tag{3.70}$$

where $d_{\text{nom,i}}$ is the nominal delay of the gate, P is the global value of a process parameter P, and P_i represents the random variation in the process parameter for gate i, and a_i and b_i are fitting parameters. The variance of node delay can then be expressed as

$$Var[d_i] = a_i^2\, Var[P_i] + b_i^2\, Var[P] \tag{3.71}$$

where variance of P_i represents intra-die random variations and variance of P represents inter-die variations. In addition, P_i and P are assumed to be independent Gaussian RVs with equal variance. Without loss of generality, we can assume that P_i and P are standard Gaussian RVs with zero mean and unit variance. The correlation between node delays can now be expressed as

$$corr[d_i, d_j] = \frac{cov(d_i, d_j)}{\sqrt{Var[d_i]Var[d_j]}} = \frac{b_i b_j}{\sqrt{(a_i^2 + b_i^2)(a_j^2 + b_j^2)}}. \tag{3.72}$$

The term

$$f_i^{\text{node}} = \frac{b_i}{\sqrt{a_i^2 + b_i^2}} \tag{3.73}$$

is defined to be the *chip-to-node* correlation of node i. Note that the correlation between the delay of two nodes can be estimated by taking the product of their node-to-chip correlations.

The delay of any path p can be expressed as sum of node delays, and their correlation (or path correlation) can be expressed as

$$corr\left[\sum_{i=1}^{m} d_i, \sum_{j=1}^{n} d_j\right] \tag{3.74}$$

$$= \frac{\sum_{i=1}^{m} b_i \sum_{j=1}^{n} b_j}{\sqrt{\left(\sum_{i=1}^{m} a_i^2 + (\sum_{i=1}^{m} b_i)^2\right)\left(\sum_{j=1}^{n} a_j^2 + \left(\sum_{j=1}^{n} b_j\right)^2\right)}}.$$

As in the case of node delay we can define a *chip-to-path* correlation as

$$f_p^{\text{path}} = \frac{\sum_{i=1}^{m} b_i}{\sqrt{\sum_{i=1}^{m} a_i^2 + (\sum_{i=1}^{m} b_i)^2}}. \tag{3.75}$$

As we perform a traversal of path p assume that the path that includes the first k nodes has a chip-to-path correlation of f_k and let the next node have a delay of $d_{\text{nom,k}+1} + a_{k+1}P_{k+1} + b_{k+1}P$. Then, after some algebraic steps we can write the chip-to-path correlation of the path that includes the next node as

$$f_{k+1} = \sqrt{\frac{1+\alpha}{f_k^2 + \alpha + \beta}} \tag{3.76}$$

where

$$\begin{aligned} \alpha &= 2\frac{b_{k+1}}{h(k)} + \frac{b_{k+1}^2}{h^2(k)} \\ \beta &= \frac{a_{k+1}^2}{h^2(k)} \\ h(k) &= \sum_{i=1}^{k} b_i. \end{aligned} \tag{3.77}$$

Thus while traversing a DAG we only need to propagate the chip-to-path correlation and $h(k)$ through the nodes. Since computation of the bounds requires us to estimate the maximum and minimum path correlations, we must compute the minimum and maximum chip-to-path correlation at each

PO of the DAG. The product of the two largest and two smallest chip-to-path correlation coefficient gives the largest and smallest path correlation, respectively.

3.4 Parameter-Space Techniques

In the previous section we discussed techniques to estimate timing yield using the pdf of maximum arrival time at the output of a design. These methods are also known as *performance-space methods*, since we map the impact of variations in process parameters to variations in performance. The timing yield calculation given the distribution of the maximum arrival time at the output is fairly simple, however generating the distribution itself is a complicated task. Another class of timing analysis techniques known as *parameter-space methods* is the focus of this section. These methods find a region in the parameter-space that represents the feasible region in terms of a timing constraint on the design. The pdf of the process parameter is then integrated over the feasible region, which is much more complicated than the hypercubic feasible region in performance-space. However, this approach to estimate the probability that the design satisfies the timing constraint deals with a distribution that is fairly simple and in most cases assumed to be multinormal. Monte Carlo based integration techniques can be used to estimate the timing yield by computing the surface integral of the feasible region [48]. However, these approaches have high computational requirements that quickly become unreasonable when intra-die variations are considered along with spatial correlations and reconvergence.

3.4.1 Parallelepiped Method

Two different parameter-space approaches that provide reasonable tradeoffs between runtime complexity and accuracy were proposed in [65]. The methods are path-based timing approaches based on linear models for gate delay and slew as a function of variations in process parameters and are amenable to any arbitrary distribution of the underlying process parameters. The subset of paths to be statistically analyzed are selected based on results from a nominal static timing analysis engine. Statistical timing analysis is then performed on each of the n paths to estimate the slack, which is expressed as

$$s_i = s_{\text{nom, i}} + \sum_{j=1}^{p} \alpha_{ij} \Delta P_j \tag{3.78}$$

where $s_{\text{nom, i}}$ is the slack of path i under nominal conditions, ΔP_j represents the variation in the jth of the p process parameters and α_{ij} is the sensitivity of the slack of path i to variations in process parameter j. If a positive slack d_0 is

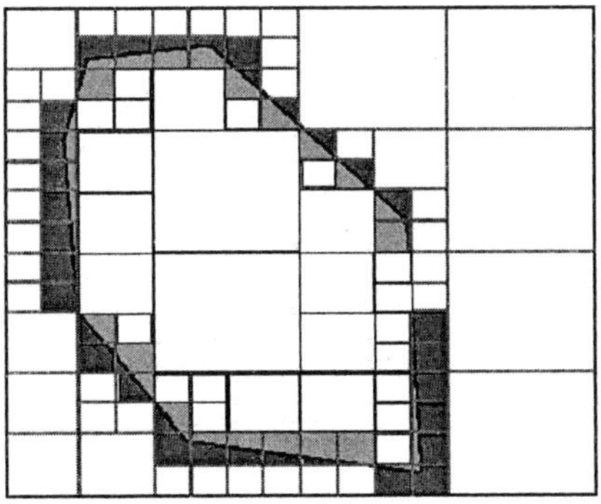

Fig. 3.17. The parallelepiped technique to approximate the integral in parameter-space. The squares in the center and the extreme right are completely feasible or infeasible are not subdivided further. (©2005 IEEE)

desired for each of the paths, then the feasible region in the performance-space of path slacks can be expressed as

$$Y_{\text{perf}} = \{\mathbf{s}|s_i \geq d_0,\ i = 1, 2, \ldots, n\} \tag{3.79}$$

where $\mathbf{s}$ is a vector of path slacks. Note that the above region defines a n-dimensional hypercube in s-space. An equivalent set (in the sense that a sample in the corresponding space has the same probability of lying within these sets) Y_{param} in the parameter-space can be defined as

$$Y_{\text{param}} = \{\mathbf{\Delta P}|s_{\text{nom, i}} + \sum_{j=1}^{p} \alpha_{ij}\Delta P_j \geq d_0,\ i = 1, 2, \ldots, n\} \tag{3.80}$$

where $\mathbf{\Delta P}$ is a p-dimensional vector of variation in the parameter-space. Note that each of the above n equations defines a hyperplane in the parameter-space and the feasible region is a convex polyhedron analogous to the feasible region in linear programming problems. Generating each of the corner points of this polyhedron is a computationally complicated task and thus determining the feasible region Y_{param} is not straightforward. One of the techniques proposed in [65] is based on determining the feasible region by recursively dividing the complete parameter-space into smaller *parallelepipeds* [36].

A parallelepiped in three dimensions is a *prism* whose sides are all parallelograms, and it is a convex object. The timing analysis procedure is based on the fact that if a set of points satisfy a set of linear constraints, then so will any point generated using a convex combination of the initial set of points. The complete set of points that can be generated using convex combinations of points is also known as the *convex hull* of the set of points. Since a parallelepiped is a convex object, if we find that each of its vertices are feasible in terms of the constraints expressed in (3.80), then the entire parallelepiped is feasible. If a parallelepiped has some vertices that are feasible and some that are infeasible, then the parallelepiped is divided into smaller parallelepipeds (by dividing it in half along each axis). This procedure is recursively continued until a parallelepiped is generated such that all its vertices are either feasible or infeasible. At the end of this procedure a weighted sum of the parallelepipeds based on their location is calculated to estimate the timing yield. This procedure is illustrated in Fig. 3.17, where the squares (a parallelepiped in 2-dimensions) in the center (right boundary) of the figure are not sub-divided since all their vertices are feasible (infeasible). However, any square that intersects the boundary of the convex region is shown to be further divided until a square is generated with all its vertices either inside or outside the convex region.

In practice, the initial bounding box as shown in Fig. 3.17 can be generated using a hypercube of size which has dimensions of $\pm 4\sigma$. Since the probability that a given process parameter lies outside this bound is very small, even if the space is feasible in terms of performance the weighting coefficient of a region outside the bounding box would be sufficiently small to make its contribution insignificant. In addition, a limit is imposed on the number of recursion levels used in dividing a parallelepiped and a lower bound on the yield is obtained by counting squares at the lowest recursion level only if all its vertices are feasible. If only some of the vertices are feasible at this point in recursion then the entire region is assumed to be infeasible.

This approach has a complexity that grows exponentially with maximum recursion depth R and the dimension of the parameter-space p. This follows from the fact that through the course of the algorithm we generate a 2^p-ary tree. Thus we need to perform $O(2^{pR})$ statistical timing analysis checks. If the statistical timing analysis check has a complexity that grows as the product of the number of paths and process parameters, then the overall complexity of this approach is $npO(2^{pR})$.

3.4.2 Ellipsoid Method

Another approach to determine a lower bound to the region shown in Fig. 3.17 determines the volume of the largest ellipsoid that can be inscribed in the feasible region expressed in (3.80). Then we can integrate the probability distribution of the process parameters over this space, which can be easily characterized, instead of the complete feasible region defined by the convex polyhedron. This is illustrated figuratively in Fig. 3.18. Let us rewrite the original constraint set (3.80) as

$$Y_{\text{param}} = \{\mathbf{\Delta P} | \mathbf{A}^T \mathbf{\Delta P} \geq \mathbf{b_0},\ i = 1, 2, \ldots, n\} \tag{3.81}$$

where $\mathbf{b_0} = \text{vec}(d_1 - s_{\text{nom},1}, \ldots)$ (vec is an operator that converts a set of n numbers to a n-dimensional column vector), and $\mathbf{A}$ is a $P \times n$ matrix with

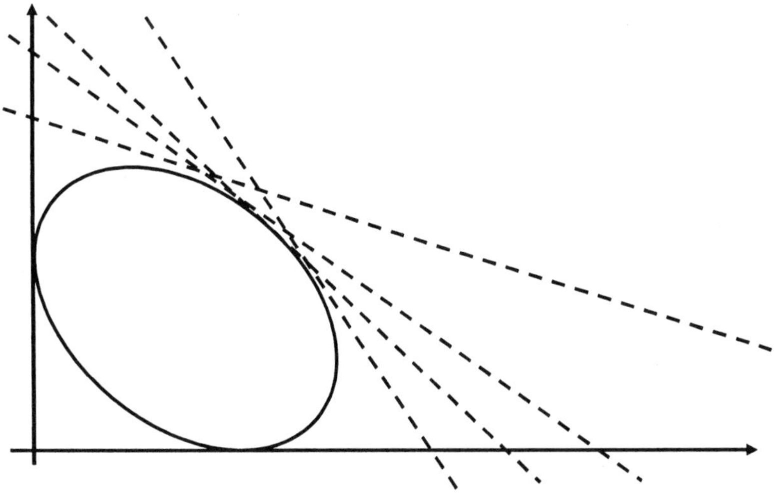

Fig. 3.18. The ellipsoid technique uses an ellipse to approximate the feasible region and provides a simple representation of the region over which the integration is performed. (©2005 IEEE)

$A_{ij} = \alpha_{ji}$. The set of points enclosed by an ellipsoid can be expressed as a matrix operation on the set of points within the unit sphere as

$$E = \{\mathbf{By} + \mathbf{d} \mid \|\mathbf{y}\| \leq 1\} \tag{3.82}$$

where **B** is a symmetric positive-definite matrix. The volume of this matrix is represented by the determinant of matrix **B** (det(**B**)). Therefore, our goal is to find the matrix **B** with the largest determinant such that all points in the set E (3.82) satisfy (3.81). This constraint can be represented as

$$\mathbf{A}^T(\mathbf{By} + \mathbf{d}) \leq \mathbf{b_0}, \ \|y\| \leq 1. \tag{3.83}$$

Writing out the above set of equations component-wise we obtain

$$\mathbf{A}_i^T\mathbf{By} + \mathbf{A}_i^T\mathbf{d} \leq b_{0i}, \ i = 1, \ldots, n, \ \ \|y\| \leq 1 \tag{3.84}$$

which can be simplified using the *Cauchy-Schwarz inequality* as

$$\|\mathbf{BA}_i^T\| + \mathbf{A}_i^T\mathbf{d} \leq b_{0i}, \ i = 1, \ldots, n. \tag{3.85}$$

Now, we can write the problem of embedding the ellipse with the largest volume within the region defined by the constraint (3.85) as

$$\text{Max}: \ \log\det\mathbf{B}$$

$$
\begin{aligned}
s.t.\ \ & \|\mathbf{B}\mathbf{A}_i^T\| + \mathbf{A}_i^T\mathbf{d} \le b_{0i},\ \ i = 1,\ldots,n \\
& \mathbf{B} \succ 0 \\
& \mathbf{B} = \mathbf{B}^T
\end{aligned}
\tag{3.86}
$$

where $B \succ 0$ implies that $\mathbf{B}$ is required to be positive-definite and the objective function still maximizes determinant $\det \mathbf{B}$ since log is a monotonic operator. The above problem is a convex optimization problem and can be solved using standard non-linear convex optimization techniques. However, efficient primal-dual interior point methods are available [144][157] that solve the problem in comparatively few iterations. These techniques are based on the ideas of interior-point methods used to solve linear and semi-definite programming problems. In fact, semi-definite problems are a special class of the above optimization problem in (3.86) [144].

3.4.3 Case-File Based Models for Statistical Timing

The previous two approaches seek a computationally efficient approach to define the feasible region based on parallelepipeds and ellipsoids. Now, we discuss the ideas presented in [96], which seek to define bounds on the yield of a simplified set of paths and develops methods upon which yield based case-files can be defined, guaranteeing a given yield if the design meets timing constraints using the developed case-file. The approach seeks to avoid the problems associated with handling spatially correlated variations. In the previous section we looked at techniques that used PCA to simplify the correlation structure. However, generating the correlation structure from process data is time consuming and complicated and in many cases the information is not available during the design phase. Moreover, the correlation structure can be process-dependent, which has a direct bearing on the complexity of timing analysis. As discussed, PCA has a computational complexity that increases very strongly with increasing grid-size, however it was argued that since PCA is a one-time investment it does not increase the complexity of SSTA. If the correlation changes significantly due to changes in layout then the above argument falls through and we need to perform PCA after changes have been made to the layout. The approach in [96] shows that it is sufficient to know the variance of process parameters and the number of principal components to perform parametric-space timing analysis. The number of principal components can be estimated by identifying the number of basic independent physical processing steps that lead to systematic process variations. Alternatively, this number can be estimated by measuring the yield of a test-chip fabricated using the same process and using the proposed model to back-calculate the number of principal components, which can then be used for other designs.

Let us consider a process parameter X and define the parametric yield as

$$
Y(x) = \mathcal{P}(X(i) \le x,\ i = 1, 2, \ldots, n) \tag{3.87}
$$

where n is the number of basic structures in the design for which X is defined. Let us decompose $X(i)$ as

$$X(i) = X_{\text{nom}} + X_{\text{inter}} + X_{\text{intra}}(i) \tag{3.88}$$

where X_{nom} is the nominal value of the parameter X, X_{inter} is the inter-die variation which is the same for all $i's$ and $X_{\text{intra}}(i)$ is the intra-die variation which is different for each structure i in the design. In addition, the intra-die variation has a systematic component that depends on the spatial co-ordinates (x_i, y_i) of i and a random component. This can be expressed as

$$X_{\text{intra}}(i) = X_{\text{sys}}(x_i, y_i) + X_{\text{rand}}(i). \tag{3.89}$$

The overall variance in parameter X for the ith instance can then be expressed as

$$\sigma^2(i) = \sigma^2_{\text{inter}} + \sigma^2_{\text{sys}}(x_i, y_i) + \sigma^2_{\text{rand}}(i). \tag{3.90}$$

The systematic component can be expressed using PCA as

$$X_{\text{sys}}(x_i, y_i) = \sum_{j=1}^{p} a_{ij} Z_j \tag{3.91}$$

where Z_j are Gaussian RVs with zero mean and unit variance. Note that the only parameters we are interested in are p, which is the number of PCA components, and $\sum_j a_{ij}^2$ which is the variance of the spatially correlated component of variation in parameter $X(i)$. Using the above modeling approach we rewrite (3.87) as

$$Y(x) = \mathcal{P}(X_{\text{nom}} + X_{\text{inter}} + X_{\text{sys}}(x_i, y_i) + X_{\text{rand}}(i) \leq x,\ i = 1, 2, \ldots, n). \tag{3.92}$$

Let us normalize the RV representing inter-die variations and define $z_0 = X_{\text{inter}}/\sigma_{\text{inter}}$, and consider an event A that z_0 is u standard deviations away from its mean value. Now using *Bayes' Theorem*, which states that

$$\mathcal{P}(\mathcal{B}) = \sum_{x=-\infty}^{\infty} \mathcal{P}(\mathcal{B}|A = u)\mathcal{P}(A = u), \tag{3.93}$$

we can write

$$Y(x) = \int_{-\infty}^{\infty} \mathcal{P}(\ X_{\text{sys}}(x_i, y_i) + X_{\text{rand}}(i) \leq x - X_{\text{nom}} - \sigma_{\text{intra}} u,\ i = 1, 2, \ldots, n)\phi(u)\,\mathrm{d}u \tag{3.94}$$

where ϕ is the distribution of z_0, which is the standard Gaussian pdf. The probability term in the above expression depends on the intra-die component of variation alone. Now, let us define

$$X_{\max} = \max_i (X_{\text{sys}}(x_i, y_i) + X_{\text{rand}}(i)) \tag{3.95}$$

and let the cdf of $X_{\max}$ be $V_{\max}$, then we can rewrite (3.94) as

$$\begin{aligned} Y(x) &= \int_{-\infty}^{\infty} V_{\max}(x - X_{\text{nom}} - \sigma_{\text{intra}} u)\phi(u)\,\mathrm{d}u \\ &= E\left[V_{\max}(x - X_{\text{nom}} - \sigma_{\text{intra}} z_0)\right]. \end{aligned} \tag{3.96}$$

Now consider the case when $p = 1$ and note that this corresponds to the case when systematic intra-die variations are perfectly correlated across the die. This can be expected to be true for gate-length variations in very small designs. In this case we can rewrite (3.96) as

$$V_{\max}(a) = \int_{-\infty}^{\infty} \prod_{i=1}^{n} \mathcal{P}(X_{\text{rand}}(i) \le a - \sigma_{\text{sys}}(x_i, y_i)v)\phi(v)\,\mathrm{dv} \tag{3.97}$$

where v is a Gaussian RV with zero mean and unit variance. The probability in the above expression can be written as the cdf of a standard Gaussian RV Φ as

$$V_{\max}(a) = \int_{-\infty}^{\infty} \prod_{i=1}^{n} \Phi\left(\frac{a - \sigma_{\text{sys}}(x_i, y_i)v}{\sigma_{\text{rand}}(i)}\right)\phi(v)\,\mathrm{dv}. \tag{3.98}$$

Since Φ is a monotonic function we can lower bound $V_{\max}(a)$ as

$$\begin{aligned} V_{\max}(a) \ge& \int_{-\infty}^{0} \prod_{i=1}^{n} \Phi\left(\frac{a - v\min_i(\sigma_{\text{sys}}(x_i, y_i))}{\sigma_{\text{rand}}(i)}\right)\phi(v)\,\mathrm{dv} \\ &+ \int_{0}^{\infty} \prod_{i=1}^{n} \Phi\left(\frac{a - v\max_i(\sigma_{\text{sys}}(x_i, y_i))}{\sigma_{\text{rand}}(i)}\right)\phi(v)\,\mathrm{dv}. \end{aligned} \tag{3.99}$$

Similarly, taking the maximum and minimum value of the random variation, we can rewrite (3.99) when $a \ge 0$ as

$$\begin{aligned} V_{\max}(a) \ge& \int_{-\infty}^{0} \Phi^n\left(\frac{a - v\min_i(\sigma_{\text{sys}}(x_i, y_i))}{\max_i(\sigma_{\text{rand}}(i))}\right)\phi(v)\,\mathrm{dv} \\ &+ \int_{0}^{a/\max_i(\sigma_{\text{sys}}(x_i, y_i))} \Phi\left(\frac{a - v\max_i(\sigma_{\text{sys}}(x_i, y_i))}{\max_i(\sigma_{\text{rand}}(i))}\right)\phi(v)\,\mathrm{dv} \\ &+ \int_{a/\max_i(\sigma_{\text{sys}}(x_i, y_i))}^{\infty} \Phi\left(\frac{a - v\max_i(\sigma_{\text{sys}}(x_i, y_i))}{\min_i(\sigma_{\text{rand}}(i))}\right)\phi(v)\,\mathrm{dv} \end{aligned} \tag{3.100}$$

and when $a \le 0$ as

$$V_{\max}(a) \geq \int_{-\infty}^{a/\min_i(\sigma_{\text{sys}}(x_i,\,y_i))} \Phi^n\left(\frac{a - v\min_i(\sigma_{\text{sys}}(x_i,\,y_i))}{\max_i(\sigma_{\text{rand}}(i))}\right)\phi(v)\,\mathrm{dv}$$
$$+ \int_{a/\min_i(\sigma_{\text{sys}}(x_i,\,y_i))}^{0} \Phi\left(\frac{a - v\min_i(\sigma_{\text{sys}}(x_i,\,y_i))}{\min_i(\sigma_{\text{rand}}(i))}\right)\phi(v)\,\mathrm{dv}$$
$$+ \int_{0}^{\infty} \Phi\left(\frac{a - v\max_i(\sigma_{\text{sys}}(x_i,\,y_i))}{\min_i(\sigma_{\text{rand}}(i))}\right)\phi(v)\,\mathrm{dv}. \quad (3.101)$$

The bound developed above holds with equality if $\sigma_{\text{inter}} = \sigma_{\text{sys}} = \sigma_{\text{rand}} = \sigma$, in which case the expression for $V_{\max}(a)$ simplifies to

$$V_{\max}(a) = E\left[\Phi^n\left(\frac{a}{\sigma} - z_1\right)\right]. \quad (3.102)$$

Using (3.102) we can lower bound the yield as

$$Y(x) = E\left[\Phi^n\left(\frac{x - X_{\text{nom}}}{\sigma} - z_0 - z_1\right)\right]. \quad (3.103)$$

Now let us consider the case where $p > 1$. In this case we can write (3.96) as

$$V_{\max}(a) = \mathcal{P}\left(\sum_{j=1}^{p} a_{ij}z_j + X_{\text{rand}}(i) \leq a,\, i = 1, 2, \ldots, n\right)$$
$$= \int_{z_1=-\infty}^{\infty} \cdots \int_{z_2=-\infty}^{\infty} \left(\prod_{i=1}^{n} P_i(a)\right)\phi(z_1)\cdots\phi(z_p)\mathrm{d}z_1\cdots\mathrm{d}z_p \quad (3.104)$$

where

$$P_i(a) = \mathcal{P}\left(X_{\text{rand}}(i) \leq a - \sum_{j=1}^{p} a_{ij}z_j\right). \quad (3.105)$$

Using *Cauchy's inequality*

$$\sum_{j=1}^{n} a_{ij}z_j \leq \sqrt{\sum_{j=1}^{n} a_{ij}^2}\sqrt{\sum_{j=1}^{n} z_j^2} \quad (3.106)$$

we can develop a bound on $P_i(a)$ as

$$P_i(a) = \mathcal{P}\left(\sum_{j=1}^{p} a_{ij}z_j \leq a - X_{\text{rand}}(i)\right)$$

$$\geq \mathcal{P}\left(\sqrt{\sum_{j=1}^{n} a_{ij}^2}\sqrt{\sum_{j=1}^{n} z_j^2} \leq a - X_{\text{rand}}(i)\right) \tag{3.107}$$

$$= \mathcal{P}\left(X_{\text{rand}}(i) \leq a - \sigma_{\text{sys}}(x_i, y_i)\sqrt{\sum_{j=1}^{n} z_j^2}\right)$$

$$= \Phi\left(\frac{a - \sigma_{\text{sys}}(x_i, y_i)\sqrt{\sum_{j=1}^{n} z_j^2}}{\sigma_{\text{rand}}(i)}\right).$$

Through the use of Cauchy's inequality, [96] develops a bound for the yield that is independent of the $a'_{ij}s$ themselves, and is only dependent on the overall variance. Now, using the above expression $V_{\max}$ can be bounded as

$$V_{\max}(a) \geq E\left[\prod_{i=1}^{n} \Phi\left(\frac{a - \sigma_{\text{sys}}(x_i, y_i)\sqrt{\sum_{j=1}^{n} z_j^2}}{\sigma_{\text{rand}}(i)}\right)\right]$$

$$\geq E\left[\Phi^n\left(\frac{a - \max_i(\sigma_{\text{sys}}(x_i, y_i))\sqrt{\sum_{j=1}^{n} z_j^2}}{\sigma_{\text{rand}}(i)}\right)\right]. \tag{3.108}$$

From the theory of probability distributions we know that the squared sum of p independent Gaussian RVs has a *chi-square* distribution with p degrees of freedom, which is symbolically represented as χ_p^2. Thus, we can rewrite (3.108) as

$$V_{\max}(a) \geq \int_0^{\infty} \Phi^n\left(\frac{a - \max_i(\sigma_{\text{sys}}(x_i, y_i))\sqrt{u}}{\sigma_{\text{rand}}(i)}\right) f_{\chi_p^2}(u)\, du. \tag{3.109}$$

As in the case of $p = 1$, we can develop bounds for the cases when a is either positive or negative using the maximum and minimum values of the variance of the random component.

Let us now integrate the ideas developed above with a path-based timing analysis technique. Consider a set of N critical paths and assume that these paths are node-and-edge disjoint. This assumption effectively makes the delay distribution of each path independent in terms of correlations arising due to reconvergence, and the only correlation results from correlated variations in process parameters. For this simplified network, we can express the timing yield as

$$Y(t_0) = \mathcal{P}(D_j < t_0,\, j = 1, \ldots, N). \tag{3.110}$$

Now, we can develop bounds for the yield expression in (3.110) using the expressions developed above. However, as in the case of electromigration, as the number of paths tends to infinity the yield of a design goes to zero. To resolve this contradiction, [96] proposes to use a truncated normal distribution for process parameters and derives bounds on the yield of designs which is found to be independent of the number of paths n. Let us represent this bound as $Y_0(x)$. In the context of (3.110), if the desired yield is y we can write

$$t_0 = Y_0^{-1}(y). \tag{3.111}$$

If we assume that each of the paths is an M-gate path and that gate delay can be expressed as

$$D_i = \sum_{i=1}^{k} \alpha_i P_i \tag{3.112}$$

which represents the gate delay as a function of k process parameters (P_k) and $\alpha's$ capture the sensitivity of gate delay to each of the process parameters, then each gate delay should satisfy the constraint that

$$\frac{t_0}{M} \leq \sum_{i=1}^{k} \alpha_i P_i. \tag{3.113}$$

This implies that we can construct a worst-case delay model that can be used to perform traditional timing analysis on the design while guaranteeing that the desired timing yield is achieved. If we assume that the worst-case file is developed with equal margins for all process parameters, then we have the condition that

$$\frac{P_1}{\sigma_{P_1}} = \frac{P_2}{\sigma_{P_2}} = \cdots = \frac{P_k}{\sigma_{P_k}} = \Delta. \tag{3.114}$$

Combining (3.113) and (3.114) we get

$$\Delta = \frac{t_0}{M \sum_{i=1}^{k} \alpha_i \sigma_{P_i}} \tag{3.115}$$

which gives the point at which worst-case files should be developed to guarantee a desired yield y for the circuit. It is important to note that such a case-file can only be developed for cases where all critical paths have similar logical depths, and is not applicable to all DAG topologies.

3.5 Bayesian Networks

The Bayesian Network based approach was proposed in [19] and computes the exact pdf under the assumption that node delays are independent. Since the delay at the inputs of a gate are correlated due to reconvergent fanouts, it is not possible to compute the distribution of delay at the output node of the

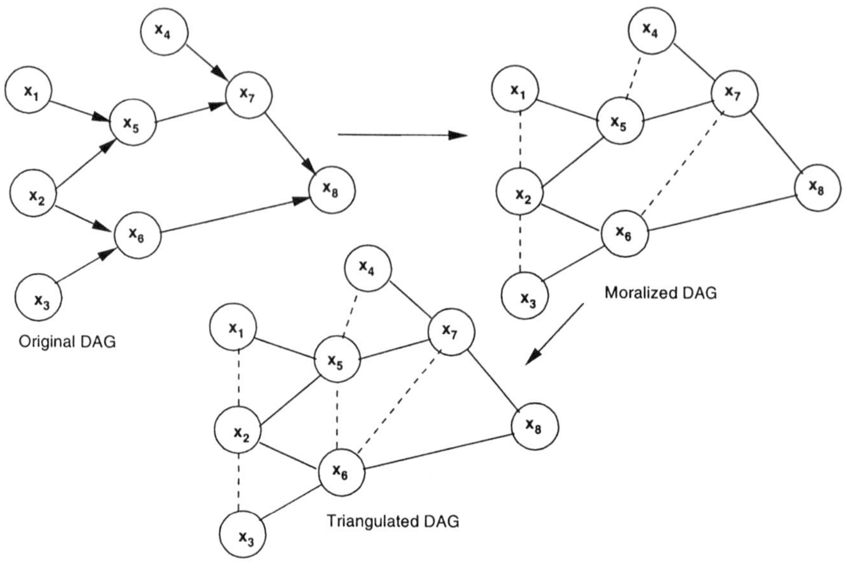

Fig. 3.19. A Bayesian Network for timing analysis is generated by moralizing and then triangulating the DAG [19]. (©2005 IEEE)

DAG based only on the individual distribution of delay at the inputs of the node. Developing an expression for the joint pdf in terms of the gate delay pdfs of all nodes in the circuit is also computationally infeasible. Although the approach based on Bayesian networks has exponential complexity, the complexity grows exponentially with the size of the largest clique in the circuit and not the circuit size itself. A *clique* is a subset of nodes in the circuit, such that each pair of nodes are connected by an edge. The size of the largest clique in the circuit grows much more slowly than circuit size and results in better performance using this approach. Now let us define a *Bayesian network.*

Definition 3.12. *A Bayesian network is a set of variables and a set of directed edges between the variables that form a DAG. Each variable A has a finite number of mutually exclusive states that it can take and if* $B_1, \ldots, B_n$ *are its predecessor nodes, then a conditional probability distribution* $\mathcal{P}(A|B_1, \ldots, B_n)$ *is associated with each node.*

The approach is based on breaking down the computation of the complete joint distribution of delay of each node in a circuit to smaller factors. Let us consider a DAG as shown in Fig. 3.19. The probability distribution of the delay of node 8 (X_8) can be represented as

$$\mathcal{P}(X_8) = \sum_{X_1,\ldots,X_7} \mathcal{P}(X_1, X_2, \ldots, X_8). \tag{3.116}$$

However, computing the complete joint distribution function of X_8 is computationally very expensive. If the DAG has n nodes with each node taking m discrete values, then the overall number of computation steps required in (3.116) is $O(m^n)$. However, the above expression can be simplified using Bayes' Theorem to

$$\mathcal{P}(X_8) = \sum_{X_1,\ldots,X_7} \mathcal{P}(X_8|X_1, X_2, \ldots, X_7)P(X_1, X_2, \ldots, X_7). \tag{3.117}$$

Since X_7 and X_6 are the only predecessor nodes of X_8, we have

$$\mathcal{P}(X_8|X_1, X_2, \ldots, X_7) = \mathcal{P}(X_8|X_6, X_7). \tag{3.118}$$

Using the above relation we can simplify (3.116), and then using expressions of the form (3.118) for each node we can finally write (3.116) as

$$\begin{aligned}\mathcal{P}(X_8) = \sum_{X_6,X_7} \mathcal{P}(X_8|X_6, X_7) \sum_{X_5}\sum_{X_4} \mathcal{P}(X_7|X_5, X_4) \sum_{X_2} \mathcal{P}(X_2) \\ \sum_{X_1} \mathcal{P}(X_5|X_2, X_1)\mathcal{P}(X_1) \sum_{X_3} \mathcal{P}(X_6|X_3, X_2)\mathcal{P}(X_3).\end{aligned} \tag{3.119}$$

Using the above equation the joint probability distribution can be broken down so that we do not need to compute the joint distribution of more than three variables at a time. The process of breaking down the joint distribution into factors is performed by first changing the DAG into a graph by removing the directionality with each edge. The graph is then *moralized* by connecting the predecessors of each node by an edge, since the delay pdf of a node can be completely determined by the joint distribution of the delay of the inputs. The next step involves *triangulation* of the graph to remove all chordless cycles of length greater than three. Thus, Bayesian networks ensure a partitioning of the initial DAG such that, by partitioning the circuit into cliques, the delay pdf of a node in the graph can be found by computing the joint distribution of nodes within a clique.

The cliques are then arranged within a *clique tree*, using techniques described in [40][64][110] as shown in Fig. 3.20. The joint probability distribution of X_1 and X_2 is then passed to clique C_1 to obtain the joint distribution of X_1, X_2 and X_5 as

$$\mathcal{P}(X_1, X_2, X_5) = \mathcal{P}(X_5|X_1, X_2)\mathcal{P}(X_1, X_2). \tag{3.120}$$

The same procedure is repeated for clique C_2. The computation for clique C_3 is performed as

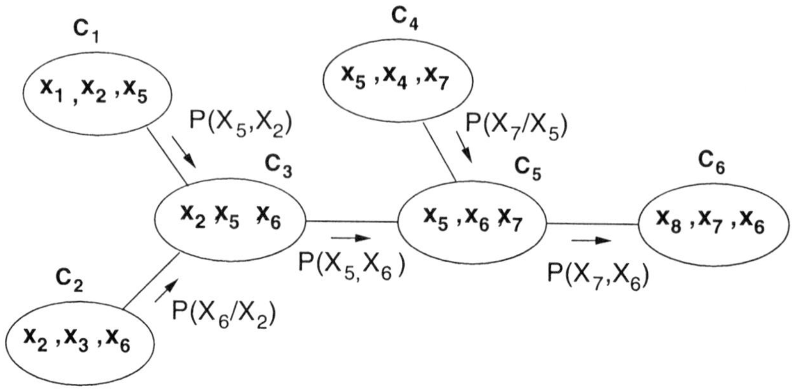

Fig. 3.20. A clique tree that can be traversed to completely determine the distribution of delay at the output node [19]. (©2005 IEEE)

Table 3.2. The number of nodes, edges and cliques in the DAG for the ISCAS'85 benchmark circuits, along with the size of the largest clique. The size of the largest clique is observed to grow much more slowly than the size of the circuit [19].

Circuit	Nodes	Edges	Max Clique	Cliques (%)
c17	11	12	4	8
c432	196	336	38	150
c499	243	408	32	183
c880	443	729	53	305
c1355	587	1064	49	402
c1908	913	1497	67	678
c2670	1426	2075	89	1084
c3540	1719	2936	189	1195
c5315	2485	4386	139	1701
c7552	3719	6144	77	2593

$$\begin{aligned}\mathcal{P}(X_2, X_5, X_6) &= \mathcal{P}(X_6|X_2)\mathcal{P}(X_5, X_2) \\ &= \frac{1}{\mathcal{P}(X_2)} \sum_{X_3} \mathcal{P}(X_2, X_3, X_6) \sum_{X_1} \mathcal{P}(X_1, X_2, X_5).\end{aligned} \tag{3.121}$$

All the computations as shown in Fig. 3.20 can be performed similarly to find the distribution for X_8. Using this approach the complexity of delay pdf computation for the output node can be reduced from $O(m^n)$ to $O(m^c)$, where c is the size of the largest clique in the graph. Table 3.2 shows the size of the largest clique for the ISCAS'85 [23] benchmark circuits. The size of the largest clique is much smaller than the circuit itself, and grows at a much smaller rate as circuit size increases.

The specification of the Bayesian network involves the computation of conditional probabilities for all combinations of the states of inputs and outputs. To reduce the computational overhead involved, [19] proposes to use the following techniques:

1) Fanin reduction: The number of conditional probabilities that need to be evaluated grows exponentially with the number of inputs of a gate. Thus, a large reduction in computation can be achieved by breaking down a large fanin gate into two stages, each with a smaller number of fanins.

2) Series reduction: Before constructing the Bayesian network, all series edges in a graph are merged into a single edge using the procedure described in Sec. 3.2.

3) Input reduction: If a node has two or more inputs, and the earliest arrival time at one of the nodes is greater than the latest arrival time at another node, then the connection to the latter node can be broken since it does not impact the pdf of delay at the output.

4

Statistical Power Analysis

The two main components of power dissipation are dynamic and static power dissipation. Dynamic power dissipation corresponds to power dissipated during the switching of nodes in a circuit and is spent in charging capacitances associated with the transistors and wires. On the other hand, static power dissipation corresponds to power dissipation due to the continuous flow of currents through the devices even in steady-state, when the logic states are not changing. In this chapter, we will develop techniques to statistically analyze different components of power. Let us first review some of the basics of power dissipation.

The dynamic power dissipation is given by the well-known equation

$$P_{Dyn} = V_{dd}^2 f \sum_{gates} C_{gate} P_{switch}, \tag{4.1}$$

where the summation is over all gates in the design. C_{gate} is the capacitance of a gate, V_{dd} is the supply voltage, f is the frequency of operation and P_{switch} is the switching probability of the gate. Here we have neglected the short-circuit component of dynamic power dissipation, which is due to the current that flows from the power supply to the ground when the devices are switching and both the pull-up and pull-down network of a gate are conducting. This component of power dissipation is generally small and can be safely neglected. However, it is important to note that if a design is highly unoptimized and has large transition times, then the short-circuit power dissipation can form a significant fraction of the total power dissipation.

Leakage power has grown with process scaling process to contribute a significant fraction of the total power budget. A study from Intel Corporation shows that leakage power will contribute approximately 50% of the total power dissipation in the 90 nm technology node. The prominence of leakage currents (I_{off}) in modern integrated circuits (ICs) has been spurred by the continued scaling of subthreshold voltage (V_{th}) and gate oxide thickness (T_{ox}). In addition, both subthreshold and gate leakage currents are known to be highly sensitive to process variations due to its exponential dependence on many of the key process parameters. Hence, it is critical to analyze leakage power statistically. The focus of this chapter will be to develop techniques which enable efficient and accurate statistical analysis of leakage power.

4.1 Overview

Leakage currents can arise due to varying phenomena. Reference [28] lists eight different mechanisms of leakage current. Not all these components of leakage are important during normal modes of operation, and subthreshold leakage (I_{sub}) and gate leakage (I_{gate}) currents are the most significant components of leakage current. In future technologies, band-to-band tunneling (BTBT) [121] leakage is expected to increase considerably and will form another major component of leakage power dissipation.

The exponential relationship dependence of I_{sub} on V_{th} and I_{gate} on T_{ox} is central to the problem of both leakage analysis and optimization. This continued scaling in V_{th} and T_{ox} in scaled technologies in order to maintain good device switching speeds at low supply voltages, has been the reason for large leakage currents. With the proliferation of portable applications that spend significant time in standby mode, large I_{off} values become a critical roadblock to improved battery lifetimes [62]. Thus, leakage power minimization has become a key objective and a number of methods for leakage reduction have been proposed for standby mode and during run-time [76],[66],[95],[128],[151],[57],[68],[149],[148],[107].

In addition to the rapid growth of I_{off} with each technology generation fluctuations of I_{off} from die to die or even gate to gate have also increased. This is especially true for subthreshold leakage currents, since controlling V_{th} is made more difficult in nanometer scale MOSFETs by Drain-Induced Barrier Lowering (DIBL) and discrete dopant effects [13]. While DIBL has been a problem since channel lengths first reached submicron dimensions, it is exacerbated in sub-100 nm devices by fundamental scaling limitations on oxide thickness (T_{ox}). Reductions in T_{ox} have kept DIBL at reasonable levels since the gate could also be more strongly coupled to the channel. Discrete dopant effects are important only in very narrow devices at advanced technologies but lead to potentially large random fluctuations in channel doping levels, and hence, V_{th}. In a projected 50 nm technology, the V_{th} 3σ uncertainty due to discrete dopant effects is expected to be comparable to the magnitude of the nominal V_{th} itself [13]. For T_{ox} values below 1.5 nm, gate oxide leakage effects become significant and limit the scalability of T_{ox}. Although gate oxide thickness is generally well controlled in a process, the strong exponential dependence of I_{gate} on T_{ox} causes large variations in I_{gate} due to small variations. The BTBT component of leakage, expected to become a major contributor in future generations, is also exponentially sensitive to variation in channel doping.

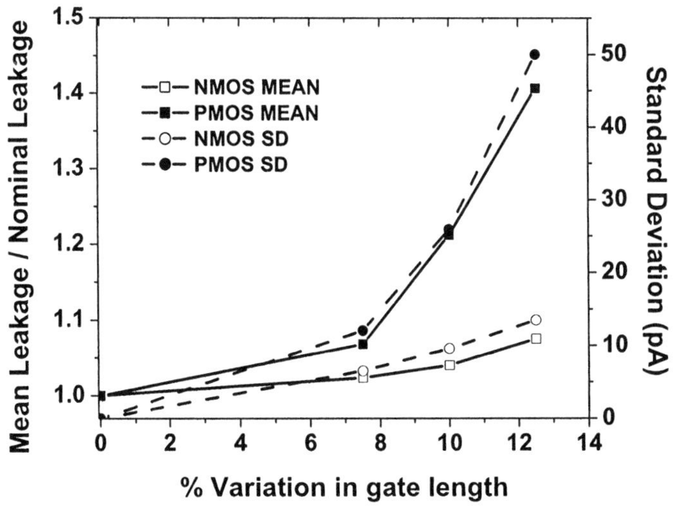

Fig. 4.1. Dependence of mean and standard deviation of leakage current on 3σ variation in gate-length.

With the growing uncertainty in process parameters, estimation of I_{off} for a device becomes difficult, making the use of traditional delay-oriented corner models for leakage analysis impractical [98]. Worst-case model files can easily exhibit 10–100X larger I_{off} than a nominal device, which leads to excessive guard-banding and overly conservative design practices. However, ignoring I_{off} variability altogether is also not an option since a small number of very leaky devices can easily dominate the total static power consumption. Figure 4.1 shows that the average leakage can be much larger (30% for PMOS with a 3σ variation in gate-length of 12.5%) than the nominal leakage due to the exponential dependence of current on the gate-length. This observation also invalidates the use of nominal device model files for even typical dies. The results also show that the degradation of PMOS leakage current with variations in the gate-length is much worse than the NMOS counterpart with the same degree of gate-length variation. This is due to the fact that DIBL effects in PMOS devices are typically worse than in NMOS devices [142].

Monte Carlo (MC) simulations provide a method to analyze the effect of process variations. However, MC techniques are very expensive in terms of time complexity and cannot be used to efficiently guide leakage optimization. Hence, an analytical approach to leakage current estimation is very useful to enable the prediction of leakage power in a design before it has been fabricated [69]. In this chapter, we first discuss leakage models and then discuss techniques to estimate the mean and variances of different components of power dissipation at the chip-level proposed in [69]. We then discuss an approach to estimate the mean and variances of leakage currents for a design described at the gate-level. A technique proposed to estimate the complete probability distribution function (pdf) of subthreshold leakage using empirical subthreshold equations considering both intra-die and inter-die components of variation is then presented. Finally, techniques to estimate the impact of environmental parameters on leakage currents proposed in [136] are discussed.

4.2 Leakage Models

We first discuss the traditional device equations that are used to model various components of leakage currents that we will be using throughout this chapter. The subthreshold leakage current is the current that flows between the source and drain of a device when the device is turned off. The charge transport occurs through diffusion along the surface of the device and is expressed as

$$I_{sub} = I_0 \exp\left(\frac{V_{gs} - V_{th}}{nV_T}\right)\left(1 - \exp\left(\frac{-V_{ds}}{V_T}\right)\right) \tag{4.2}$$

where

$$I_0 = \mu_0 C_{ox}(W/L_{eff})V_T^2(n-1) \tag{4.3}$$

where C_{ox} is the gate oxide capacitance, $V_T = KT/q$ is the thermal voltage, V_{th} is the threshold voltage of the device, and n is the subthreshold swing coefficient. The threshold voltage of a device depends on the source-to-body voltage V_{sb} and the drain-to-source voltage V_{ds} of the device due to body and DIBL effects, respectively, and can be expressed as

$$V_{th} = V_{fb} + |2\phi_p| + \frac{\lambda_b}{C_{ox}}\sqrt{2qN_{ch}\epsilon_s\left(|2\phi_p| + V_{sb}\right)} - \lambda_d V_{ds} \tag{4.4}$$

where V_{fb} is the flat-band voltage, ϕ_p is the surface potential, λ_b is the body effect factor, q is the charge of an electron, N_{ch} is the channel doping concentration, ϵ_s is the permittivity of Silicon, and λ_d is the DIBL coefficient.

The gate leakage current results from the tunneling of electrons (holes) from the substrate to the gate of a NMOS (PMOS) device. As shown in Fig. 4.2, the gate tunneling current is composed of several components. I_{gso} and I_{gdo} are the leakage currents that flow through the gate-to-S/D extension overlap regions, and I_{gc} is the gate-to-inverted channel tunneling current. A fraction of I_{gc} flows to the source (I_{gcs}) and the drain (I_{gcd}) [25]. The key dependency of gate leakage on process parameters can be expressed as [28]:

$$I_{gate} = WA_g\left(\frac{V_{dd}}{T_{ox}}\right)^2 \exp\left(-B_g\frac{T_{ox}}{V_{dd}}\right) \tag{4.5}$$

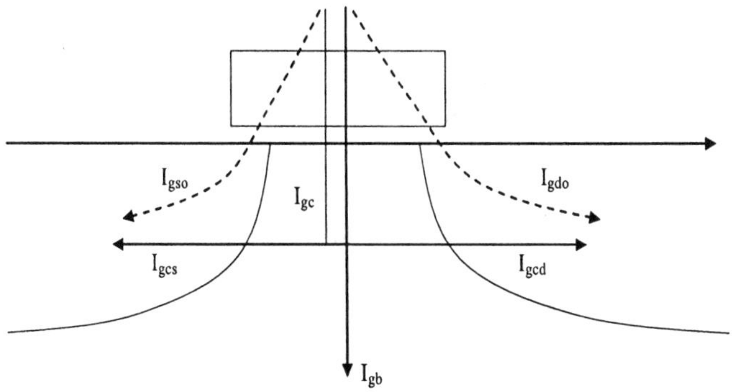

Fig. 4.2. Components of tunneling gate current.

where T_{ox} is the oxide thickness, W is the device width, and A_g and B_g are process dependent physical parameter. The equation shows that gate leakage is a strong function of gate oxide thickness.

The BTBT component of leakage of a device in the off state can be expressed as [93]:

$$I_{BTBT} = \sum_{\text{side, bottom}} W L_k A \frac{\xi_k}{E_g^{1/2}} V_{dd} \exp\left(\frac{-BE_g^{3/2}}{\xi_k}\right) \tag{4.6}$$

where $L_{\text{side}} = X_j$ the junction depth, $L_{\text{bottom}} = L_{SDE} + L_d$ (L_d being the length of the junction), ξ_{side} and ξ_{bottom} are the electric fields at the side and bottom junctions, A and B are physical parameters, and E_g is the band-gap voltage. The dominant component of BTBT leakage comes from the *side* component since the doping concentration is strongest at the sides of the junction, the above expression can be simplified as

$$I_{BTBT} = W X_j A \frac{\xi}{E_g^{1/2}} V_{dd} \exp\left(\frac{-BE_g^{3/2}}{\xi}\right) \tag{4.7}$$

where

$$\xi = \sqrt{\frac{2qN_{halo}N_{sd}}{\epsilon_{Si}(N_{halo} + N_{sd})}\left(V_{dd} + \frac{KT}{q}\ln\left(\frac{N_{halo}N_{sd}}{n_i^2}\right)\right)} \tag{4.8}$$

where N_{halo} is the halo doping concentration, N_{sd} is the source/drain doping concentration, and n_i is the intrinsic doping concentration.

These expressions define the dependence of various components of leakage currents on the device characteristics. We will discuss approaches to use these equations directly for statistical analysis, however it will become obvious as we go through the next few chapters that it is better to simplify these equations. This is achieved by capturing the strong sensitivities from the above expressions and using empirical expressions to model the leakage current components.

4.3 High-Level Statistical Analysis

High-level statistical analysis techniques are useful to estimate performance parameters of a design when detailed information about the design is not available. Early in the design stage detailed gate-level information is not available and only parameters such as the total device width and the relative fraction of on/off devices in a design are available. Considering 4.1 we note that due to the linear dependence of dynamic power dissipation on gate-length and gate width, any given variation in these process parameters results in a similar variation in dynamic power. However, leakage components are exponentially related to process parameters and small variations in these process parameters result in large variations in the leakage current itself.

Since leakage currents have very wide distributions, using worst-case models can result in huge overestimation of leakage. In addition, as noted before the nominal values of process parameters do not correspond to the average value of leakage currents. Such information can, therefore, become crucial in allowing the designers to make critical changes regarding leakage power dissipation early in the design process. For example, such information can point designers to sections of the design where excessive leakage power is consumed and specialized leakage reduction techniques can be utilized to control the leakage power dissipation for those sections of the design.

To estimate the impact of the within-die component of variation in leakage power, we discuss the approach proposed in [69]. The variations in the process parameters are assumed to be normally distributed. Given an estimate of the total device width in a design the average subthreshold leakage current considering variations can be expressed as

$$E\left[I_{sub}\right] = I_0 w \frac{1}{\sigma\sqrt{2\pi}} \int_{x_{min}}^{x_{max}} \exp\left(\frac{-(-x-\mu)^2}{2\sigma^2}\right) \exp\left(\frac{\mu - x}{a}\right) \mathrm{d}x \tag{4.9}$$

where I_0 is the nominal subthreshold leakage current per unit width, w is the total device width, μ and σ are the mean and standard deviation, respectively, of the process parameter represented as x [69]. The above equation can be used to consider the impact of variations in both gate-length or threshold voltage. In the case where x represents gate-length, a captures the relationship between gate-length and subthreshold leakage whose numerical value can be estimated using SPICE simulations. If the parameter x represents V_{th} then a will be nV_T as in (4.2). The second exponential term in the above equation thus captures the dependence of subthreshold leakage current on the process parameter being considered. The first exponential term in the above equation represents the Gaussian distribution of x and provides the fraction of device width that can be expected to be associated with a given value of the process parameter x. Equation (4.9) can then be re-written as

$$\begin{aligned} E[I_{sub}] &= I_0 w \frac{1}{\sigma\sqrt{2\pi}} \exp\left(\frac{\sigma^2}{2\lambda^2}\right) \int_{x_{min}}^{x_{max}} \exp\left(\frac{-(x-\mu)^2}{2\sigma^2} + \frac{\mu - x}{\lambda} + \frac{-\sigma^2}{2\lambda^2}\right) \mathrm{d}x \\ &= I_0 w \frac{1}{\sigma\sqrt{2\pi}} \exp\left(\frac{\sigma^2}{2\lambda^2}\right) \int_{x_{min}}^{x_{max}} \exp\left[\left(\frac{x-\mu}{\sqrt{2}\sigma} + \frac{\sigma}{\sqrt{2}\lambda}\right)^2\right] \mathrm{d}x. \qquad (4.10) \end{aligned}$$

The integral in the above equation is then rewritten using the transformation

$$t = \left[\frac{x-\mu}{\sqrt{2}\sigma} + \frac{\sigma}{\sqrt{2}\lambda}\right] \Rightarrow \mathrm{d}x = \sqrt{2}\sigma \mathrm{d}t \qquad (4.11)$$

which simplifies the integral to

$$\begin{aligned} E[I_{sub}] &= I_0 w \frac{1}{\sqrt{\pi}} \exp\left(\frac{\sigma^2}{2\lambda^2}\right) \int_{\frac{x_{min}-\mu}{\sqrt{2}\sigma}+\frac{\sigma}{\sqrt{2}\lambda}}^{\frac{x_{max}-\mu}{\sqrt{2}\sigma}+\frac{\sigma}{\sqrt{2}\lambda}} \exp\left(-t^2\right) \mathrm{d}t \\ &= \frac{I_0 w}{2} \exp\left(\frac{\sigma^2}{2\lambda^2}\right) \\ &\quad \left[\operatorname{erf}\left(\frac{x_{max}-\mu}{\sqrt{2}\sigma} + \frac{\sigma}{\sqrt{2}\lambda}\right) + \operatorname{erf}\left(\frac{\mu - x_{min}}{\sqrt{2}\sigma} - \frac{\sigma}{\sqrt{2}\lambda}\right)\right] \qquad (4.12) \end{aligned}$$

where erf is the error function. The algebraic details regarding the derivation of (4.12) can be found in [69]. Note that when $x \gg 1$ then $erf(x) \rightarrow 1$ and both terms in the error function in (4.12) become much greater than one. Therefore, the final simplified expression for leakage can be written as

$$E[I_{sub}] = I_0 w \exp\left(\frac{\sigma^2}{2\lambda^2}\right) \qquad (4.13)$$

The results in [69] present leakage power measurement data for 960 samples of a 180 nm 32-bit microprocessor. Upper bound of leakage current was estimated by assuming that all gates are operating at their worst-case corner, while the lower bound was estimated by using nominal values for all process parameters. Results show that for most of the samples, using a lower bound as an estimate underestimates the leakage by as much as 6.5X and using an upper bound results in overestimation by 1.5X. The technique discussed above shows a good correlation with data and the calculated leakage is within 20% of the measured value for more than 50% of the samples, as compared to 11% and 0.2% when upper and lower bounds are used as estimates.

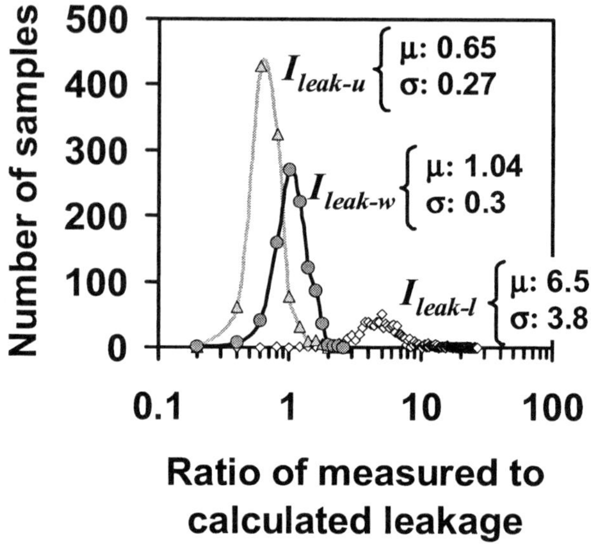

Fig. 4.3. Ratio of measured to calculated subthreshold leakage current distribution for the upper and lower bound analysis techniques described [69]. (©2005 IEEE)

In general, The technique can be extended to estimate other components of leakage power as well. The key step in this technique was to simplify the subthreshold leakage model to only have an exponential dependence on the process parameter under consideration (if the dependence is expressed linearly, as in the case of device width, then the problem boils down to the situation in dynamic power estimation). Though the simple model is inexact, as can be easily seen by looking at the expressions in (4.2) and (4.5), reasonable accuracy can be achieved by using a fitting parameter (a in this case). Models which satisfy this criterion have been developed for gate leakage [78] and are found to provide good fidelity and accuracy. The approach outlined above can then be directly extended to consider the impact of variations in gate leakage. Similar simplifications of leakage expressions will form the cornerstone of many of the techniques that we will discuss in later sections.

4.4 Gate-Level Statistical Analysis

In this section we will develop techniques to analyze the power dissipation of a gate-level design. As compared to the previous section, the approaches discussed in this section will be concerned with the estimation of leakage currents for individual gates and then the summation of these estimates to calculate the overall leakage of a design. In this section, we will first develop a general technique that has been used to estimate the mean and variance of leakage currents [129] and present the application of this technique to estimate the mean and variance of various leakage current components. Later in the section we will discuss the techniques to estimate the complete pdf of subthreshold leakage current considering both intra-die and inter-die variations [117]. The technique used to integrate intra-die and inter-die variation is fairly general and can be easily incorporated into other approaches, that consider only intra-die variation, to consider both components of variation as well.

4.4.1 Dynamic Power

If we consider the equation of dynamic power dissipation (4.1) it is evident that process variation result in variation in dynamic power only through variation in the switched capacitance, as the switched capacitance varies linearly with gate-length and gate widths. Furthermore, the variation in dynamic power dissipation is much smaller compared to variations in leakage power which depends exponentially on a number of process parameters, as we saw earlier. The impact of variations in gate-length or gate width on dynamic power dissipation can be easily estimated since the dynamic power dissipation is a weighted sum of the individual random variables (RVs) (representing gate-length or gate width). The gate-lengths are assumed to come from a multinormal probability distribution, and can be mapped to a linear combination of independent RVs using principal component analysis as discussed in Chap. 2. Therefore, the sum can be represented as another Gaussian RV. The parameters of the Gaussian RV are estimated using the following property of Gaussian RVs.

Let X_i be an independent Gaussian RVs with mean μ_i and standard deviation σ_i and let Y be a linear combination of the X_i's which is expressed as:

$$Y = \sum_i a_i X_i + b \tag{4.14}$$

then the mean and variance of Y can be expressed as:

$$\begin{aligned} \mu_Y &= \sum_i a_i \mu_i + b \\ \sigma_Y &= \sqrt{\sum_i a_i^2 \sigma_i^2}. \end{aligned} \tag{4.15}$$

On the other hand, dynamic power dissipation has an inverse dependence on T_{ox} and, hence it is not as straightforward to consider this impact. It is important to note that the T_{ox} variations are generally very well controlled as compared to the variations in gate-length and gate width, and the variations in T_{ox} can generally be ignored while considering dynamic power. Also note that variation in process parameters results in a similar variation in dynamic power in terms of the ratio of SD and mean. Hence, variations in dynamic power dissipation have not been a pressing concern, compared to the variation exhibited by leakage power.

4.4.2 Leakage Power

The first step to estimate the overall pdf of leakage power dissipation is to estimate the mean and variance of different components of power dissipation. As in the previous section, we assume that the variations in the process parameters are normally distributed.

Estimating Parameters of the Distribution

Considering only intra-die variations, let $g(x)$ represent the dependence of some component of leakage current on a process parameter x which is assumed to be distributed according to the distribution function $f(x)$. The mean or the expected value of $g(x)$ can then be expressed as

$$E\left[g(x)\right] = \int_{-\infty}^{\infty} g(x) f(x) \mathrm{d}x \tag{4.16}$$

using the Law Of The Unconscious Statistician (LOTUS). Then using Taylor's formula, $g(x)$ can be expanded around the mean value of x and (4.16) can be re-written as

$$E\left[g(x)\right] = \int_{-\infty}^{\infty} \left(\sum_{n=0}^{\infty} \frac{g^n(\eta)}{n!} (x-\eta)^n \right) f(x) \mathrm{d}x \tag{4.17}$$

$$= \sum_{n=0}^{\infty} \frac{g^n(\eta)}{n!} \int_{-\infty}^{\infty} (x-\eta)^n f(x) \mathrm{d}x \tag{4.18}$$

where η is the expected value of $f(x)$. The term within the integral in (4.18) corresponds to the central moments of $f(x)$. If $f(x)$ is assumed to be Gaussian then only the terms corresponding to even values of n contribute to the sum.

Since variations in process parameters are generally within a range of 10–30% of the mean value, we can assume that x is concentrated around its mean value. Using this assumption, we can neglect higher order terms in (4.18). Note that improvements in accuracy can only be obtained by considering two additional higher order derivatives,. Each such addition in (4.18) provides two orders of improvement in accuracy. A similar approach can be used to estimate higher order moments of $g(x)$, where instead of taking the Taylor's expansion of $g(x)$, the appropriate function of $g(x)$ is used.

Having established the general approach to estimate the parameters of any source of power dissipation, let us now consider the case of subthreshold leakage in a single device which is turned off. In the off-state, the gate-to-source voltage (V_{gs}) of a device is zero and the drain-to-source voltage (V_{ds}) is V_{dd}, which results in the simplified expression for subthreshold leakage as

$$I_{sub} = I_0 \left(1 - \exp\left(\frac{-V_{dd}}{V_T}\right)\right) \exp\left(\frac{-V_{th}}{nV_T}\right) = I_0' \exp\left(\frac{-V_{th}}{nV_T}\right). \tag{4.19}$$

. Equation (4.19) shows that the subthreshold leakage current is a function of the threshold voltage, drawn dimensions, and gate-oxide thickness. Hence, to estimate the variation in subthreshold leakage, the dependence of threshold voltage on other process parameters needs to be established.

The approach developed in [129] uses simplified expressions for the body effect and DIBL coefficient to estimate this dependence. If we consider variations in gate-length alone, the variation in threshold voltage can be expressed as

$$\frac{\Delta V_{th}}{\Delta L} = \frac{\partial V_{th}}{\partial \lambda_d}\frac{\mathrm{d}\lambda_d}{\mathrm{d}L} + \frac{\partial V_{th}}{\partial \lambda_b}\frac{\mathrm{d}\lambda_b}{\mathrm{d}L} = K_L^{V_{th}}. \tag{4.20}$$

The dependence of the body effect coefficient λ_b on the process parameters is expressed as

$$\lambda_b = 1 - \left(\sqrt{1 + \frac{2W}{X_j}} - 1\right)\frac{X_j}{L} \tag{4.21}$$

On the other hand, the physical dependence of λ_d on process parameters is much more complicated and empirical expression developed in [97] can be used. Using (4.20), and isolating the terms in (4.19) which depend on gate-length we obtain:

$$I_{sub} = \frac{I_0' L_{nom}}{L} \exp\left(\frac{-V_{th} + K_L^{V_{th}}(L - L_{nom})}{nV_T}\right) = \frac{K_0}{L}\mathrm{e}^{K_1 L}. \tag{4.22}$$

Hence, using $g_L(x) = \mathrm{e}^{K_1 x}/x$ in (4.16) we can estimate the mean of leakage power considering variations in gate-length. Expressions similar to (4.20) are developed in [129] to estimate the variations in subthreshold leakage due to variations in other process parameters. This approach to estimate the dependence of threshold voltage on process parameters neglects all second-order effects in the body-effect coefficient and DIBL and, therefore, results in inaccuracies. An improvement to this approach was recently proposed in [156]. In this work the dependence of $\mathrm{d}V_{th}/\mathrm{d}L$ in (4.20) on gate-length is captured by calculating its average value over the range of variations in gate-length ($\pm 3\sigma$).

Having analyzed the simplified problem of evaluating the mean and variance of an individual device, let us now consider the case when a set of devices are connected in parallel or series, which is used in the construction of most gates. Since the different devices in a gate are in close proximity of each other, the RV defining the variations in these gates can be assumed to be perfectly correlated. We now consider the case where there are n devices in parallel. In this case, the overall leakage $I_{sub,parr}$ can be expressed as

$$I_{sub,parr} = \sum_{i=1}^{n} I_{sub,i} \tag{4.23}$$

where $I_{sub,i}$ is the individual leakage of each device. The evaluation of the mean and variance of the set of parallel devices now becomes straightforward and can be expressed as

$$\begin{aligned} \mu\left[I_{sub,parr}\right] &= \sum_{i=1}^{n} \mu\left[I_{sub,i}\right] \\ Var\left[I_{sub,parr}\right] &= \sum_{i=1}^{n} \sigma^2\left[I_{sub,i}\right]. \end{aligned} \tag{4.24}$$

The case of series-connected devices is much more complicated since no accurate and simple expression is known to exactly estimate the leakage current through a stack of off devices. Various approaches for the analysis of leakage current in stacks have been developed [33], [56]. The approach in [56] is simple but is not found to provide accurate results. The approach in [33] is more accurate and general in the sense that it can model stacks of arbitrary length. Based on this approach, statistical models for the leakage current of a stack with two off transistors can be obtained with some minor assumptions. The approach can in theory be extended to stacks with a larger number of off transistors, although the complexity increases rapidly with the stack length. For the purpose of analysis, [33] assumes that the on transistors in a stack behave as short circuits which is true except for the case when the top-most transistor of the stack is on and induces a V_{th} drop in the voltage seen by the rest of the stack. Barring this case, the source-drain voltage of the lower off transistor (since we have only two off transistors in the case) is given by [33],

$$V_{ds} = \frac{nV_T}{1 + 2\lambda_d + \lambda_b} \ln\left[\exp\left(\frac{\lambda_d V_{dd}}{nV_T}\right) + 1\right]. \tag{4.25}$$

The leakage current expression (4.2) for a stack can then be simplified to

$$I_{sub,series} = I_0 \exp\left(\frac{-V_{th}}{nV_T}\right) - I_0 \exp\left(\frac{-(V_{th} + nV_{ds})}{nV_T}\right) \tag{4.26}$$

which expresses the leakage current as a difference of two exponential terms. To estimate the mean and SD of this expression we utilize the fact that

$$E[X - Y] = E[X] - E[Y]$$
$$E[(X - Y)^2] = E[X^2] + E[Y^2] - 2E[XY] \tag{4.27}$$

where X and Y are any two RVs. Note that, under the assumption that variation in V_{ds} are small, the terms whose expected values need to be evaluated are in the same form as (4.19) and the same steps as discussed above can be repeated. Again, we notice that approximations need to be made while performing statistical analysis, if physical models for leakage currents are used, which can result in inaccuracies.

The general approach outlined in (4.16)–(4.18) has also been used to estimate the mean and variance of gate leakage and band-to-band tunneling (BTBT) leakage [121] currents in [94]. Having established techniques to estimate the mean and variance of different components of leakage power dissipation, we now discuss techniques to estimate the complete pdf of leakage currents. First, we develop a new empirical model to obtain the desired accuracy with certain key characteristics which simplifies the estimation of the complete pdf while providing reasonable accuracy.

Simultaneous variation of multiple parameters

The approach discussed above can be easily extended to consider simultaneous variations in process parameters. The problem of evaluating the leakage in this case is simplified by our earlier approximation of linearizing the effect of the change in threshold voltage with the process parameters (4.22). Under these assumptions the expression of the subthreshold current can be expressed in the form of (4.22) as

$$I_{sub} = Kg_1(x)g_2(y) \tag{4.28}$$

where x and y are two different process parameters. Ideally, the choice of process parameters which are used in the analysis should be made such that the parameters are independent of each other. This implies that parameters such as gate-length and threshold voltage should not be used as process parameters since variations in gate-length result in variations in threshold voltage. A better approach in this case is to consider variations in gate-length, channel doping and gate oxide thickness, which can be assumed to have independent variations. Under this assumption, we have a product of two independent RV in (4.28), which can be handled using the fact that the expectation of the product of two independent RVs is the same as the product of their expectations. Using this fact, we can estimate the mean and standard deviation of the leakage current in terms of the functions in (4.28) as

$$\mu[I_{sub}] = E[Kg_1(x)g_2(y)] = KE[g_1(x)]E[g_2(y)]$$
$$\sigma[I_{sub}] = \sqrt{K^2(E[g_1^2(x)]E[g_2^2(y)] - E^2[g_1(x)]E^2[g_2(y)])}. \quad (4.29)$$

The terms in the right hand side of (4.29) can be evaluated using the technique developed to estimate moments of leakage currents considering variations in a single parameter.

The same ideas can also be extended to estimate the parameters of total leakage [94]. The total leakage can be expressed as:

$$I_{leak} = I_{gdo} + I_{sub} \quad (4.30)$$

Then, the mean and SD of total leakage can be obtained using the first two moments, which are expressed as:

$$E[I_{leak}] = E[I_{sub}] + E[I_{gdo}]$$
$$E[I_{leak}^2] = E[I_{sub}^2] + E[I_{gdo}^2] + 2E[I_{sub}I_{gdo}]. \quad (4.31)$$

The cross product term has the same exponential form as the other components of leakage currents and is treated similarly, by writing it as a product of functions of single process parameters.

The results obtained in [94] are shown in Tables 4.1 and 4.2 where V_{fb} is the flat-band voltage, N_{dep} is the channel doping concentration and N_{halo} is the halo doping concentration. The results show that Monte Carlo and analytical results track very well except for cases when the variations are larger than 20%, where the SD shows significant error. This could result from the fact that only the first few terms are retained in the Taylor's expansion in (4.18) and higher order terms become important when variations in process parameters are larger. As expected, gate leakage shows the strongest sensitivity to variations in T_{ox} due to its strong exponential relationship. Subthreshold leakage shows strong sensitivity to variations in L, T_{ox}, and V_{fb}, but is much less sensitive to variations in doping concentrations, corroborating the results found in [129]. Generally, variations in T_{ox} are very well controlled and most of the variations in I_{sub} result from variations in gate-length and threshold voltage.

Estimating the probability density function

We begin by describing the method to compute an analytical expression for the pdf of subthreshold leakage for an individual device using simplified empirical models. First, the dependence of I_{sub} on L is characterized by the function h such that $I_{sub} = h(L)$, which is then used to determine the inverse function $g(I_{sub})$, that expresses L as a function of $I_{sub} : L = h^{-1}(I_{sub}) = g(I_{sub})$. In order to compute the pdf of the leakage, it is essential that:

Table 4.1. Comparison of the analytical approach with Monte Carlo simulations in estimating the impact of process variation on I_{gdo} (Nominal value is 8.43nA) [94].

Parameter Varied	Variation (3σ)	Mean(nA) Monte Carlo	Mean(nA) Analytical	SD(nA) Monte Carlo	SD(nA) Analytical
V_{dd}	20%	8.54	8.56	1.71	1.68
T_{ox}	10%	9.30	9.32	4.43	4.29
T_{ox}	10%	12.50	12.55	14.20	11.00
L_{SDE}	20%	8.44	8.43	0.56	0.56
W	10%	8.43	8.43	0.28	0.28
All	10%	9.40	9.36	4.55	4.46
All	20%	12.89	12.74	14.82	11.80

Table 4.2. Comparison of the analytical approach with Monte Carlo simulations in estimating the impact of process variation on I_{sub} (Nominal value is 3.72nA) [94].

Parameter Varied	Variation (3σ)	Mean(nA) Monte Carlo	Mean(nA) Analytical	SD(nA) Monte Carlo	SD(nA) Analytical
V_{fb}	10%	7.09	6.08	11.70	7.54
V_{dd}	20%	3.74	3.74	0.46	0.43
N_{pocket}	20%	4.44	4.45	2.91	2.62
N_{dep}	20%	3.78	3.79	0.72	0.72
L	20%	6.97	6.27	13.62	9.45
T_{ox}	20%	4.51	4.54	3.17	2.88
W	10%	3.73	3.73	0.38	0.38
All	10%	9.11	8.38	19.18	11.66
All	20% (V_{fb}=10%)	17.55	15.00	61.43	38.00

(1) the function g is a closed-form expression, and
(2) the function h is differentiable over the given range of currents.

Unfortunately, the complexity of the relationship between leakage current and channel length (i.e., the function $h(L)$) does not allow for the derivation of $g(I_{sub})$ such that it satisfies these two conditions. Therefore, an approximate empirical fit has to be used for the function $h(L)$ so that the required inverse function can be computed while maintaining good accuracy. Given the closed form expression of $g(I_{sub})$ and the pdf of $L = f_x(L)$, we can express the pdf of I_{sub} using the above expressions as [109]:

$$f_y(I_{sub}) = \frac{f_x(g(I_{sub}))}{h'(L)} \tag{4.32}$$

where f_y is the pdf of I_{sub}. In this analysis, we assume that the drawn gate-length has a Gaussian distribution with a fixed mean μ and standard deviation σ. Using these assumptions the pdf of I_{sub} can be written as follows:

$$f_y(I_{sub}) = \left(\frac{1}{h'(L)}\right)\left(\frac{1}{\sigma\sqrt{2\pi}}\right)\exp\left(\frac{-(g(I_{sub})-\mu)^2}{2\sigma^2}\right). \tag{4.33}$$

Finally, to calculate the mean and standard deviation of the leakage current distribution of a gate, we perform numerical integration of $f_y(I_{sub})$ over the given range of leakage currents:

$$\mu[I_{sub}] = E[I_{sub}] = \sum_i I_{sub} f_y(I_{sub})$$

$$\sigma[I_{sub}] = \sqrt{\left(\sum_i I_{sub}^2 f_y(I_{sub}) - \left(\sum_i I_{sub} f_y(I_{sub})\right)^2\right)}. \tag{4.34}$$

The next step is to compute $f_y(I_{sub})$ in more detail for a single device. We initially discuss the approach for a single device and then extend the approach for a stack of two or more transistors.

Single Transistor Stacks (Inverters)

Based on the BSIM3 device model, the subthreshold current through a device can be expressed as (4.2). The term $(1 - \exp(-V_{ds}/V_T))$ can be neglected for an inverter since $V_{ds} = V_{dd}$ is much greater than the thermal voltage V_T. We also set $V_{gs} = 0$ since the source nodes of either device in an inverter are tied directly to a supply rail or to the ground rail. V_{th} is the threshold voltage and is given by (4.4). These equations in principle enable us to calculate the mean and standard deviation (SD) using the device model-files for a given technology. However, analytical expressions for leakage current based on these parameters are found to fit very poorly even for 180 nm technologies. In particular, nebulous definitions for the values for technology constants such as N_{sub} and X_j produce large errors in the analytical current expressions. Furthermore, the models for body-effect coefficient and particularly DIBL coefficient used in the previous section to estimate the mean and variance of subthreshold leakage are inadequate and can produce unrealistically small values for these parameters resulting in large errors in the values for leakage currents.

Note that the actual BSIM3 model used to compute leakage current in SPICE simulations is much more complex than these simplified expressions. In addition, the constraints placed on functions g and h necessitates the use of further simplifications to derive a suitable analytical expression for current in terms of drawn gate-length. Figure 4.4 shows that a simplified BSIM3

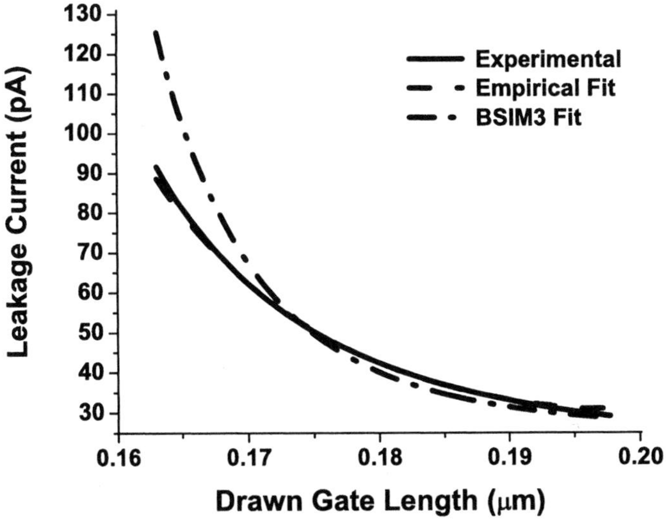

Fig. 4.4. Comparison of the BSIM3 fit and analytical fit for $h(L)$ with results from SPICE.

model vastly overestimates the leakage current for devices with gate-lengths that deviate by more than 5% from the nominal value. Since these conditions correspond to the devices that contribute a large portion of leakage current, the resulting pdf will be skewed to the right, rendering the simplified BSIM3 fit unacceptable. Therefore, we use a new empirical model to express leakage current I as a function of L. This empirical model is expressed as

$$I_{sub} = q_1 \exp\left(q_2 L + q_3 L^2\right) = h(L) \tag{4.35}$$

This expression circumvents the use of V_{th} as an intermediate variable in expressing the current as a function of the gate-length. However, it maintains the general form of the BSIM3 model and has the following properties:

(1) It preserves the exponential dependency of I on L.
(2) It is easily invertible (as shown below).
(3) It yields closed form expressions for both I and L.
(4) It accurately fits currents for both individual NMOS/PMOS as well as transistor stacks.

Figure 4.4 also shows the comparison between the values for leakage current obtained from SPICE simulations and the values obtained from both the simplified BSIM3 fit and the empirical fit for a single stacked device (4.35) for a 10% variation in gate-length. From the plot it can be seen that the empirical model provides a much better fit over a wide range of channel lengths.

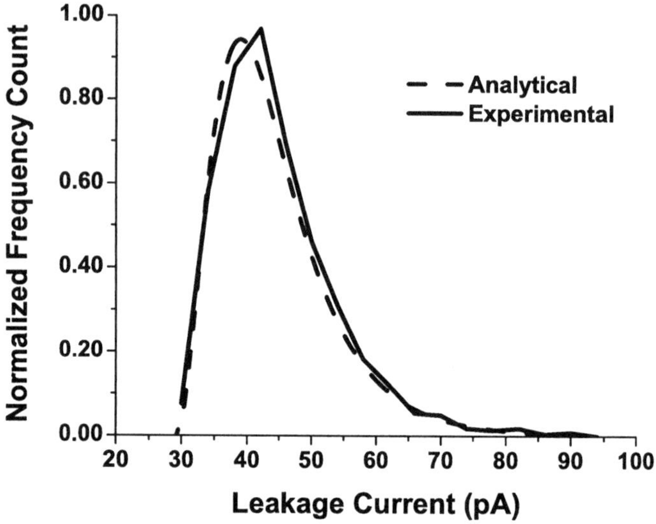

Fig. 4.5. Comparison of the SPICE pdf with the analytical pdf

Equation 4.35 is a simple exponential quadratic equation that can be inverted to obtain an analytical expression for L as follows:

$$L = \frac{1}{2q_3}\left(-q_2 + \sqrt{q_2^2 - 4q_3 \ln\left(\frac{q_1}{I_{sub}}\right)}\right) = g(I_{sub}) \qquad (4.36)$$

Using the expressions (4.33) and (4.34) with the functions g and h as specified by (4.35) and (4.36), we can obtain the pdf of I. Figure 4.5 presents the comparison between the pdf obtained from SPICE simulations and the pdf obtained analytically for a single stacked device with 10% 3σ variation in gate-length. The plots of the pdfs, including the tail portion, match well and have a lognormal shape.

Series-Connected Devices (Stacks)

In the case of a stack of transistors, the gate-length variation impacts the leakage current of the bottom transistor in the stack in two ways:

1) gate-length variation of the bottom transistor directly modulates its threshold voltage.

2) gate-length variation of the top transistor indirectly affects the leakage of the bottom transistor by altering the voltage drop across the top transistors of the stack.

Hence, the analytical expression of current as a function of gate-length is more complex for a stack of multiple transistors. Since the devices in a stack are placed close together in the layout, the simplifying assumption can be made that their gate-length variations are perfectly correlated. Similar empirical expressions for stacks of two and three transistors are also derived, and the method can be extended to stacks of arbitrary length in a straightforward manner.

In an inverter the term $(1 - \exp(-V_{ds}/V_T))$ in (4.2) was neglected since the drain-source voltage V_{ds} in the leaking device was much greater than the thermal voltage V_T. However, for a device with stacks of two or three transistors, the value of the intermediate node voltage (V_{ds2} and V_{ds3}) is much lower. The empirical model (4.35) is sufficiently general enough to model the leakage current in stacked circuits using the same general form. The current can be empirically modeled with a new set of fitting parameters in (4.35). Naturally, this set of constants is different for different stack depths and also for NMOS and PMOS devices, since the drain-source voltages differ in these situations. Equation (4.36) is then solved using the suitable coefficients in the quadratic expression to obtain the value of channel length as a function of I_{sub}. The pdf for stacked devices can be similarly determined.

As discussed in the previous section, the on transistors in a stack can be approximated as a short circuit except for the case in which the on transistor is at the top (bottom) of a NMOS (PMOS) stack. This effect can be modeled by estimating the leakage current under the assumption that the V_{th} drop is a constant value that corresponds to the nominal V_{th} of the device. This allows us to use the same models for stacks of transistors with an effectively reduced power supply voltage.

Leakage Distribution of Circuit Blocks

Having developed a methodology to accurately predict the pdf of a single gate, we will discuss the approach to estimate the leakage current distribution of circuit blocks considering within-die variations. For now, we assume that process parameters in different gates are independent of each other and hence uncorrelated. Since the distribution of the leakage currents of a single gate is close to lognormal the leakage current for a circuit block as a whole is a sum of lognormals. Thus, to find the distribution of the total leakage current, given k lognormal RVs we need to find the distribution of the sum S given as

$$S = X_1 + X_2 + \cdots + X_k = \mathrm{e}^{Y_1} + \mathrm{e}^{Y_2} + \cdots + \mathrm{e}^{Y_k} \tag{4.37}$$

where $X_1, \ldots, X_k$ are independent lognormal RVs. Sums of lognormals, assuming independence, can be well approximated by another lognormal RV [16]. Various approaches are known to estimate the parameters of the final lognormal used to approximate the sum. As shown in [16] the simple *Wilkinson's approximation* [127] is more accurate as compared to other complex approaches for our range of interest in the cumulative distribution function (cdf) of leakage current. In Wilkinson's approximation the sum of the mean and variance of the individual gate leakage current distributions, $X_1, X_2, \ldots, X_k$ is matched with the first two moments of S, which gives

$$\begin{aligned} \mu[S] &= E[S] = \mu_1 + \mu_2 + \cdots + \mu_k \\ \sigma[S] &= \sqrt{\sigma_1^2 + \sigma_2^2 + \cdots + \sigma_k^2} \end{aligned} \tag{4.38}$$

where the $\mu's$ and $\sigma's$ are the means and standard deviations of the leakage currents of the individual gates. To express the resulting pdf of the circuit subthreshold leakage current as a lognormal, we note that the lognormal pdf is given as

$$f(x) = \frac{1}{x\beta\sqrt{2\pi}} \exp\left(\frac{-(\ln(x) - \alpha)^2}{2\beta^2}\right) \tag{4.39}$$

where α and β are the parameters of the lognormal distribution. If $Y(\mu, \sigma)$ is a Gaussian random variable and the corresponding lognormal X is related to Y as $X = \exp(Y)$, then the parameters of the lognormal are the mean and variance of the corresponding Gaussian distribution. We can compute these parameters based on the mean and variance of the lognormal. The mean and variance of the lognormal can be expressed as a function of its parameters as:

$$\begin{aligned} E[X] &= \exp\left(\alpha + \beta^2/2\right) \\ Var[X] &= \exp\left(2(\alpha + \beta^2/2)\right) - \exp(2\alpha + \beta^2). \end{aligned} \tag{4.40}$$

Equation (4.40) can be solved for α and β (the mean and variance of the Gaussian that are the parameters of the lognormal) in terms of the mean and variance of the lognormal as:

$$\begin{aligned} \alpha &= \frac{1}{2} \ln\left(\frac{E^4[X]}{E^2[X] + Var[X]}\right) \\ \beta^2 &= \ln\left(\frac{Var[X] + E^2[X]}{E^2[X]}\right). \end{aligned} \tag{4.41}$$

The parameters of the lognormal can then obtained using (4.41), which completely determines the pdf of the leakage current of the circuit block considering uncorrelated within-die variations. Note that for large circuit blocks the leakage current distribution will approach a Gaussian due to the *central limit theorem* [109]. On the other hand, as shown in [88], both S (4.37) as well as the log of S can be approximated by a lognormal when a large number of independent lognormals are summed. Thus, for large k, the shape of a lognormal distribution tends towards the shape of a Gaussian distribution [47], and using a lognormal distribution to approximate sums of lognormals is justified.

Table 4.3. Comparison of subthreshold leakage estimated obtained using the analytical approach with Monte Carlo simulations considering intra-die and inter-die variations in gate-length.

Circuit	Mean(nA) Monte Carlo	Mean(nA) Analytical	Error(%)	SD(pA) Monte Carlo	SD(pA) Analytical	Error(%)
c17	0.2	0.3	8.3	36.0	37.0	2.8
c432	7.1	7.2	1.4	190.0	210.0	10.5
c499	19.0	20.0	5.3	280.0	330.0	17.9
c880	17.0	17.0	0.0	280.0	330.0	17.9
c1355	21.0	22.0	4.8	320.0	370.0	15.6
c1908	16.0	17.0	6.3	260.0	300.0	15.4
c2670	32.0	33.0	3.1	350.0	410.0	17.1
c3540	39.0	40.0	2.6	420.0	480.0	14.3
c6288	120.0	120.0	0.0	900.0	1010.0	12.2

Table 4.3 compares the results obtained using the approach discussed above as compared to Monte Carlo simulations for the ISCAS'85 benchmark circuits [23], for a 3σ variation in gate-length of 10%. The Monte Carlo mean and SD are estimated using a random input vector for each circuit. The results show that the average error in estimating the mean over all the circuits is 3.5% with a maximum error of 8.3%. The average error in the SD is 13.7% with a maximum error of 17.9%.

Accounting for Inter-Die and Intra-Die Variations

As discussed in Chap. 1 process variation can be classified into inter-die variation and intra-die variations. Intra-die variation refers to variations within a particular circuit block or chip, whereas inter-die variations occur as fluctuations from one die to the next. The drawn gate-length of a transistor i is expressed as an algebraic sum of the nominal gate-length (L_{nominal}), the intra-die variation (ΔL_{intra}) and the inter-die variation (ΔL_{inter}). Consequently, the total variance is also a sum of the inter-die and intra-die variances:

$$\begin{aligned} L_{\text{gate, i}} &= L_{\text{nominal}} + \Delta L_{\text{inter}} + \Delta L_{\text{intra, i}} \\ \sigma^2_{\text{gate, i}} &= \sigma^2_{\text{inter}} + \sigma^2_{\text{intra, i}} \end{aligned} \tag{4.42}$$

where σ_{inter} and σ_{intra} are the SD's of the inter-die and intra-die variations in gate-length, respectively. Note that in the (4.42), the RV ΔL_{inter} is shared by all devices in a design (creating correlation between their leakage currents), whereas the random variables $\Delta L'_{\text{intra}}s$ assigned to each of the transistors are independent (reducing the correlation of their leakage currents). ΔL_{inter} can

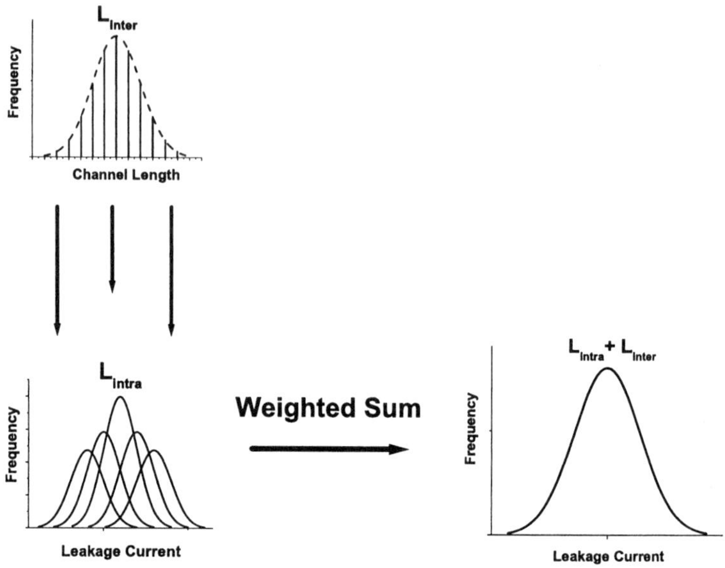

Fig. 4.6. Probability density functions for (a) Channel length considering only intra-die variation with its discrete sampling (b) Leakage current pdf corresponding to each discrete sample point in (a) considering intra-die gate-length variation (c) Leakage current considering both inter- and intra-die variation.

also be interpreted as variations in the mean gate-length of different samples of a chip, whereas ΔL_{intra} represents variations in gate-length of individual devices from this mean value. Note that the above approach estimates the pdf of subthreshold leakage of a circuit block considering only intra-die variations given the mean value of gate-length. Thus we can utilize Bayes' Theorem [109], which states that the probability of an arbitrary event $\mathcal{A}$, can be expressed as

$$\mathcal{P}(\mathcal{A}) = \sum_{x=-\infty}^{\infty} \mathcal{P}(\mathcal{A}|X = x)\mathcal{P}(X = x) \tag{4.43}$$

where X is a RV with a pdf $f(x)$. If X represents the RV associated with inter-die variations then the probability of the event that the subthreshold leakage of a circuit block considering both components of variations lies within a given range, can be estimated using (4.43). The term associated with the conditional probability in (4.43) corresponds to the evaluation of the pdf discussed when only intra-die variations are considered given a mean value of that variation.

To compute the total leakage, accounting for both types of gate-length variation, the pdf of L_{inter} can be discretized as shown in Fig. 4.6(a). For each

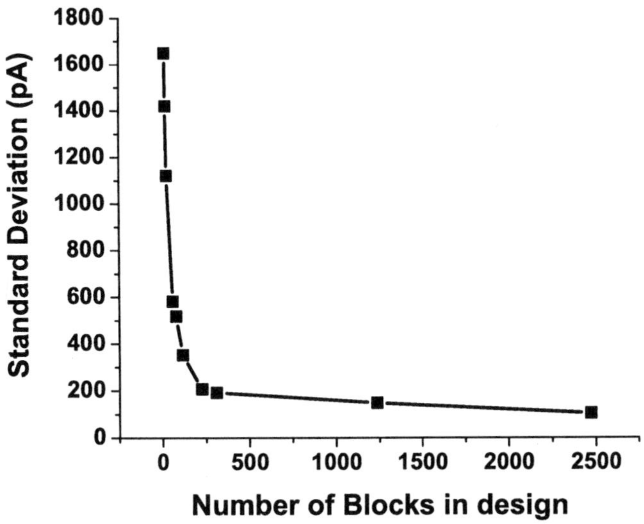

Fig. 4.7. The relation between the standard deviation of total leakage current in a chip and the number of blocks that constitute the chip sheds insight on the validity of gate independence assumptions.

discrete point $L_{\text{inter, j}}$ on the pdf of L_{inter}, consider the intra-die variation of the channel length as a normally distributed pdf, whose mean is $L_{\text{inter, j}}$ and standard deviation is σ_{intra}. Corresponding to this distribution of channel length, we obtain a pdf of the leakage current for the circuit using the approach outlined in (4.32)–(4.41). Thus, we obtain a family of these pdfs of leakage current as shown in Fig. 4.6(b), where each pdf is associated with a conditional probability that corresponds to the pdf value of $L_{\text{inter, j}}$ on the pdf of L_{inter}. To obtain the pdf of leakage current considering both inter- and intra-die variation we form a weighted sum of the family of pdfs using (4.43). This can be expressed as

$$\mathcal{P}(I_{sub} < i < I_{sub} + \Delta I) = \sum_{j=1}^{n} \Big(\mathcal{P}_{\text{intra, j}}(I_{sub} < i < I_{sub} + \Delta I) \\ \mathcal{P}_{\text{inter}}(L_{\text{inter, j}}) \Big) \tag{4.44}$$

where $\mathcal{P}_{\text{inter}}(L_{\text{inter, j}})$ is the probability of occurrence of jth point in the set of n discrete points selected on the inter-die pdf. P_{intra} is calculated based on the lognormal distribution of the leakage current corresponding to the jth point, $L_{\text{inter, j}}$ on the L_{inter} pdf.

Intra-chip variations often exhibit spatial correlation such that devices that are closer to one another have a higher probability of being alike than devices that are far apart. In our analysis so far, we have assumed that the intra-die gate-length variation expressed by the random variables ΔL_{intra} assigned to each gate is independent. However, spatial correlation will result in dependence of these random variables. Hence, we examine the impact of such correlation on the statistical leakage estimation using Monte Carlo simulation.

For simplicity, we model the effect of spatial correlation using clusters of gates in a circuit, such that ΔL_{intra} of gates within a cluster are perfectly correlated, while ΔL_{intra} of gates between different clusters are independent. Large cluster sizes therefore reflect a stronger spatial correlation of intra-die gate-length variation while small cluster sizes reflect a weak spatial correlation. Figure 4.7, shows that the standard deviation of subthreshold leakage current for a design as a function of the number of clusters in the design. As the number of clusters is decreased, the size of each individual cluster increases, representing a stronger spatial correlation. From the plot, we see that, due to the averaging effect of a large number of uncorrelated variables, the variability in leakage current converges to a relatively small value as the number of clusters is increased. For designs with 250 or more clusters, the standard deviation has largely converged, and the impact of spatial correlation can be ignored. In other words, we can approximate the case having 250 gate clusters with perfectly correlated intra-die gate-length variation within each cluster, with the case where all gates are considered to have independent intra-die gate-length variation (as assumed in the analysis in this section).

In typical process technologies, spatial correlation drops off sharply for distances greater than 0.1 mm. Hence, even for a small design with a die area of $2.5\,\text{mm}^2$, the number of independent gate clusters is sufficient to perform statistical leakage current analysis assuming independence of intra-die gate-length variation. Since most practical designs are significantly larger than $2.5\,\text{mm}^2$, spatial correlation does not pose a significant issue for statistical leakage current estimation for such designs. In Chap. 5 we will consider an approach to consider the impact of these correlated variations for small designs based on principal components analysis.

Figure 4.8 shows the impact of varying the distribution of inter-die process variation on the pdf of the leakage current while keeping the standard deviation of the total gate-length σ_{total}=15% of the mean. The figure shows that when inter-die process variation is increased (and consequently the intra-die variation is decreased), the pdf tends to a lognormal shape. Note that for the case where there is no intra-die process variation, all gate-lengths on a single die will be at their nominal values. Hence, the pdf of this leakage current due to inter-die process variation alone should be similar to the pdf of the leakage current of a single gate which, as we know, can be closely approximated by a lognormal. The figure suggests that, since leakage current is well characterized in terms of the I_{DDQ} values across die, the shape of this leakage current pdf

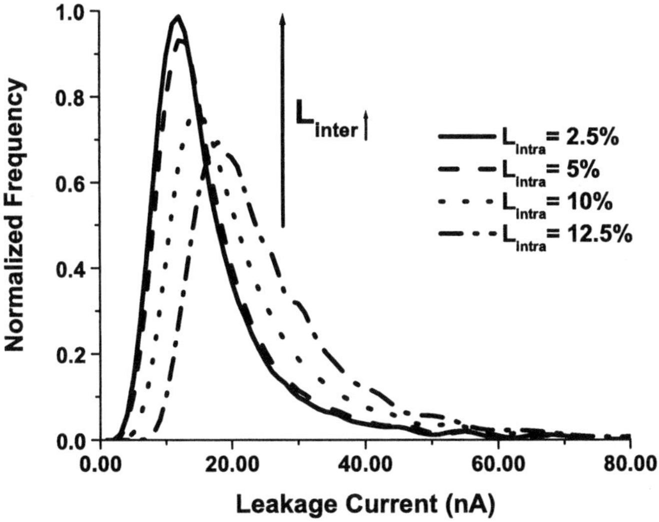

Fig. 4.8. Probability density functions of leakage current for different contributions of inter- and intra-die process variation. The total variation is 15%.

Table 4.4. Comparison of subthreshold leakage estimated obtained using the analytical approach with Monte Carlo simulations considering intra-die variations in gate-length.

Circuit	Mean(nA) Monte Carlo	Mean(nA) Analytical	Error(%)	SD(pA) Monte Carlo	SD(pA) Analytical	Error(%)
c17	0.4	0.4	0.0	0.5	0.4	20.0
c432	10.0	10.0	0.0	9.2	7.6	17.4
c499	28.0	27.0	3.6	24.1	19.5	19.1
c880	24.6	23.9	2.8	21.2	17.4	17.9
c1355	32.2	30.6	5.0	30.2	23.9	20.9
c1908	23.6	23.3	1.3	21.9	17.5	20.1
c2670	48.2	45.4	5.8	41.3	33.7	18.4
c3540	57.5	54.5	5.2	47.4	38.2	19.4
c6288	186.7	175.4	6.1	183.5	152.0	17.2

can be a useful way to estimate the contribution of the inter-die or intra-die component to the total process variation.

Table 4.4 compares the results of the analytical approach to Monte Carlo simulation considering both intra- and inter-die variation. The table lists the data for the case where intra-die and inter-die standard deviation have been assumed to be 10% and 11% of mean, respectively, which make up a total standard deviation of 15% variation based on (4.42). As can be seen, the error in the estimated mean is always within 6.1% and that for the standard deviation within 21%. When comparing the median and the 95th/99th percentile points estimated using the traditional approach to the statistical approach, we can see that the traditional approach significantly overestimates the leakage for higher confidence points since all the devices are assumed to be operating at a pessimistic corner point. Since the relationship between the gate-length and leakage current is monotonic, the median point as estimated by the traditional analysis is very close to the nominal leakage current.

4.4.3 Temperature and Power Supply Variations

To this point in the chapter we have been concerned with the impact of variations in process parameters on power dissipation. In this section, we will consider variation in power supply and temperature. As discussed in Chap. 1, these variations are fundamentally different from process variations and a completely different set of techniques needs to be used to consider their impact.

If we consider the expressions for dynamic and leakage power(subthreshold and gate), we can note that only subthreshold leakage is dependent on variations in temperature. As shown in [136] subthreshold leakage has a superlinear dependency on temperature, and a change in temperature of 30°C can affect leakage by as much as 30%. Variations in power supply have strong quadratic and cubic impact [76] on dynamic and leakage power, respectively. In this section, we will discuss a technique proposed in [136] to estimate dynamic and subthreshold leakage power while considering variations in power supply and temperature, which can be easily mapped to consider variations in other components of leakage current as well. This work proposed the first approach to consider realistic variations in supply voltage and temperature which are strongly influenced by the power grid decoupling capacitor locations [32] and the profile of the currents drawn by the transistors. In addition, these variations demonstrate strong locality and linear approximations of the temperature and power supply variations over a chip results in large inaccuracies.

Variations in supply voltage and temperature cause variations in the currents drawn from the power grid by the active devices that impacts the amount of power dissipated in this region as well. This, in turn, affects the supply voltage (through IR-drop etc.), and the temperature (increased power dissipation results in a higher temperature). Thus, a solution to this problem involves an iterative solution of a nonlinear set of equations. Therefore, we need an efficient tool capable of performing full-chip power grid and thermal analysis. Generally, a set of nonlinear equations is solved using an iterative Newton-Raphson technique [105], which become impractical for current VLSI designs. In this work an iteration-based approach as outlined in Fig. 4.9 is used to

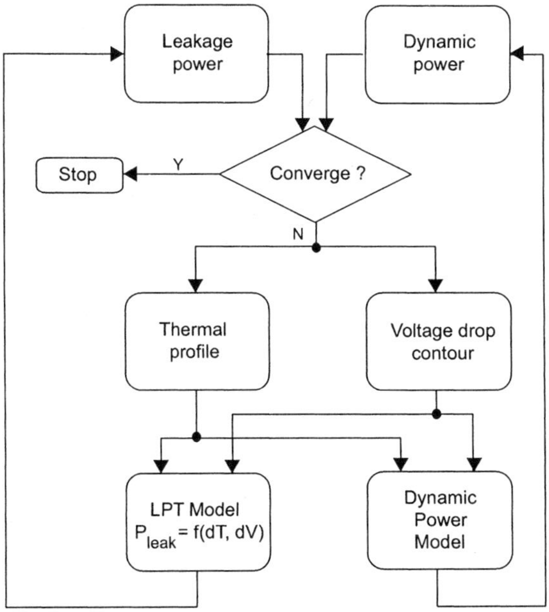

Fig. 4.9. Iterative flow used for leakage estimation under power supply and temperature variations [136]. (©2005 IEEE)

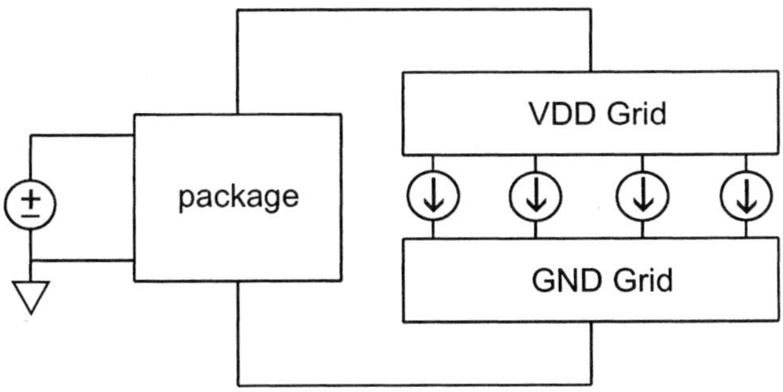

Fig. 4.10. Model for the complete power supply network [136]. (©2005 IEEE)

improve efficiency. This technique is built upon an efficient temperature and power grid simulator which act as inputs to the leakage (LPT) and dynamic power models. This loop is then repeated until convergence is achieved. We will first discuss the analysis techniques for the power grid and temperature simulation techniques used in [136] and then discuss the leakage and dynamic power models.

Most chip-level power grid techniques decouple the large power grid network from the nonlinear devices which are connected to this network. Assuming a perfect power supply grid, the current used by the nonlinear devices are estimated. This current should consist of both the leakage currents as well as

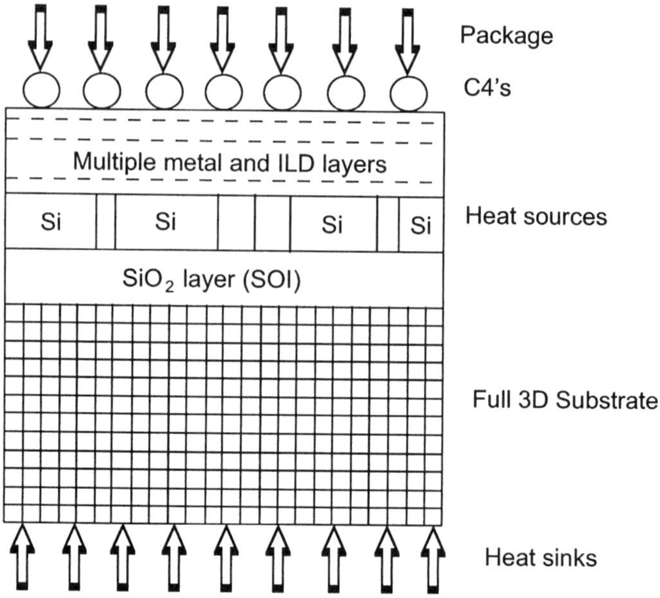

Fig. 4.11. Thermal model of a chip [136]. The package and heat sinks are assumed to be thermally ideal with constant temperature. (©2005 IEEE)

the current required for switching. These current profiles are then modeled as idealized current sources and connected to the resistive power grid network to complete the power supply network. In general the power grid network is modeled as a resistive mesh with layers of metals being connected through resistive vias. The decoupling capacitors act as capacitances between the power and ground networks and the top metal layer is connected to the ideal voltage regulators through resistive and inductive elements.

In particular, leakage current estimation only requires a DC solution of the power grid network. In this case the inductive and capacitive elements are replaced by shorts and opens, respectively, and the entire network becomes a large linear network of resistances. This is represented in Fig. 4.10 where the *VDD-grid* and *GND-grid* are resistive networks which are connected to the *package* which is again modeled as a resistive network. A typical power grid can consist of millions of nodes and specialized techniques are required to solve these systems with reasonable memory and run-time requirements. The implementation in [136] uses an iterative algebraic multi-grid AMG solver solver [135]. The technique simplifies the problem by initially coarsening the power grid which maps the problem to a smaller power grid. The solution obtained using the coarser grid is (using direct solution of the matrix equations) is then mapped back to the original power grid using interpolation techniques.

The decoupling technique used to simplify the problem of analyzing the power grid is also utilized to obtain a thermal solution of the chip. The full-chip thermal model is shown in Fig. 4.11 which includes the silicon substrate, the package and the heat sinks. The thermal profile in a general 3D medium satisfies the heat conduction equation

$$\rho c_p \frac{\partial T(x,y,z,t)}{\partial t} = \nabla[k(x,y,z,T)\nabla T(x,y,z,t)] + g(x,y,z,t) \tag{4.45}$$

subject to the boundary condition

$$k(x,y,z,T)\frac{\partial T(x,y,z,t)}{\partial n_i} + h_i T(x,y,z,T) = f_i(x,y,z) \tag{4.46}$$

where T is the temperature, g is the power density of heat sources (which in our case would be the power density of devices at the silicon surface), k is the thermal conductivity, ρ is the material density, c_p is the specific heat capacity, h_i is the heat transfer coefficient on the boundary, f_i is a function of the position and n_i is the unit vector normal to the surface element i. Under normal operating conditions the thermal conductivity can be assumed to be independent of position and temperature. In addition, under steady state conditions the differentials with respect to time drop-out which simplifies (4.45) to

$$k\nabla^2 T(x,y,z) + g(x,y,z) = 0. \tag{4.47}$$

Depending on the packaging type, which determines the positions of the heat sinks, different forms of boundary conditions need to be enforced which can be obtained from (4.48).

$$k(x,y,z,T)\frac{\partial T(x,y,z,t)}{\partial n_i} + h_i T(x,y,z,t) = f_i(x,y,z) \tag{4.48}$$

The above partial differential equation (PDE) in (4.47) is solved using standard finite-difference techniques. The method requires the domain of interest to be replaced by a grid. At each grid point each term in the partial differential is replaced by a difference formula which may include the values of T at that and neighboring grid points. The thermal resistance of each of the 3D grid cube of dimensions $(\mathrm{d}x, \mathrm{d}y, \mathrm{d}z)$ to the flow of heat in the direction x is expressed as

$$R_i = \frac{\mathrm{d}x}{k\mathrm{d}y\mathrm{d}z}. \tag{4.49}$$

To obtain the thermal resistance at the convective boundary, $\mathrm{d}x/k$ in (4.49) is replaced by the heat transfer coefficient. By substituting the difference formula and the discretized thermal resistances into the PDE, a difference equation is obtained which is solved to obtain the solution to the original PDE. Again, as in the power supply analysis case, the solution of the complete set of difference equations involves a huge number of nodes, and an AMG based solver is used.

Table 4.5. Comparison of various leakage estimation scenarios (Initial leakage estimate is 9.6W) [136].

Variations Considered	ΔV (mV)	ΔT (^oC)	Total Leakage (W)
Voltage Temperature	min: -4 max: -184	min: -4.2 max: 25.3	7.75
Voltage	min: -4 max: -184		7.77
Temperature		min: -4.2 max: 25.3	9.63
Uniform	-120	0	5.31

Using the simulation techniques described above the temperature and power supply map of the entire silicon surface can be obtained. The next step in the power estimation approach, as outlined in Fig. 4.9, is to calculate the change in power dissipations based on the new temperature and supply voltages estimated in the simulation step. An empirical second-order polynomial model is used, where the coefficients are obtained using regression analysis, and has the form:

$$\frac{I_{leak}(\Delta T, \Delta V)}{I_{leak}(0,0)} = 1 + a_1 \Delta T + a_2 (\Delta T)^2 + b_1 \Delta V + b_2 (\Delta V)^2 + c_2 \Delta T \Delta V \quad (4.50)$$

The values of these coefficients are found to have very small variations from one standard cell to the other. Dynamic power is assumed to be independent of temperature variations and has a simple quadratic dependence on power supply variations. The results obtained using the above approach show that the leakage power is more strongly affected by power supply variations as compared to temperature variations. After one iteration of the approach, the leakage power of a design becomes less than the initial value due to the correlation in power supply and temperature variations. Table 4.5 lists the leakage estimate after one iteration. The uniform variation refers to uniform 10% V_{dd} drop and a uniform 85^oC temperature. This simple assumption results in a 30% underestimation in leakage. In addition, most of the correction in leakage from the initial estimate is found to happen in the first iteration. For the case of the design used in Table 4.5 the first iteration provides a 19.2% reduction in leakage. Further iterations only result in a change in 0.5%. Thus, using only one iteration is sufficient to provide reasonable accuracy in the leakage power estimate.

In this chapter, we have discussed techniques to analyze various leakage power components with variations in process and environmental parameters. In the next chapter, we will use the ideas developed in this chapter and in

Chap. 3 to estimate the true parametric yield of a design given both power and performance constraints.

5

Yield Analysis

As we have seen in previous chapters, variations have a tremendous impact on both power and performance of current integrated circuit (IC) designs. In particular, leakage power which has grown to contribute a significant fraction of total power and is also known to be highly susceptible to process variations due to its exponential dependence on threshold voltage [28]. In [20], a 20X variation in leakage power for 30% delay variation between fast and slow dies was reported. Both the variation in leakage power and delay affect the

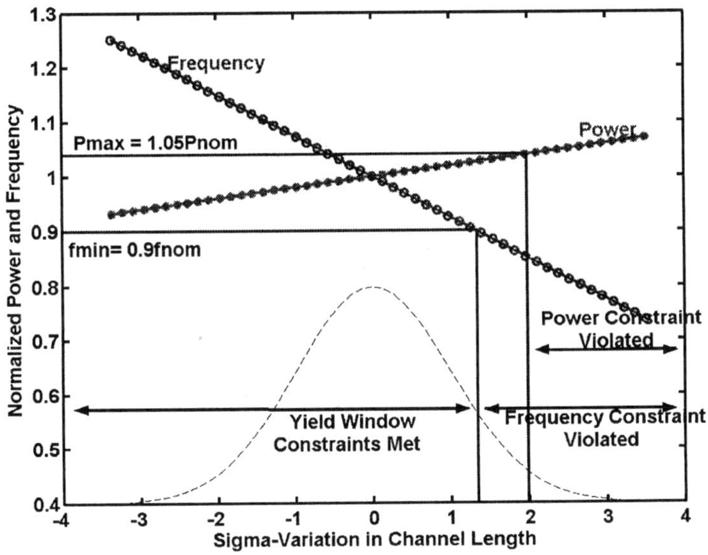

Fig. 5.1. In older technologies, power and performance resulted in a one-sided constraint on the feasible region for yield. (©2005 IEEE)

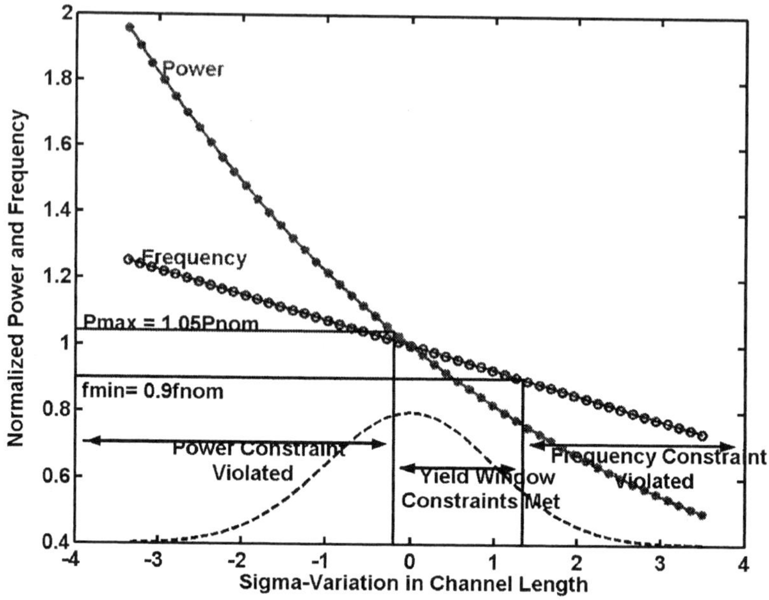

Fig. 5.2. In current technologies, the yield window is constrained from one side by a performance constraint and power on the other. (©2005 IEEE)

number of dies that meet the specifications, and therefore affect yield. In this chapter, we will discuss yield analysis techniques, which consider both delay and leakage variability. We will be specifically concerned with the strong inverse correlation between power and delay and the impact this has on the fraction of dies that satisfy both timing and power constraints. Figure 5.1 shows the situation where total power is dominated by dynamic power [116]. This causes parts that have a lower delay (smaller gate length) to have a lower total power dissipation as well. However, with increasing leakage power the situation changes, as illustrated in Fig. 5.2. Samples of a design that can operate at higher frequencies now dissipate more power as well becuase the feasible region in terms of yield in now constrained from both sides. This results in a significant loss in parametric yield.

This change in yield loss can be easily captured by considering the correlation in power and performance, which has changed from being positive in dynamic power dominated systems to negative in leakage power dominated scenarios. This correlation results in most of the fastest chips in a lot to have unacceptable leakage and vice versa and results in the two-sided constraint on yield. To demonstrate the importance of power-delay correlation, Table 5.1 shows yield for varying values of correlation factors at which simple expressions for yield can be obtained. $F(x)$ represents the the cdf function of a Gaussian RV. The yields are estimated for delay constraints of D standard deviations (SD) from the mean at a fixed power constraint P. The results in

Table 5.1. Estimated yield for different values of correlation coefficient. Power constraint is set at 1.5X the nominal leakage power.

	Estimated Yield		
	Corr=-1.0	Corr=0.0	Corr=1.0
Yield Expression	max $(\Phi(D)+\Phi(P),0)$	$(0.5+\Phi(D))*$ $(0.5+\Phi(P))$	0.5 $\Phi(\min(D,P)l$
D=-1	0.000	0.095	0.159
D=0	0.100	0.300	0.500
D=1	0.441	0.505	0.600
D=2	0.577	0.586	0.600
D=3	0.599	0.599	0.600

Table 5.1 clearly show that the correlation of power and delay has a strong impact on parametric yield, particularly for mid- to high-performance speed bins.

This yield loss will worsen in future technologies due to increasing process variations and the continued significance of leakage power. Another troublesome observation is that increased variation not only results in a larger spread of leakage power but also in higher average leakage power. Additionally, most current optimization approaches do not consider process variations and are unaware of their impact on yield. These approaches invariably result in the formation of a timing wall and result in yield loss due to increased susceptibility to process variations [14].

In the last two chapters we have looked at a number of techniques to perform statistical timing or power analysis. However, these analysis approaches neglect the correlation of power and performance. Hence, performing optimization based on these analysis methodologies can potentially harm overall parametric yield. In particular, timing yield optimization using a statistical timing analyzer will result in yield loss due to the power constraint while power minimization techniques will harm timing-based yield. Hence in this chapter we will discuss true yield estimation approaches, which consider both power and performance. We will look at optimization in more detail in the next chapter.

Recently, [118] presented a chip-level approach to estimate the yield in separate frequency bins given a power constraint. This high-level approach is based on global circuit parameters such as total device width on a chip. Since it does not use circuit specific information from a gate level netlist, it is difficult to use for optimization of gate-level parameters, such as the threshold voltage and sizes of individual gates. However, it is able to provide insight into the achievable parametric yield early in design cycle and can be crucial in making alterations in the design early in the design cycle to achieve better yield. This approach will be discussed in Sec. 5.1.

Another important requirement for an accurate yield estimation approach is to consider all classes of variations which have significantly different impact on delay [21] and power [117], as discussed in Chap. 3 and Chap. 4. Process variations are typically classified into inter-die and intra-die components. Intra-die variations are further classified as having correlated and random components. Traditionally, inter-die variations have been the dominant source of variations but with process scaling, the random and correlated components of intra-die variations now exceed inter-die variations [44]. The relative magnitude of these components of variation also depends on the process parameter being considered. For example, gate length variations are generally considered to have roughly comparable random and correlated components whereas gate length-independent threshold voltage is commonly assumed to vary randomly due to random dopant fluctuations [124]. The approach proposed in [130] considers all sources of variations and performs gate level yield analysis. This will be the focus of our discussion in Sec. 5.2. Finally, in Sec. 5.3, we will consider the sensitivity of parametric yield to the supply voltage [116], and develop a yield estimation approach by mapping back the feasible region from the power-performance space to the space of process parameters.

5.1 High-Level Yield Estimation

The computation of a high-level estimate for yield which was proposed in [118] and is based on developing expressions for the total leakage of a design, considering both subthreshold and gate leakage. Both inter-die and intra-die variability in gate length, threshold voltage and oxide thickness is considered. The expressions for leakage are developed in terms of the global or the inter-die variability, which has a given fixed value for a particular sample of a design. Since the model is developed for full-chip yield estimation, the contribution of the correlated component of intra-die variability can be safely neglected. This follows from our discussion in Sec. 4.4.2 where we found, that for a design of reasonable size, the impact of correlation on leakage variance is minimal.

In addition, based on simulations performed using Berkeley predictive technology models (BPTM) and industry data showing the relative impact of inter- and intra-die variability, the authors argue that chip-performance is dictated by global variability in gate length. Based on these observations, different frequency bins are mapped to a feasible global gate length fluctuation, which is then used to estimate the fraction of chips that meet the leakage power dissipation constraint.

5.1.1 Leakage Analysis

Let us first consider the analytical model used to estimate the leakage current of a given design, which is expressed as a sum of the subthreshold and gate leakage current

$$I_{\text{tot}} = I_{\text{sub}} + I_{\text{gate}}. \tag{5.1}$$

Both components of leakage current are expressed as a product of the nominal leakage and a function that captures the variation in leakage from the nominal value based on variations in process parameters. Let us represent this variation in process parameters as a vector $\mathbf{\Delta P}$, using which we can write leakage current as:

$$I_{\text{leak}} = I_{\text{leak,nom}} f(\mathbf{\Delta P}) \tag{5.2}$$

where $I_{\text{leak,nom}}$ represents the leakage under nominal conditions. As we saw in Chap. 4, using analytical leakage expressions based on BSIM device models results in extremely complicated expressions for the non-linear function f, which makes further statistical analysis cumbersome. To simplify the problem, a carefully selected empirical equation is used to capture the nature of f that provides sufficient accuracy and ease of analysis. The variation in process parameters $\mathbf{\Delta P}$ is decomposed into an intra-die and inter-die component, which are referred to as local ($\mathbf{\Delta P_l}$) and global ($\mathbf{\Delta P_g}$) variations, respectively. Thus, the random variable (RV) corresponding to the total variation is expressed as a sum

$$\mathbf{\Delta P} = \mathbf{\Delta P_l} + \mathbf{\Delta P_g} \tag{5.3}$$

where the sum of the variances of the global and local variations gives the overall variance in the process parameter. Now, let us consider the choice of f for each of the components of leakage and perform statistical analysis to estimate the leakage distribution using these expressions for f.

Subthreshold Leakage

To capture the dependence of subthreshold leakage on variations in process parameters using a functional form, we note that it is exponentially dependent on threshold voltage. However, the threshold voltage is itself related to a number of physical parameters through complex device phenomena. A number of second order effects such as DIBL, narrow width effect and other short channel effects play a significant role in determining the subthreshold leakage current. Considering the three process parameters of interest (L_{eff}, V_{th} and T_{ox}), subthreshold leakage is most strongly influenced by variations in channel length. Channel length independent V_{th} variation arises mostly due to random dopant variations and has a significant role in leakage variability as well. However, T_{ox} is a comparatively well controlled process parameter and has a much smaller influence on subthreshold leakage [129], given the much smaller sensitivity of subthreshold leakage to gate oxide thickness. Based on these observations we can capture the variation in subthreshold leakage on process parameters as

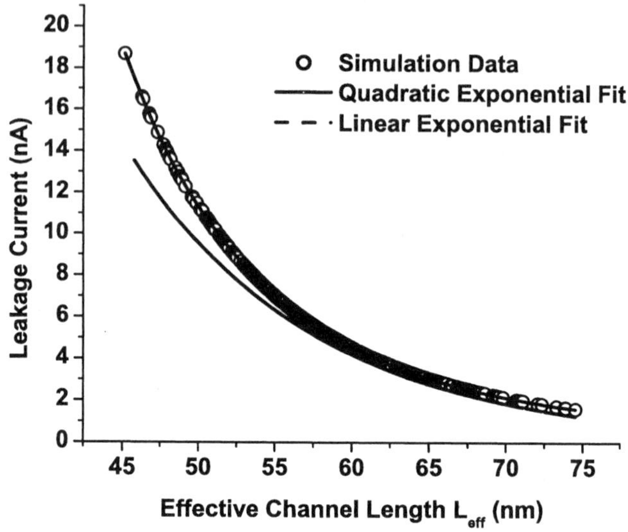

Fig. 5.3. Comparison of quadratic and linear exponential fit of leakage with effective channel length. Nominal L_{eff}=60 nm.

$$f(\mathbf{\Delta P}) = f_1(\Delta L_{\text{eff}}) f_2(\Delta V_{th}) \tag{5.4}$$

where ΔL_{eff} captures the dependence of subthreshold leakage on L_{eff} and the associated influence on threshold voltage, and ΔV_{th} captures the variation in subthreshold leakage resulting from doping concentration variations. For better accuracy, f_1 is assumed to be an exponential of a quadratic function of gate length (as in Chap. 4). This accuracy improvement is much more significant in sub-100nm technologies as shown in Fig. 5.3 which compares a quadratic and linear exponential fitting function with SPICE data obtained for 60 nm devices using BPTM models. As shown in the figure, a linear exponential is not able to accurately model the leakage value for low values of gate lengths that have the maximum subthreshold leakage. Therefore, using a linear exponential will result in an underestimation of mean leakage.

On the other hand, a linear exponential is found to provide reasonable accuracy while considering doping concentration variations. Thus, we can write

$$f(\mathbf{\Delta P}) = \exp\left(-\frac{L + c_2 L^2}{c_1}\right) \exp\left(-\frac{c_3 V}{c_1}\right) \tag{5.5}$$

where c_1, c_2 and c_3 are fitting parameters which can be obtained using SPICE simulations. Using (5.5) and (5.2), we can finally write

$$I_{\text{sub}} = I_{\text{sub,nom}} \exp\left(-\frac{L + c_2 L^2 + c_3 V}{c_1}\right). \tag{5.6}$$

Decomposing the variability into global and local components as

$$L = L_g + L_l \qquad V = V_g + V_l \tag{5.7}$$

we can rewrite (5.6) as

$$I_{\text{sub}} = I_{\text{sub,nom}} \exp\left(-\frac{(L_g + L_l) + c_2(L_g + L_l)^2 + c_3(V_g + V_l)}{c_1}\right) \tag{5.8}$$

$$= I_{\text{sub,nom}} \exp\left(-\frac{L_g + c_2 L_g^2 + c_3 V_g}{c_1}\right) \exp\left(-\frac{L_l + \lambda_2 L_l^2 + \lambda_3 V_l}{\lambda_1}\right) \tag{5.9}$$

where the λ_i's are assumed to be in the same ratio as the c_i's and are related as

$$\frac{c_i}{\lambda_i} = 1 + 2c_2 L_g. \tag{5.10}$$

The above relation can be easily obtained by matching the two right hand sides in (5.8). We now calculate the expected value of leakage for a given sample of a design that correspond to a fixed global variability in process parameters. Based on our discussion regarding correlations in leakage variability, we can assume that the RVs, which correspond to local variability for each gate, are mutually independent. Recall that the central limit theorem states that the sum of a large number of independent RVs

$$x = x_1 + x + 2 + \cdots + x_n \tag{5.11}$$

converges in distribution to a Gaussian distribution with the following parameters

$$\mu_x = \mu_{x_1} + \mu_{x+2} + \cdots + \mu_{x_n} \tag{5.12}$$

$$\sigma_x^2 = \sigma_{x_1}^2 + \sigma_{x_2}^2 + \cdots + \sigma_{x_n}^2 \tag{5.13}$$

where μ_i and σ_i correspond to the mean and standard deviation of RV i, respectively. Note that if the above expression is not dominated by one RV, and there are a large number of a RVs with comparable mean and variance, then the ratio $\mu_x/\sigma_x \rightarrow 0$ as the number of summed RVs increases. Therefore, if we sum a large number of similar RVs, the final distribution can be approximated as a single value which corresponds to the sum of the mean of individual RVs. Now taking the expectation over the local variability in (5.8), we can write

$$I_{\text{sub}} \approx I_{\text{sub,nom}} \exp\left(-\frac{L_g + c_2 L_g^2 + c_3 V_g}{c_1}\right) E\left[\exp\left(-\frac{L_l + \lambda_2 L_l^2 + \lambda_3 V_l}{\lambda_1}\right)\right]. \tag{5.14}$$

Since the RV V_l captures the variation in threshold voltage that results from random doping fluctuations, it is assumed to be statistically independent of L_l, which gives:

$$I_{\text{sub}} \approx I_{\text{sub,nom}} \exp\left(-\frac{L_g + c_2 L_g^2 + c_3 V_g}{c_1}\right) E\left[\exp\left(-\frac{L_l + \lambda_2 L_l^2}{\lambda_1}\right)\right] E\left[\exp\left(-\frac{\lambda_3 V_l}{c_1}\right)\right]. \tag{5.15}$$

The expected value of a lognormal RV $Y = e^X$, where X is Gaussian RV with mean μ_x and sigma σ_x, can be expressed as

$$\mu_y = \exp\left(\mu_x + \frac{\sigma_x^2}{2}\right) \tag{5.16}$$

Using the above expression and the fact that any linear multiple of a Gaussian RV is Gaussian, we can write:

$$E\left[\exp\left(-\frac{X}{a}\right)\right] = \exp\left(-\frac{\mu_x}{a} + \frac{\sigma_x^2}{2a^2}\right) \tag{5.17}$$

$$Var\left[\exp\left(-\frac{X}{a}\right)\right] = E\left[\exp\left(-\frac{X}{2a}\right)\right] - E^2\left[-\frac{X}{a}\right]. \tag{5.18}$$

To handle the squared exponential term, we need to estimate the mean and variance of RVs of the form $Z = e^{(-X + a_2 X^2)/a_1}$, which can be obtained in closed form for the case when X is a zero mean Gaussian RV with standard deviation σ_x, as

$$E[Z] = \left(\sqrt{1 + \frac{2a_2}{a_1}\sigma_x^2}\right)^{-1} \exp\left(\frac{\sigma_x^2}{2a_1^2 + 4\sigma_x^2 a_1 a_2}\right)$$
$$Var[Z] = E\left[\exp\left(\frac{-X + a_2 X^2}{(a_1/2)}\right)\right] - E^2[Z]. \tag{5.19}$$

Using the above expressions, (5.15) can be rewritten as

$$I_{\text{sub}} \approx I_{\text{sub, g}} S_L S_V \tag{5.20}$$

where

$$
\begin{aligned}
I_{\text{sub,g}} &= I_{\text{sub, nom}} \exp\left(-\frac{L_g + c_2 L_g^2 + c_3 V_g}{c_1}\right) \\
S_L &= \left(\sqrt{1 + \frac{2\lambda_2}{\lambda_1}\sigma_{L_l}^2}\right)^{-1} \exp\left(\frac{\sigma_{L_l}^2}{2a_1^2 + 4\sigma_{L_l}^2 \lambda_1 \lambda_2}\right) \\
S_V &= \exp\left(\frac{\lambda_3^2 \sigma_{V_l}^2}{2\lambda_1^2}\right).
\end{aligned}
\tag{5.21}
$$

To calculate the full-chip subthreshold leakage, the above expression is evaluated separately for NMOS and PMOS devices and then multiplied by the effective PMOS and NMOS device width, respectively. Effective width is the actual device width scaled by the percentage of devices that are expected to be non-conducting on average, and the appropriate scale factor which captures the stacking effect that reduces subthreshold leakage when devices connected in series are simultaneously non-conducting. Finally, we can write the full chip subthreshold leakage (for a given fixed global variation) as

$$
I_{\text{sub}} \approx \left(\sum_{d \in N} \frac{W_d}{q^N}\right) I_{\text{sub, g}}^N S_L^N S_V^N + \left(\sum_{d \in P} \frac{W_d}{q^P}\right) I_{\text{sub, g}}^P S_L^P S_V^P \tag{5.22}
$$

where N, P represent the set of NMOS and PMOS devices, respectively, W_d represents the device width, q represents the scaling factor based on the probability of the transistor being off and the number of off transistors in series. The scale factor q can be different for NMOS and PMOS devices which is represented by the superscripts N and P, and S_L and S_V are as expressed in (5.21), and are calculated separately for NMOS and PMOS devices.

Gate Leakage

Gate leakage is known to be extremely sensitive to variations in T_{ox} and hence any variation in gate leakage resulting from variations in gate length variations can be safely ignored. The strong sensitivity of gate leakage results from a strong exponential dependence of gate leakage current on gate oxide thickness. Moreover, variations in gate length have a linear dependency on gate leakage and do not affect the mean gate leakage current. As we saw in subthreshold leakage, discussed above, we can use the central limit theorem to approximate the leakage current by its mean value.

We first approximate $f(\mathbf{\Delta P})$ as

$$
f(\mathbf{\Delta P}) = I_{\text{gate, nom}} \exp\left(-\frac{T}{a}\right) \tag{5.23}
$$

where T corresponds to the variation in gate oxide thickness, and the fitting parameter a can be obtained using SPICE simulations. Again, decomposing the total variation into local and global components we can write

$$I_{\text{gate}} = I_{\text{gate, nom}} \exp\left(-\frac{T_g}{a}\right) \exp\left(-\frac{T_l}{a}\right) \tag{5.24}$$

where $I_{\text{gate, nom}}$ is the nominal gate leakage current, T_g is the global fluctuation in gate oxide thickness which is constant for a given sample of the design and T_l is the local variability in gate oxide thickness. Using the same arguments as in the case of subthreshold leakage, we again approximate the local variability as a scaling factor, and write

$$I_{\text{gate}} \approx E[I_{\text{gate}}] = I_{\text{gate, g}} S_T \tag{5.25}$$

where

$$\begin{aligned} I_{\text{gate, g}} &= I_{\text{gate, nom}} \exp\left(-\frac{T_g}{a}\right) \\ S_T &= \exp\left(\frac{\sigma_{T_l}^2}{2a^2}\right). \end{aligned} \tag{5.26}$$

The full-chip gate leakage can now be written as

$$I_{\text{gate}} \approx \left(\sum_{d \in N} \frac{W_d}{p^N}\right) I_{\text{gate, g}}^N S_T^N + \left(\sum_{d \in P} \frac{W_d}{p^P}\right) I_{\text{gate, g}}^P S_T^P \tag{5.27}$$

where the summation is across PMOS and NMOS devices. We use the scale factor p for gate leakage instead of q for subthreshold leakage in (5.22).

The total leakage is obtained using (5.1) by summing the expressions for subthreshold leakage (5.22) and gate leakage (5.27). Note that we have considered only two kinds of devices (PMOS and NMOS) in the above equations. In the case where we have devices with different nominal threshold voltages or gate oxide thicknesses, we will have additional terms which account for these devices.

Table 5.2 compares the leakage estimated using the above analytical technique with Monte Carlo methods. We consider three different cases: 1) without any variability, 2) with only global variability, and 3) with both local and global variability. The results are generated using 60 nm BPTM devices. The middle columns list the amount of variations in each of three process parameters considered. The results show that the error in the analytical approach is always less than 5% as compared to Monte Carlo and points to good accuracy of the proposed leakage analysis methodology. Furthermore, the table also shows that when within die variability is considered, the leakage of the design increases by a further 15%, which results from the exponential dependence of leakage currents on process parameters.

Table 5.2. Comparison of the analytical approach with SPICE based Monte Carlo simulations.

Case	Parameter sigma (σ) values			Mean Leakage (μA)	
	(L_g, L_l)	(V_g, V_l)	(T_g, T_l)	Experimental	Analytical
No Variation	(0,0)	(0,0)	(0,0)	14.97	15.22
Only die-to-die	(-1,0)	(-1,0)	(-1,0)	20.82	21.32
No Variation	$(-1, \pm 3)$	$(-1, \pm 3)$	$(-1, \pm 3)$	24.01	24.95

5.1.2 Frequency Binning

Parametric yield analysis is performed by frequency binning, in which samples are analyzed for their maximum operating frequency and placed into a frequency bin that corresponds to the measured performance. However, if the performance is below a lower limit, the sample is discarded as being useless. In addition, a power constraint is imposed on each of the frequency bins. Chip samples that dissipate more power than a given value are also discarded because they exceed the heat dissipation capacity of the heat removal system. This limit may also be imposed by the kind of package used for the design.

With continued technology scaling, subthreshold leakage has grown to contribute a significant fraction of the total power budget, which correlates negatively with circuit delay, and high performance chips are frequently found to have power which is higher than the imposed constraint. This is known as the two-sided constraint on the yield of current designs. We will look at this issue in more detail in the following section.

Circuit performance is a function of all three process parameters we used for leakage analysis. However, of the three, gate length variation is found to have the strongest influence on circuit performance. This is illustrated in Fig. 5.4 which shows the variation in the delay of a 17-stage ring oscillator in a 100 nm process for varying amount of variation in global values of process parameters. Variation in gate length can be seen to have the strongest influence, while variation in threshold voltage and gate oxide thickness have minimal impact on performance. This results from the fact that a smaller gate oxide thickness or threshold voltage increases the drive strength of a gate which reduces delay. However, it also increases the capacitive loading of the previous gate, which increases delay. The same dependence holds true for local variability and therefore, we can neglect the impact of variations in threshold voltage and oxide thickness, both global and local, on circuit performance.

In Chap. 3, we found that local variations in gate length have a strong influence on the mean circuit delay while global variations result in an increased variance of circuit delay. However, this work is based on the assumption that local variation in gate length do not have a strong influence on delay. An approach that considers this impact will be discussed in the next section. Thus,

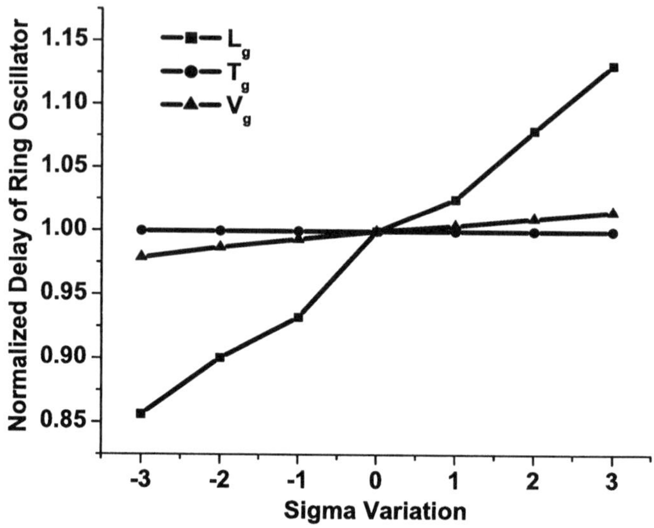

Fig. 5.4. Comparison of the influence of parameter variation on ring oscillator delay.

the performance of a design can now be approximated as a function of global variations in gate length alone.

5.1.3 Yield Computation

We will now discuss a technique to estimate the yield of a design in each of the frequency bins based on our discussion of leakage variability and frequency binning. Using the mapping between performance and global gate length fluctuation, we can estimate the range of the parameter L_g that corresponds to each of the performance bins. We assume that the frequency bin is represented by the smallest value of L_g in this range, for the purpose of leakage computations. Using this value of global gate length variability, we can rewrite the leakage equation for samples within this frequency bin as

$$I_{\text{sub}} = A_s^N \exp\left(-\frac{c_3^N V_g}{c_1^N}\right) + A_s^P \exp\left(-\frac{c_3^P V_g}{c_1^P}\right) \tag{5.28}$$

$$I_{\text{gate}} = A_g^N \exp\left(-\frac{T_g}{a_1^N}\right) + A_g^P \exp\left(-\frac{T_g}{a_1^P}\right) \tag{5.29}$$

where expressions for A_s and A_g can be easily obtained using parameters in (5.21), (5.22) and (5.27). Note that total leakage can now be expressed as a sum of four RVs, where each RV has a lognormal distribution. Since sums of correlated or uncorrelated lognormal RVs can be accurately approximated by another lognormal, we can express the distribution of total leakage as a lognormal. Let us represent the total leakage as

$$I_{\text{tot}} = \mathrm{e}^{X_1} + \mathrm{e}^{X_2} + \mathrm{e}^{X_3} + \mathrm{e}^{X_4} = Y_1 + Y_2 + Y_3 + Y_4 \approx Z \tag{5.30}$$

where X_i's are Gaussian RVs and Y_i's are lognormal RVs. To approximate the sum in the equation above as another lognormal RV, we can use Wilkinson's method, which is based on matching the first two moments of I_{tot} with the moments of Z. Note, in contrast to the approach discussed in Chap. 4, where we confronted uncorrelated RVs, we are dealing with correlated RVs in (5.27). This does not overly complicate matters since the expected value remains the same as in the case of uncorrelated RVs and is expressed as

$$E\left[I_{\text{tot}}\right] = E\left[Y_1\right] + E\left[Y_2\right] + E\left[Y_3\right] + E\left[Y_4\right] \tag{5.31}$$

which can be easily evaluated using (5.17). When we perform variance computation, we need to compute the expected value of the product of $Y_i's$. We can simplify this computation since the product of two lognormal RVs is another lognormal RV:

$$\begin{aligned} E\left[I_{\text{tot}}^2\right] &= \sum_{i,j=1}^{4} E\left[Y_i Y_j\right] \\ &= \sum_{i,j=1}^{4} E\left[\mathrm{e}^{X_i + X_j}\right]. \end{aligned} \tag{5.32}$$

The above expectation can now be calculated by estimating the moments of $X_i + X_j$, which is straightforward given the mean, variance and correlation of X_i and X_j. Having estimated the mean $\mu_{I_{\text{tot}}}$ and variance $\mu_{I_{\text{tot}}}$ of total leakage, we then calculate the parameters that define the lognormal RV I_{tot}. These parameters correspond to the mean and variance of the Gaussian RV associated with this lognormal RV, which were shown in Chap. 4 to be

$$\mu_{N,\, I_{\text{tot}}} = \frac{1}{2} \log\left(\frac{\mu_{I_{\text{tot}}}^4}{\mu_{I_{\text{tot}}}^2 + \sigma_{I_{\text{tot}}}^2}\right) \tag{5.33}$$

$$\sigma_{N,\, I_{\text{tot}}}^2 = \log\left(1 + \frac{\sigma_{I_{\text{tot}}}^2}{\mu_{I_{\text{tot}}}^2}\right). \tag{5.34}$$

The quantile numbers for the lognormal distribution can also be expressed in terms of the Gaussian cdf (or the error function), since an exponential transformation is monotonic. The cdf of a lognormal RV is expressed as

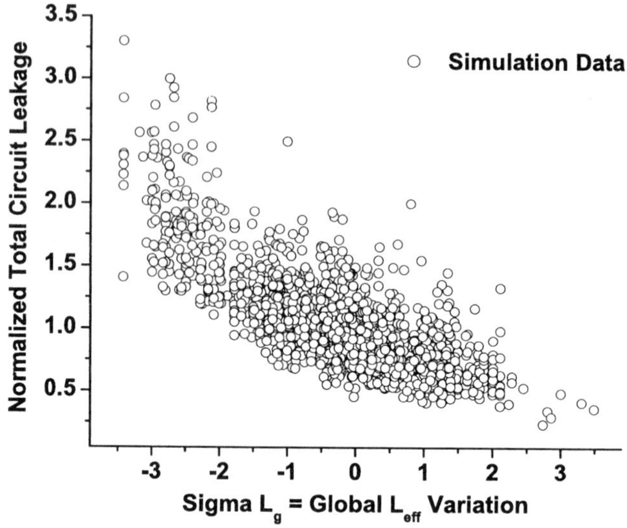

Fig. 5.5. Scatter plot showing the fluctuation in total circuit leakage power.

$$\mathrm{CDF}\left(I_{\text{total}}\right)=\frac{1}{2}\left(1+\operatorname{erf}\left(\frac{\log(I_{\text{tot}})-\mu_{N,\,I_{\text{tot}}}}{\sqrt{2}\sigma_{N,\,I_{\text{tot}}}}\right)\right) \tag{5.35}$$

where erf refers to the error function. Using the above equation, the leakage constraint for a given desired yield as well as the fraction of chips that satisfy a given power constraint can be estimated.

Figure 5.5 shows the scatter plot of leakage, generated using SPICE simulations, for 2000 samples of a circuit. The x-axis corresponds to the variation in the global gate length (which is a first-order approximation of performance) in terms of the number of sigma deviations from the nominal value. If all samples are considered, we see that we get an overall spread of 14X in leakage currents. In addition, for a given value of L_g we observe a distribution in leakage, which in the case of $L_g = 0$ has a spread of 3X. This *local distribution* for a given L_g value results from variation in the global variations in threshold voltage and gate oxide thickness, which have a strong influence on the overall leakage power but a weak impact on performance. The band structure does not result from local variability since local variability in process parameters acts as a scaling parameter for large designs and does not result in a distribution for leakage. The variation in circuit leakage for a given value of L_g will result in a fraction of chips having power dissipation levels which are higher than the imposed constraint and will result in reduced yield.

In Fig. 5.6, we superimpose analytically generated contours on top of the previous figure. These contour plots are generated by calculating $\mu_{N,\,I_{\text{tot}}}$ and

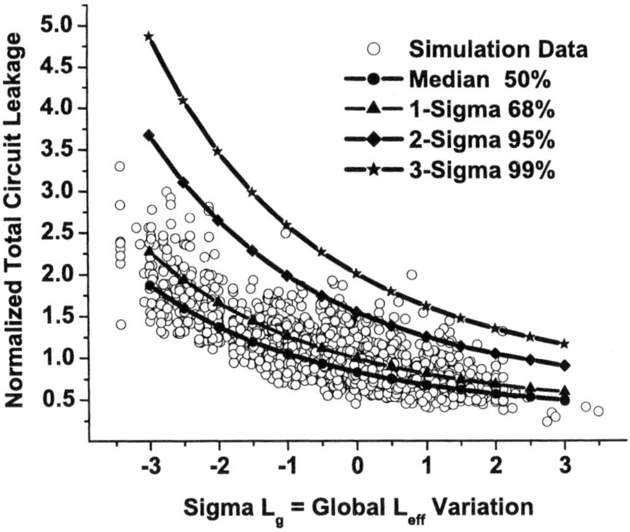

Fig. 5.6. Scatter plot of leakage with superimposed analytically computed contour lines.

$\sigma_{N, I_{\text{tot}}}$ for each value of L_g, and then using (5.35) to generate the leakage power number that corresponds to a given percentile point on the local distribution. The figure shows that for values of gate length much smaller or much larger than nominal, most samples are within a 1σ range. However, for nominal values of gate length, a significant fraction of samples result in leakage which lies outside the 1σ range. This can be understood from the fact that global variability in L_g, V_g and T_g are independent and the probability that all these process parameters are at their process corners is much smaller than the probability that L_g lies close to its nominal value and V_g and T_g are at their corners. This implies that for frequency bins that correspond to nominal performance, there is a potential for a much larger loss in yield if leakage power constraints are defined in terms of sigma deviations from the nominal leakage power for that particular bin. However, the same power constraint is generally imposed on all frequency bins, and given the wedge shaped distribution for the complete scatter plot, we can expect the lowest yield for the highest performance bin.

Since L_g directly corresponds to performance, we can partition the scatter plot as shown in Fig. 5.7, where the vertical lines now correspond to performance bins. Note that when assuming a frequency constraint which corresponds to the performance obtained at a global gate length variability of 1σ, we throw away all samples of a design that lie to the right of this region in this scatter plot. The power constraint imposed on all performance bins is that

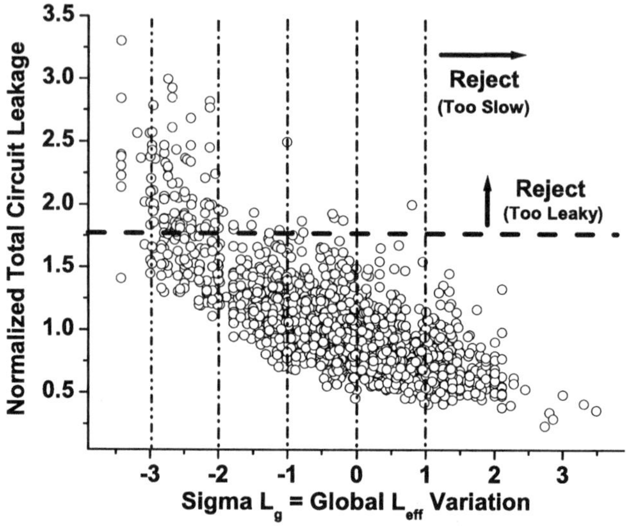

Fig. 5.7. Scatter plot showing power and performance constraints for different performance bins.

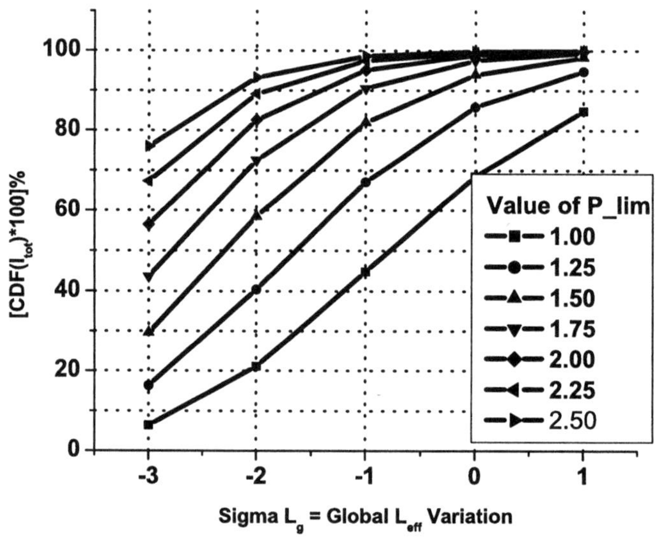

Fig. 5.8. Analytically computed yield for each frequency bin for different power constraints.

the power is less than 1.75X the nominal leakage power. Given this constraint, we see that samples that lie above this line in the scatter plot correspond to samples of the design that are too leaky and are hence discarded. Figure 5.8 plots the yield in each of the performance bins for different power constraints. We see that most samples in the lowest performance bin satisfy the power constraint. However, a significant fraction of the samples in the high performance bins fail to meet the power constraint. For the power constraint of 1.75X, we find that 27.4% of the samples are lost in the highest performance bin ($L_g < -3\sigma$). This has significant impact on earned revenues, since high performance are generally sold at a much larger profit margin.

In this section, we looked at a high-level approach to consider the impact of the correlated variation in performance and leakage power. In the next section, we will develop a gate-level yield analysis technique that can be used to perform gate level optimization to improve the parametric yield of a design under power and performance constraints.

5.2 Gate-Level Yield Estimation

To perform gate-level yield analysis, we need to perform gate-level timing and power analysis which considers all components of variations and considers correlation between power and delay that they introduce. Since variations in dynamic power are comparatively insignificant, we will again concentrate on the variations in leakage power and subtract the nominal dynamic power dissipation from the total power budget to estimate yield using a given delay constraint and leakage power budget. In addition, both the leakage and timing analysis are performed using the same underlying RV so that the final correlation between leakage and timing can be estimated. The correlated components are modeled in this analysis approach using a principal component based approach that allows us to express the underlying variations in terms of independent Gaussian RVs. We then develop an approach to perform statistical leakage power analysis and express the circuit leakage power in terms of the same underlying process variations used to express the delay of the circuit. With increasing circuit size, the impact of the random component of variation on the variance of power reduces to zero due to the central limit theorem [109]. Thus, although the random component impacts the statistical circuit delay, the correlation between the random components of power and delay has a vanishingly small impact on the overall correlation in power and delay. We will show that even for small circuits with a few hundred gates, the random component has negligible impact on the overall variance of power. Since the correlation due to correlated process parameters is already captured, the correlation in power and delay can be computed allowing the construct of their joint probability distribution function (jpdf). Based on this jpdf of delay and power a closed-form approach is developed to estimate the yield of a design given delay and leakage constraints.

To simplify the analysis, we will consider process variations in gate length and gate length-independent threshold voltage (V_{th0}) alone, although the approach can be easily extended to consider other sources of variations. The modeling of process variations and the timing analysis is very similar to the statistical timing analysis approach described in Chap. 3 using principal components analysis. The process parameters are expressed as a sum of correlated and random components and the sum of variances of both these components provides the overall variation in the process parameter. To handle the correlated components of variations (inter-die and correlated intra-die) the overall chip area is divided into a grid as discussed in Chap. 2 (Fig. 2.6). To simplify the problem, we replace the set of correlated RVs by another set of mutually independent RVs with zero mean and unit variance using the principal components of the set of original correlated RVs. A vector of RVs $\mathbf{X}$ with a correlation matrix $\mathbf{C}$, can be expressed as a linear combination of the principal component vector $\mathbf{Y}$ as

$$\mathbf{X} = \mathbf{\Sigma} + \mathbf{D}^{1/2}\mathbf{\Lambda}^{-1}\mathbf{Y} \tag{5.36}$$

where $\mathbf{\Sigma}$ is the mean vector associated with $\mathbf{X}$, $\mathbf{D}$ is a diagonal matrix with elements being the eigenvalues of the covariance matrix of the RVs of $\mathbf{X}$ and $\mathbf{\Lambda}$ is the matrix whose columns are the eigenvectors of the covariance matrix. Since the correlation matrix of a multivariate (non-degenerate) Gaussian RV is positive-definite, all elements of $\mathbf{D}$ are positive and the square-root in (5.36) can be evaluated.

We now express the delay and leakage power of an individual gate as

$$d = d_{\text{nom}} + \sum_{i=1}^{p} \alpha_p(\Delta P_p) \tag{5.37}$$

$$l = \exp\left(V_{\text{nom}} + \sum_{i=1}^{p} \beta_p(\Delta P_p)\right) \tag{5.38}$$

where d_{nom} and $\exp(V_{\text{nom}})$ are the nominal values of delay and leakage power respectively, and the $\alpha's$ and $\beta's$ represent the sensitivities of delay and the log of leakage to the process parameters under consideration. The variable ΔP_p represents the fluctuation in the process parameters from their nominal value.

In a statistical scenario, the process parameters are modeled as RVs. If the overall circuit is partitioned using the 2-D grid, the delay of individual gates can be expressed as a function of these RVs. Using the principal component approach, the delay in (5.37) can be expressed as

$$d = d_{\text{nom}} + \sum_{i=1}^{p} \left(\alpha_p \sum_{j=1}^{n} \gamma_{ji} z_j\right) + \eta_d R \tag{5.39}$$

where z_j's are the principal components of the correlated RVs ΔP_p's in (5.37)-(5.38) and the γ_{ji}'s can be obtained from principal component analysis (5.36). $R \sim N(0,1)$ in the above equation represents the random component of variations of all the process parameters lumped into a single term that contributes a total variance of η_d^2 to the overall variance of delay. Similarly, the leakage power for an individual gate can be expressed as

$$l = \exp\left(V_{\text{nom}} + \sum_{i=1}^{p}\left(\beta_p \sum_{j=1}^{n} \gamma_{ji} z_j\right) + \eta_l R\right). \tag{5.40}$$

Now, using these expressions we will perform timing and power analysis. These canonical forms for the representation for delay and power will be maintained all through the analysis. We will also find that the loss in information due to the lumping of the random component of variation (which results in a large simplification during analysis) results in an insignificant error. We will first discuss the details of timing analysis, which will be a simple extension of the approach outlined in Chap.3, to consider the random component of variation.

5.2.1 Timing Analysis

The delay of each gate 'j' can be expressed as follows using the expression developed in the previous section as:

$$d_j = a_{j,0} + \sum_{i=1}^{n} a_{i,j} z_i + a_{n+1,j} R_j. \tag{5.41}$$

This serves as the canonical expression for delay. The mean delay is equal to the nominal delay and is expressed as $d_{j,nom}$. The principal components are represented by RVs z_i's and the RV corresponding random component is represented as R. The $a_{i,j}$'s function as scaling parameters to obtain the sensitivity of gate delay to the parameters represented by each of the RVs. Since the RVs used in the above expression are statistically independent zero mean unit variance Gaussian RVs, we can express the variance of gate delay as

$$Var(d_j) = \left(\sum_{i=1}^{n} a_{i,j}^2\right) + a_{n+1,j}^2 \tag{5.42}$$

and the covariance of delay with any one of the principal components can be obtained as

$$Cov(d_j, z_i) = E\left[d_j, z_i\right] - E\left[d_j\right] E\left[z_i\right] = a_{i,j}^2 \quad \forall i = 1, 2, \ldots, n. \tag{5.43}$$

To perform statistical timing analysis, we need to define the max and the sum operation for delay expressions. We assume the delay expression at each node in the circuit graph is represented in canonical form (5.41), and we define the sum operation for two delay expressions d_j and d_k as:

$$\mathrm{Sum}(d_j, d_k) = a_{j,nom} + a_{k,0} + \sum_{i=1}^{n} (a_{i,j} + a_{i,k})\, z_i + \sqrt{a_{n+1,j}^2 + a_{n+1,k}^2}. \quad (5.44)$$

The max operation is hard to compute accurately and results in a non-normal distribution which makes further analysis complicated. It has been argued that the max of two Gaussian RVs can be closely approximated by another Gaussian RV for the purpose of timing analysis [30][141][3]. In addition, we discussed in Chap. 3 that the delay distributions arising due to correlated reconvergent fanouts can be tightly upper bounded by assuming them to be independent. Let nodes l, m and n be related such that

$$d_l = \max(d_m, d_n). \quad (5.45)$$

Assuming that the delay at node k can again be expressed in canonical form, we can write a set of $n+2$ equation to estimate the coefficients of the delay expression at node k in terms of the coefficients at nodes m and n. This is achieved by matching the first two moments of d_k obtained using expression for the max of two Gaussian RV and matching the correlation of d_l with each of the principal components. These expressions, which were developed in [35], are discussed in Chap. 3. Using these expressions, the set of $n+2$ equations can be written as

$$\begin{aligned} a_{l,0} &= E[\max(d_m, d_n)] \\ a_{i,l} &= Cov(d_l, z_i) = Cov(\max(d_m, d_n), z_i) \ \ \forall i = 1, 2, \ldots, n \\ a_{n+1,l} &= \left(Var(\max(d_m, d_n)) - \sum_{i=1}^{n} a_{i,l}^2 \right)^{1/2} . \end{aligned} \quad (5.46)$$

To evaluate the terms in the set of equations above, we need to use expressions for the mean and variance of the canonical delay expression. We also need to estimate the covariance of two delay expressions which can be expressed as

$$Cov(d_m, d_n) = \sum_{i=1}^{n} a_{i,m} a_{i,n}. \quad (5.47)$$

By modeling the random component, the timing analysis steps are able to preserve the mean, variance and correlations, avoiding the need to scale the coefficients of the principal components to match variance, as we did in

Chap. 3, which results in the delay expressions losing their exact correlation with the principal components. For gates with more than two inputs, the technique described above is applied iteratively.

Using the technique described above, we can develop an expression for the delay of a circuit in terms of the RVs associated with process parameter variations. We now discuss the steps to perform leakage analysis, where the goal is to preserve the correlation between delay and power, which is achieved by performing a similar principal component-based analysis approach using the same underlying RVs.

5.2.2 Leakage Power Analysis

As in (5.38), leakage power is expressed as an exponential of a Gaussian RV, which is known to have a lognormal distribution. The leakage power for a complete circuit block can be expressed as a sum of correlated lognormal RVs. The authors in [1] show that this sum can be accurately approximated as another lognormal RV. It is also shown that the approximation performed using an extension of Wilkinson's method [127], which is based on matching the first two moments, provides good accuracy. Using the principal components of timing analysis, we can write the canonical form for leakage power as

$$l_j = \exp\left(b_{0,j} + \sum_{i=1}^{n} b_{i,j} z_i + b_{n+1,j} R\right) \tag{5.48}$$

where the z_i's are principal components of the RVs (used for timing analysis as well) and the coefficients b_j's can be computed using (5.36) and (5.38). Using expressions for mean and variance of lognormal RVs, the mean and variance of leakage power of gate j can be expressed as

$$E[l_j] = \exp\left(b_{0,j} + \frac{1}{2}\sum_{i=1}^{n+1} b_{i,j}^2\right) \tag{5.49}$$

$$Var(l_j) = \exp\left(2b_{0,j} + \sum_{i=1}^{n+1} b_{i,j}^2\right) - \exp\left(2b_{0,j} + \frac{1}{2}\sum_{i=1}^{n+1} a_{i,j}^2\right).$$

The correlation of the leakage of a particular gate with the lognormal RV associated with one of the principal components is expressed as:

$$E[l_j, z_i] = \exp\left(b_{0,j} + \frac{1}{2}\sum_{k=1,k\neq i}^{n+1} a_{k,j}^2 + (a_{i,j}+1)^2\right) \quad \forall j = 1, 2, \ldots, n. \tag{5.50}$$

Similarly, the correlation between the leakage currents of two gates can be expressed as:

$$E[l_m, l_n] = \exp\left(b_{0,m}b_{0,n} + \frac{1}{2}\left(\left(\sum_{i=1}^{n}(b_{i,m}+b_{i,n})^2\right) + b_{n+1,m}^2 + b_{n+1,m}^2\right)\right). \quad (5.51)$$

As compared to timing analysis, we only need to define the sum operation for leakage analysis, since the leakage for a circuit block is simply the sum of the leakage of individual gates. We again make the simplifying assumption that the sum of two lognormal RVs can be expressed as another lognormal and we express the resulting lognormal in the same canonical form (5.48). This process is then iteratively continued to sum the leakage of all gates under consideration. Note that if the random variables associated with all the gates are summed in a single step then the overall complexity of the power analysis approach becomes $O(n^2)$, where n is the number of gates in the circuit, due to the need to evaluate the complete correlation matrix. However, in the iterative summation approach, we sum two RVs of leakage in canonical form in each iterative step to obtain another RV in the same canonical form. To find the coefficients in the expression for the sum of the RVs, we match the first two moments (as in Wilkinson's method) and the correlations with the lognormal RVs associated with each of the Gaussian principal components. We outline one iterative step where we sum l_m and l_n to obtain l_l.

$$l_l = l_m + l_n \quad (5.52)$$

Each iterative step again becomes equivalent to solving a set of equations and the resulting coefficients can be expressed as

$$\begin{aligned} b_{i,l} &= \log\left(\frac{E[l_l e^{z_i}]}{E[l_l]E[e^{z_i}]}\right) \\ &= \log\left(\frac{E[l_m e^{z_i}] + E[l_n e^{z_i}]}{(E[l_m] + E[l_n])E[e^{z_i}]}\right)) \quad \forall i = 1, 2, \ldots, n. \end{aligned} \quad (5.53)$$

The remaining two coefficients are expressed as

$$b_{0,l} = \frac{1}{2}\log\left(\frac{(E[l_m]+E[l_n])^4}{(E[l_b]+E[l_c])^2 + Var(l_b) + Var(l_c) + 2Cov(l_b l_c)}\right) \quad (5.54)$$

$$b_{n+1,l} = \left[\log\left(1 + \frac{Var(l_m) + Var(l_n) + 2Cov(l_m l_n)}{(E[l_m]+E[l_n])^2}\right) - \sum_{i=1}^{n} b_{i,l}^2\right]^{1/2} \quad (5.55)$$

The expectations, variances and covariances in (5.53)-(5.54) can be evaluated using (5.49)-(5.51).

The timing and power analysis techniques outlined above can now be used to efficiently estimate the individual probability distribution functions of delay and leakage power. The correlation in delay and leakage power arising from the correlated components of variation can be estimated since the correlated variations are expressed in terms of the principal components used to develop the expressions for both delay and power.

As will be shown in the results, the dependence of the variance of leakage power on the random component is very weak. This arises due to the fact that the random component associated with each gate is independent and hence the ratio of standard deviation to mean for the sum of these independent RVs is inversely proportional to the square root of the number of RVs summed [109]. This ratio does not reduce for correlated RVs - therefore, if a large number of RVs are summed with both correlated and random components, the overall variance is dominated by the variance of the correlated component. Hence the correlation due to the random component, which is difficult to compute efficiently, is insignificant and will be neglected.

5.2.3 Yield Estimation

We now use the delay and power expressions to estimate the parametric yield of a design given leakage power and delay constraints. The parametric yield of a circuit given delay and power constraints can be expressed as

$$Y = \mathcal{P}\left(d \leq d_0, l \leq l_0\right), \tag{5.56}$$

which is the probability of the circuit delay being less than d_0, while the leakage power dissipation is less than p_0. Since delay and power are correlated, the yield cannot be simply computed by multiplying the separate probabilities. Let us refer back to the results regarding multinormal distributions discussed in Sec. 2.2.2. Under the assumption that the RVs used to represent the correlated and random variations are part of a multinormal distribution, the joint distribution of delay and the logarithm of leakage power becomes a jointly normal bivariate Gaussian distribution. We express the yield in terms of two $N(0,1)$ RVs, N_0 and N_1, which are jointly normal as:

$$Y = \mathcal{P}\left(N_0 \leq \frac{d_0 - E[d]}{\sigma_d}, N_1 \leq \frac{\log l - E[\log l]}{\sigma(\log l)}\right). \tag{5.57}$$

Since correlation does not change under a linear transformation with positive coefficients, the correlation between N_0 and N_1 remains the same as the correlation between delay and log of leakage power. The correlation coefficient of the two Gaussian RVs in the yield equation above can now be obtained using (5.47).

An approach to evaluate the probability in the above expression is to perform numerical integration of the jpdf over the desired region, but this is computationally inefficient. A look-up table based approach, though efficient,

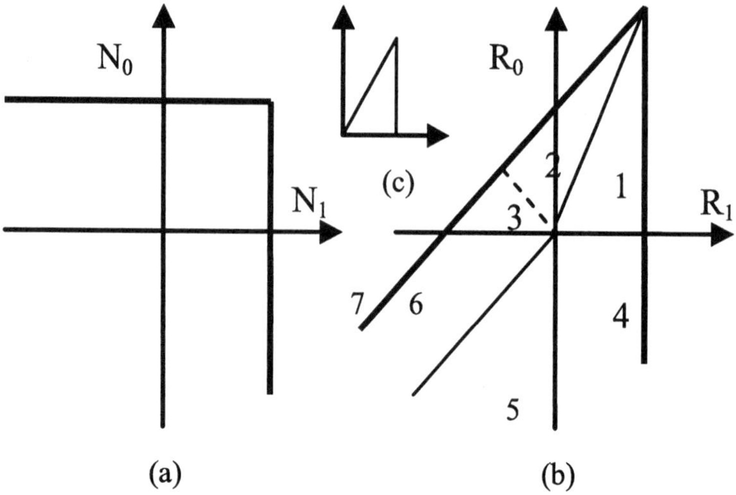

Fig. 5.9. Transformation of the feasible region from (a) to (b) under the transformation expressed in (5.58) for negative values of correlation.

involves substantial inaccuracy due to the required interpolation as noted in [24]. Hence, we adopt an analytical approach to estimate the yield which makes the approach efficient and practical within a yield optimization framework.

The feasible region defined by a pair of correlated RVs is transformed to region defined by a pair of uncorrelated RVs using the following transformation:

$$R_0 = N_0; \;\; R_1 = \left(\frac{N_1 - \rho N_0}{(1 - \rho^2)^{1/2}} \right) \tag{5.58}$$

This transformation maps the feasible region from a rectangle to a triangle as shown in Fig. 5.9 for the case where $\rho < 0$, which is the case of interest. The desired probability can be obtained by using approximate expressions developed in [24] for evaluating probabilities of uncorrelated bivariate Gaussian RVs in regions of the form shown in Fig. 5.9(c). We will also use the fact that uncorrelated RVs, that are distributed according to a joint Gaussian distribution, are also independent.

To evaluate the probability of the region shown in Fig. 5.9(b), we partition the figure as shown. The desired probability can then be expressed as a sum of the probabilities in Regions 1-6 which can be evaluated as follows:

Region 1: The region is already in the form required in Fig. 5.9(c),

Region 2: The integral of the region is circularly symmetric. Hence, if the axes are rotated such that the dotted line as shown in Fig. 5.9(b) lies along the x-axis, then Region 2 is also in the same form as Fig. 5.9(c),

Region 4: The probability in this region is

$$\mathcal{P}(R_0 \leq 0, 0 \leq R_1 \leq X) \tag{5.59}$$

where X is the point where the vertical line cuts the R_0 axis. Since R_0 and R_1 are statistically independent, this probability can be expressed as

$$\mathcal{P}(R_0 \leq 0)\mathcal{P}(0 \leq R_1 \leq X) = 0.5\Phi(X) \tag{5.60}$$

where Φ is the cdf for a Gaussian RV,

Region 3, 5, 6: The probability for this region can be expressed as

$$\mathcal{P}(R_0 \leq 0, R_1 \leq 0) + \mathcal{P}(3+6) - \mathcal{P}(6+7) \tag{5.61}$$

The first and second terms in (5.61) correspond to a region that is in the same form as Region 4 and the region for the third term is in the same form as Region 1. Thus, the desired yield expressed in (5.56) can be efficiently estimated using closed-form expressions.

In terms of computational complexity, the above approach differs from the principal components based statistical timing analysis approach discussed in Chap. 3 in the computation of an extra term associated with the random component. The overall complexity of the timing analysis remains $O(nN_g)$, where n is the number of terms in the delay expression that corresponds to the number of partitions into which the circuit is divided, and N_g is the number of gates in the circuit. The power analysis is similar and requires an additional $O(nNg)$ steps. The correlation computation requires an additional $O(n)$ steps, and the yield estimation runs in constant time. The computation of the principal components requires $O(pn^3)$ steps where p is the number of process parameters required. The cubic dependence results from the eigenvector computation required during principal component analysis. Since the principal components need to be calculated only once, it does not impact the overall complexity and hence the overall complexity of the approach is $O(nN_g)$. Thus, the runtime can be expected to increase quadratically with increase in circuit size since the size of the partitions remains the same.

Table 5.3 shows results for the ISCAS'85 [23] and MCNC benchmark circuits. The benchmark circuits are synthesized using an industrial 130 nm technology. Only channel length and gate length-independent threshold voltage variations are considered in these results. A 3σ variation of 20% of the nominal value is assumed. All variation in V_{th0} is assumed to be random (due to random dopant effects), whereas half the variation in channel length is considered to be correlated. The table compares the means and standard deviations of delay and power obtained using the proposed approach and Monte Carlo based simulations. The table also compares the coefficient of correlation of

Table 5.3. Comparison of the analytical timing and power analysis approach and Monte Carlo based simulation results

Benchmark	Analytical					Error as compared to Monte Carlo				
Circuit	Delay Mean	Delay SD	Leakage Mean	Leakage SD	Corre-lation	Delay Mean	Delay SD	Leakage Mean	Leakage SD	Corre-lation
c432	0.91	0.04	12.20	4.05	-0.91	0.5%	8.4%	0.1%	11.1%	2.0%
c499	0.89	0.03	36.14	10.29	-0.95	0.4%	11.4%	0.9%	8.8%	4.9%
c880	0.82	0.04	30.00	8.64	-0.88	1.1%	11.1%	1.1%	8.3%	4.9%
c1908	1.22	0.04	19.03	5.38	-0.95	2.1%	11.9%	1.0%	9.2%	4.0%
c2670	0.91	0.04	7.47	2.34	-0.93	1.7%	9.8%	1.7%	10.3%	0.7%
c3540	1.43	0.06	57.54	14.70	-0.74	2.0%	10.4%	1.3%	6.4%	0.9%
c5315	1.23	0.04	88.41	20.95	-0.87	1.4%	12.0%	2.7%	5.2%	4.1%
c6288	3.32	0.11	116.73	25.38	-0.79	2.6%	13.4%	3.6%	5.3%	3.2%
c7552	1.12	0.04	85.39	20.27	-0.86	0.6%	16.2%	2.5%	5.4%	8.8%
i2	0.47	0.02	4.53	1.46	-0.91	5.0%	20.1%	0.2%	10.1%	4.9%
i3	0.27	0.01	0.83	0.26	-0.89	1.0%	16.8%	1.0%	3.9%	3.6%
i4	0.38	0.02	11.34	3.73	-0.87	2.6%	20.6%	0.9%	8.1%	1.7%
i5	0.34	0.01	20.63	5.98	-0.88	1.0%	17.8%	0.0%	7.3%	8.7%
i6	0.31	0.01	13.90	4.58	-0.88	1.5%	10.4%	0.4%	9.0%	6.4%
i7	0.34	0.02	22.71	6.69	-0.65	3.1%	14.4%	0.6%	10.5%	3.6%
i8	0.52	0.02	26.71	7.66	-0.94	1.5%	18.0%	1.4%	7.9%	6.6%
i9	0.53	0.02	24.47	7.91	-0.83	1.8%	13.4%	0.6%	3.0%	0.7%
i10	1.41	0.05	49.55	12.59	-0.86	2.4%	11.1%	2.3%	7.4%	5.5%
Average						1.8%	13.7%	1.2%	7.6%	4.2%

delay and the log of leakage power which is required for yield estimation as discussed above.

The results show that the estimates obtained using the discussed approach for the values of the mean delay and leakage power are very accurate with an average error of 1.2% and 1.8% respectively. The standard deviations show an average error of 7.6% and 13.7% for power and delay respectively. Note that in general, circuits with smaller logic depth show larger error in delay compared to circuits with larger logic depths. It is also found that circuits which have a larger error have a significant component of the overall variation in delay arising from random variations. This results from the fact that the correlations in the random component are neglected, which can result in an overall smaller variance. However, the coefficient of correlation between the log of leakage power and delay shows very good match to MC results with an average error of 4.2%.

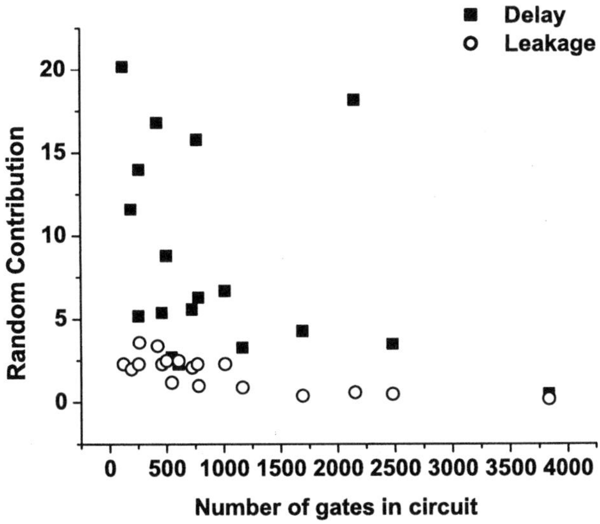

Fig. 5.10. Contribution of random variation to the variance of delay and leakage power for different circuits.

Figure 5.10 shows the contribution of the random component of variation to the variance in delay and leakage. The contribution of the random component to delay of a path is inversely proportional to the depth of the critical paths and can generally be expected to be small for circuits with large logic depths. However, with varying circuit size the contribution of random component has no consistent trends. This results from the fact that the length of the critical path is not directly proportional to circuit size. Moreover, the contribution of random variation to delay also depends on the number of critical paths in a circuit. On the other hand, the contribution of the random variations to leakage variance is typically small and can be seen to decrease with increasing circuit size. The overall contribution of the random component to the variance of leakage power is 1.8% on average with a maximum of 3.6%. This confirms the assumption that the impact of the random component of variation is negligible when estimating the correlation in power and performance.

Figure 5.11 shows a representative jpdf of the log of leakage and delay, which is a bivariate Gaussian jpdf which is described as

$$f_(X,Y)(x,y) = \frac{1}{2\pi\sqrt{1-\rho^2}} \exp\left(\frac{-1}{2(1-\rho^2)}\left[\left(\frac{x-\mu_x}{\sigma_x}\right)^2\right.\right.$$

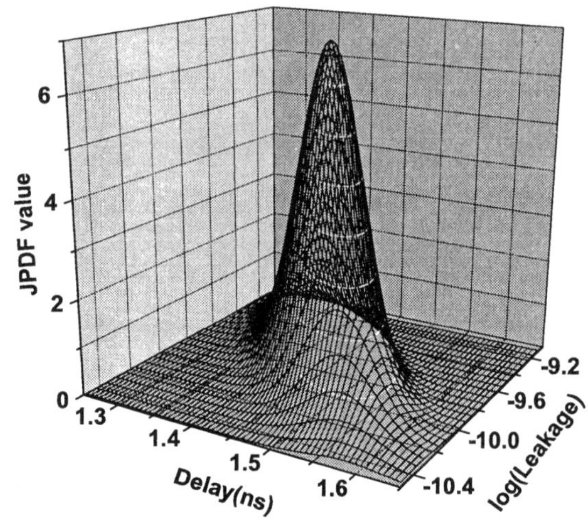

Fig. 5.11. Joint probability distribution function for the bivariate Gaussian distribution for c3540.

$$-2\rho\left(\frac{x-\mu_x}{\sigma_x}\right)\left(\frac{y-\mu_y}{\sigma_y}\right)+\left(\frac{y-\mu_y}{\sigma_y}\right)^2\Bigg]\Bigg) \quad (5.62)$$

where μ_x and σ_x are the mean and variance of RV X (which in our case represents delay) and μ_y and σ_y are the mean and variance of RV Y (which represents the log of leakage). The contours of the jpdf are ellipses with center at the mean of delay and the log of leakage, and have the equation

$$\left(\frac{x-\mu_x}{\sigma_x}\right)^2-2\rho\left(\frac{x-\mu_x}{\sigma_x}\right)\left(\frac{y-\mu_y}{\sigma_y}\right)+\left(\frac{y-\mu_y}{\sigma_y}\right)^2 = -2(1-\rho^2)\log(1-\alpha) \quad (5.63)$$

where α represents the fraction of the jpdf enclosed within the ellipse. The major-axis of the ellipse makes an angle θ with the with the x-axis that is expressed as

$$\theta=\frac{1}{2}\tan^{-1}\left(\frac{2\rho\sigma_x\sigma_y}{\sigma_x^2-\sigma_y^2}\right) \quad (5.64)$$

This angle is 135° if $\sigma_x=\sigma_y$ and $\rho<0$ (independent of the exact value of the correlation coefficient). If $\sigma_x=\sigma_y$ and $\rho=0$ then the ellipse degenerates into a circle. In addition, as the correlation is allowed to increase to an extreme

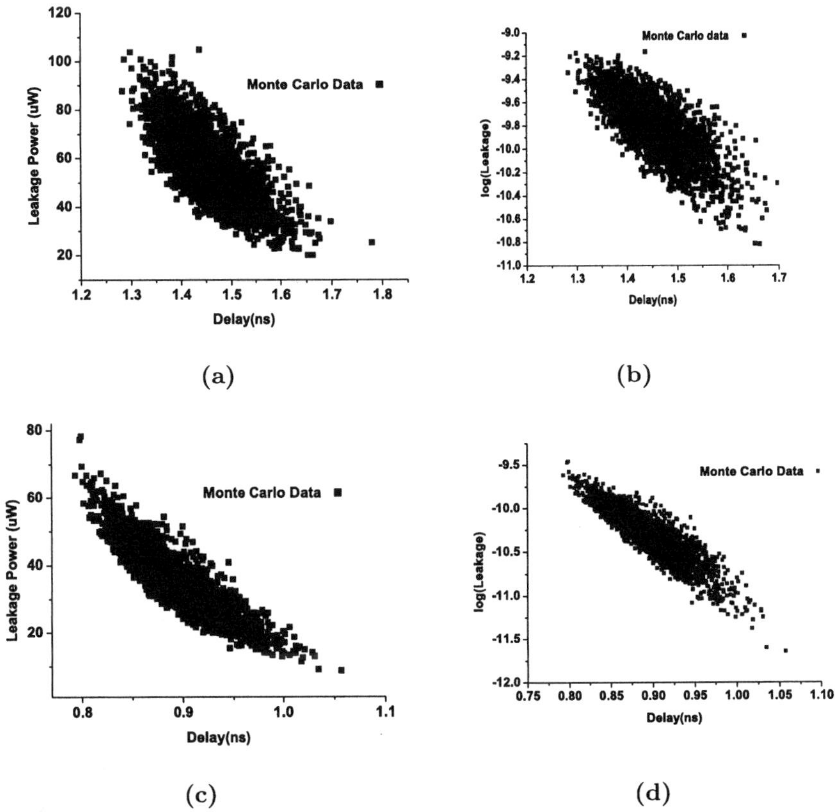

Fig. 5.12. Scatter plots obtained using MC simulations. (a) Delay and leakage for c3540 (b) Delay and log of leakage for c3540 (c) Delay and leakage for c499 (d) Delay and log of leakage c499.

value of ± 1, the contours of the jpdf concentrate around the major axis and merge into a line. Figure 5.12 shows the scatter plots obtained using MC-based simulations for two of the benchmark circuits. The scatter plots on the left show the lognormal nature of leakage power which is more evident for the case of c499 (Figure 5.12(c)) being one of the smaller circuit in the benchmark suite. Circuit c499 also shows a high correlation between delay and power which is evident from the concentration of the jpdf along the major axis of its contours. The scatter plots on the right show the Gaussian nature of the jpdf of the log of leakage and delay, since the shape of the scatter plots closely resembles an ellipse.

Table 5.4 compares the yield estimates achieved using the principal component based approach to those obtained using Monte Carlo based simulations for all benchmark circuits at different performance bins. For both bins, the

Table 5.4. Yield estimates for different frequency bins using the analytical approach and Monte Carlo based simulations.

Benchmark	Monte Carlo		Analytical		Yield Neglecting Correlation	
	$D < D_\mu$	$D_\mu < D$ $D < 1.1D_\mu$	$D < D_\mu$	$D_\mu < D$ $D < 1.1D_\mu$	$D < D_\mu$	$D_\mu < D$ $D < 1.1D_\mu$
c432	0.17	0.43	0.14	0.46	0.31	0.3
c499	0.17	0.46	0.15	0.49	0.32	0.32
c880	0.2	0.43	0.16	0.46	0.32	0.31
c1908	0.18	0.48	0.14	0.49	0.32	0.32
c2670	0.16	0.44	0.14	0.47	0.31	0.3
c3540	0.22	0.43	0.2	0.44	0.33	0.32
c5315	0.19	0.48	0.19	0.48	0.33	0.33
c6288	0.22	0.47	0.21	0.46	0.34	0.34
c7552	0.2	0.47	0.19	0.47	0.33	0.33
i2	0.17	0.4	0.14	0.45	0.31	0.3
i3	0.17	0.4	0.15	0.45	0.31	0.3
i4	0.19	0.4	0.15	0.46	0.31	0.3
i5	0.18	0.45	0.16	0.47	0.32	0.31
i6	0.2	0.39	0.15	0.45	0.31	0.3
i7	0.22	0.38	0.21	0.4	0.32	0.3
i8	0.18	0.46	0.15	0.49	0.32	0.32
i9	0.19	0.44	0.17	0.45	0.31	0.31
i10	0.2	0.46	0.18	0.47	0.33	0.32
Average	0.19	0.44	0.17	0.46	0.32	0.31

leakage power is constrained to be less than 1.1X the mean leakage value. Two different performance bins are constructed with delay confined to less than 1.0X and between 1.0X and 1.1X the mean delay. The proposed approach is seen to provide good estimates of the yield for the different frequency bins with an average error in yield of 2%. If the correlation in power and delay is ignored, the yield in the different bins can be both significantly overestimated (up to 15% in the high performance bin) and underestimated (up to 16% in the low performance bin) as shown in the last three columns of the table.

5.3 Supply Voltage Sensitivity

In the previous two sections, we presented an analysis of yield under variations in process parameters. Power supply is an additional lever that the designer can play with to influence the frequency and power of a design. In this manner, significant improvements in the shipping frequency or the yield of a design can be obtained. This idea was proposed and investigated in [116]. The approach develops a mathematical approach to analyze yield by mapping the feasible region in the power frequency space to the space of global variation in the physical parameters, gate length and threshold voltage.

As we saw in Sec. 5.1, the performance of a design can be captured reasonably accurately while considering variation in global gate length alone. The frequency can be assumed to be proportional to the saturation current of a device, which can be expressed using the alpha-power model [123] as

$$f = \frac{k\left(V_{dd} - V_{th0}\right)^{\alpha}}{V_{dd}(L + \Delta L)^2} \tag{5.65}$$

where V_{th0} is the threshold voltage, α is the velocity saturation coefficient, L is the nominal gate length of a device and ΔL represents the fluctuation in global gate length. Since variations are typically small, we can assume the frequency to be linearly dependent on the fluctuation in gate length and using Taylor's theorem, we can rewrite (5.65) as

$$f = \frac{k_f\left(V_{dd} - V_{th0}\right)^{\alpha}}{V_{dd}L^2}\left(1 + \frac{\Delta L}{L}\right)^{-2} \approx f_n\left(1 - \frac{2\Delta L}{L}\right) \tag{5.66}$$

where k_f is a proportionality constant and f_n is the frequency under nominal process conditions.

Let us now consider the influence of process parameters on the different components of power. Total power dissipation can be expressed as

$$P = P_{\mathrm{Dyn}} + P_{\mathrm{Gate}} + P_{\mathrm{Sub}} \tag{5.67}$$

where P_{Dyn} is the dynamic or switching power dissipation, P_{Gate} is the gate leakage power, and P_{Sub} refers to the subthreshold leakage power. Dynamic power can be simply expressed as

$$P_{\mathrm{Dyn}} = \alpha f C\left(1 + \frac{\Delta L}{L}\right)V_{dd}^2 = P_{\mathrm{Dyn,\,n}}\left(1 + \frac{\Delta L}{L}\right), \tag{5.68}$$

where α is the switching activity factor and f is the frequency of operation, the term $(1+\Delta L/L)$ captures the dependence of the switching capacitance on gate length to first-order and $P_{\mathrm{Dyn,\,n}}$ is the nominal dynamic power dissipation.

Gate leakage power is proportional to the device area, and is therefore directly proportional to the variation in gate length. In addition, gate leakage current is exponentially related to the power supply due to the increase in electric filed across the gate oxide. Note, that though we are not concerned with variation in V_{dd}, this dependence will play an important role when we consider the sensitivity of yield to supply voltage, which is our goal in this analysis. The gate leakage can be expressed as

$$P_{\text{Gate}} = k_{g1}(L + \Delta L)V_{dd} \exp(k_{g2}V_{dd}) = P_{\text{Gate, n}}\left(1 + \frac{\Delta L}{L}\right), \tag{5.69}$$

where K_{g1}, K_{g2} are proportionality constants and $P_{\text{Gate, n}}$ is the nominal gate leakage power.

The subthreshold leakage is exponentially proportional to the threshold voltage of the device and can be expressed as

$$P_{\text{Sub}} \approx k_{s1}V_{dd} \exp(-k_{s2}(V_{th0} + \Delta V_{th} - \eta V_{dd})) \tag{5.70}$$

where k_{s1}, k_{s2} are proportionality constants and η is the DIBL coefficient, which causes a reduction in the threshold voltage of a device with increase in drain-to-source voltage. This DIBL coefficient decreases with the increase in gate length and can be empirically expressed as

$$\eta = \eta_0 - p_1(L + \Delta L) - p_2(L + \Delta L)^2 \approx \eta_n - (p_1 + p_2 L)\Delta L \tag{5.71}$$

where η_n is the DIBL coefficient under nominal process conditions. Using (5.71), we can rewrite (5.70) as

$$\begin{aligned} P_{\text{Sub}} \approx\ & k_{s1}V_{dd} \exp(-k_{s2}(V_{th0} - \eta V_{dd})) \\ & \exp(-k_{s2}(\Delta V_{th} + (p_1 + p_2 L)\Delta L V_{dd}))) & (5.72) \\ =\ & P_{\text{Sub, n}} \exp(-k_{s2}(\Delta V_{th} + (p_1 + p_2 L)\Delta L V_{dd}))) \,. & (5.73) \end{aligned}$$

Having established the dependence of frequency and the different components of power on variation in process parameters, we now discuss the steps that are used to map the feasible region from the performance and power perspective to the feasible region in the space of variations in gate length and threshold voltage. The parametric yield is defined as the probability that the frequency is greater than a given minimum desired value f_{min} and the power dissipation is less than a given maximum value P_{max}. The constraint on frequency can be used to establish a constraint on the maximum gate length and this inequality is mathematically expressed as:

$$L_{\text{max}} \leq L + (\Delta L)_{\text{max}} = \leq L + \frac{L}{2}\left(1 - \frac{f_{\text{min}}}{f_n}\right). \tag{5.74}$$

For a given value of gate length fluctuation l, the gate leakage and dynamic power components are independent of variations in threshold voltage and their sum can be used to enforce a constraint on the subthreshold leakage power of the design. This gives,

$$P_{\text{Sub}} \leq P_{\text{max}} - (P_{\text{Dyn, n}} + P_{\text{Gate, n}})\left(1 + \frac{l}{L}\right). \tag{5.75}$$

Table 5.5. Yield estimates for using the analytical yield model and SPICE based Monte Carlo simulations [116].

Frequency Constraint	Power Constraint	Yield SPICE	Yield Analytical
0.95	1	41.16	40.02
0.95	1.1	63.18	63.00
0.95	1.2	72.30	72.00
1	1	7.64	10.06
1	1.1	30.84	31.56
1	1.2	42.70	42.73
1.05	1	0.02	0.26
1.05	1.1	3.46	4.84
1.05	1.2	14.24	12.71

Using, (5.73) we can rewrite the above inequality in terms of the minimum possible threshold voltage $V_{\min}$ as

$$\begin{aligned} V_{\min} &\geq V_{th} + (\Delta V_{th})_{\min} \\ &= V_{th0} - \frac{1}{k_{s2}} \ln\left(\frac{P_{\max} - (P_{\mathrm{Dyn,\,n}} + P_{\mathrm{Gate,\,n}})(1 + l/L)}{\exp(-k_{s2}(p_1 + 2Lp_2)V_{dd} l}\right) \end{aligned} \quad (5.76)$$

We now define the parametric yield Y in terms of the integral of a bivariate normal distribution function of gate length and threshold voltage as

$$Y = \frac{1}{2\pi\sigma_L\sigma_{V_{th}}} \int_{-\infty}^{L_{\max}} \int_{V_{\min}}^{\infty} \exp\left(-\frac{L^2}{2\sigma_L^2}\right) \exp\left(-\frac{V_{th}^2}{2\sigma_{V_{th}}^2}\right) \mathrm{d}L\mathrm{d}V_{th}. \quad (5.77)$$

where σ_L and $\sigma_{V_{th}}$ correspond to the variance of global variation in gate length and threshold voltage, respectively. The above expression can then be evaluated using numerical integration. Note, that in (5.77) we have assumed that L and V_{th} are independent RVs. However, any correlation between these two parameters can be easily considered, by integrating the appropriate bivariate distribution in the (5.77). Table 5.5 compares the yield estimate obtained using the analytical approach with the yield obtained using SPICE based Monte Carlo simulation for a 15-stage ring oscillator in a 90 nm technology with $V_{dd} = 1\,V$. It can be seen that the analytical yield estimation approach provides good accuracy.

To consider the sensitivity of yield to supply voltage, we need to consider the impact both in performance and power. Consider (5.77) and note that yield is defined by the constraint on gate length $L_{\max}$ and the constraint on

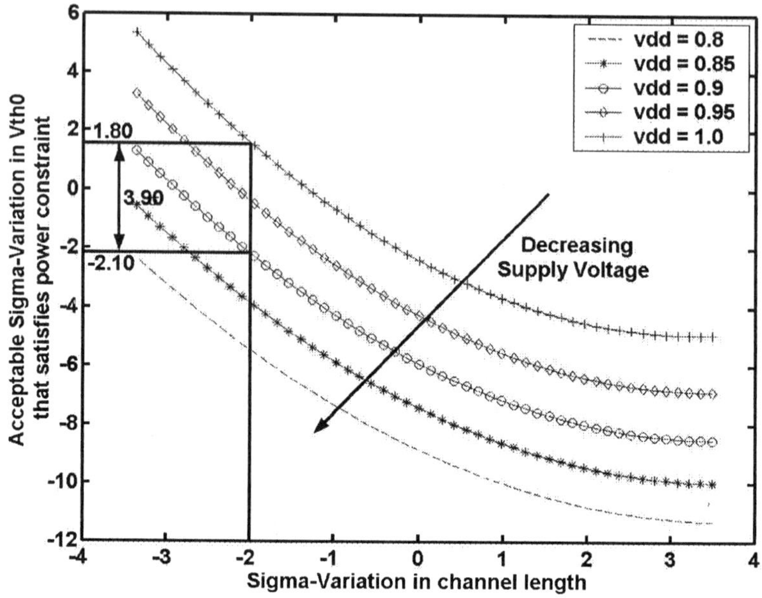

Fig. 5.13. Sensitivity of the constraint on minimum threshold voltage to changes in supply voltage [116]. (©2005 IEEE)

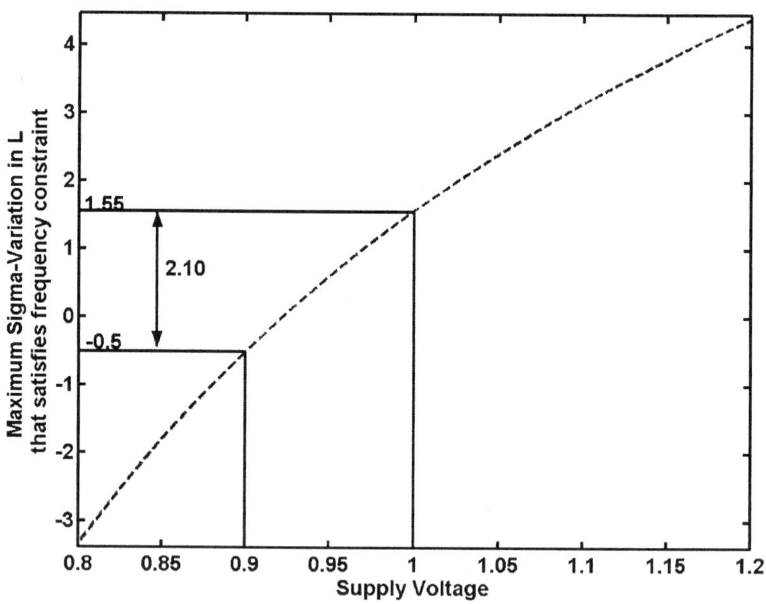

Fig. 5.14. Sensitivity of the constraint on maximum gate length to changes in supply voltage [116]. (©2005 IEEE)

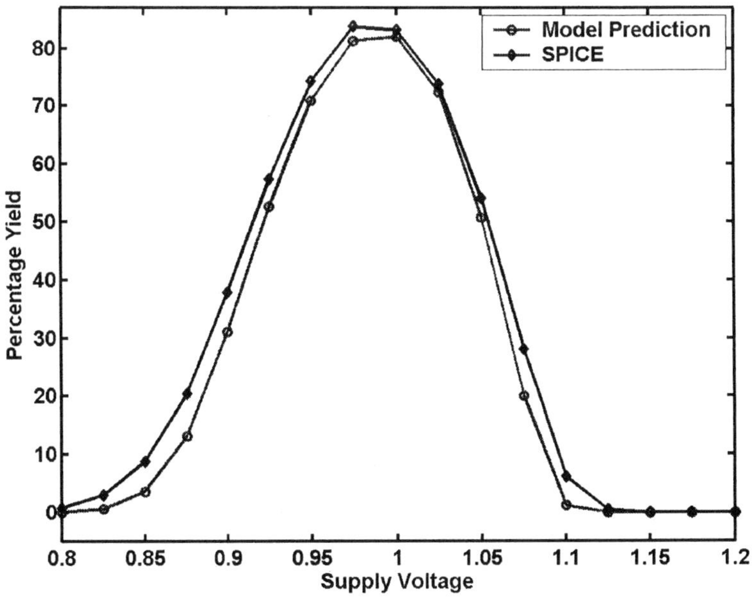

Fig. 5.15. Parametric yield as a function of the supply voltage [116]. (©2005 IEEE)

threshold voltage $V_{\min}$. Lowering the supply voltage lowers the power dissipation and results in a relaxation on the constraint on $V_{\min}$. This is illustrated graphically in Fig. 5.13 which plots the maximum possible tolerable variation in V_{th} that satisfies the power constraint that $P_{\max} \leq 1.1\,P_{\text{nom}}$, for varying supply voltages. As the supply voltage is lowered, the feasible region in terms of V_{th} increases. As an example for a global variation in gate length that corresponds to -2σ, a change in supply voltage from 1.0 V to 0.9 V results in a reduction of the feasible $V_{\min}$ from -2.1σ to 1.8σ.

On the other hand, lowering the supply voltage has a strong negative influence on the performance of the design as well. Figure 5.14 illustrates the change in the feasible region in terms of variations in gate length with varying supply voltage for a performance constraint of $f_{\min} \geq 0.95\,F_{\text{nom}}$. In this case, reducing the supply voltage from 1.0 V to 0.9 V results in a reduction of the feasible region from 1.55σ to 0.50σ. Note that the sensitivity of the feasible region to supply voltage is a strong function of the value of the supply voltage as well.

Figures 5.14 and5.15 show regions where small changes in voltage can result in either a large or a small change in the feasible region. In terms of yield, the constraint imposed on power and performance also has a strong influence on the sensitivity of yield to supply voltage. This can be understood by considering the distribution of power or performance. If the constraint is close to the nominal, then small changes in the constraint, due to changes in supply voltage will have a strong impact on the overall yield. However, as we

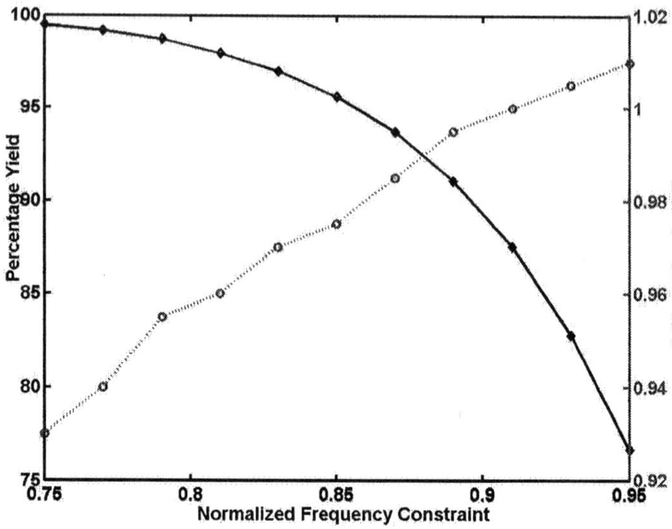

Fig. 5.16. Sensitivity of the optimized yield and the optimal supply voltage to the frequency constraint for a power constraint of 1.2X the nominal power [116]. (©2005 IEEE)

move towards the corner and the constraints become either too tight or too loose the sensitivity of yield to changes in the constraint also becomes smaller.

Figure 5.15 plots the estimated parametric yield with varying supply voltage, both using the analytical approach and using SPICE based Monte Carlo simulations. The analytical results are obtained by repeating the analysis for different supply voltage values. The plot shows good accuracy of the analytical approach and shows that the parametric yield is strongly sensitive to variations in supply voltage. A small change of even 5% in supply voltage results in a 15% degradation in yield.

The constraint imposed on $L_{\max}$ is a strong function of the frequency constraint. Hence, we would expect a large sensitivity of yield to small changes in the timing constraint. This is illustrated in Fig. 5.16, which plots the optimal power supply voltage that maximizes the yield and the maximized yield for varying frequency constraints. The figure shows that both the yield and the optimal supply voltages are strongly dependent on the yield constraint. For loose timing constraints, a lower value of power supply voltage becomes optimal, since it results in lower power dissipation as well. As the timing constraint becomes tight, the optimal power supply voltage increases to maintain good timing yield. However, this has a negative impact on power dissipation and the overall yield suffers. The same trends can also be found for the power constraint, which is shown in Fig. 5.17.

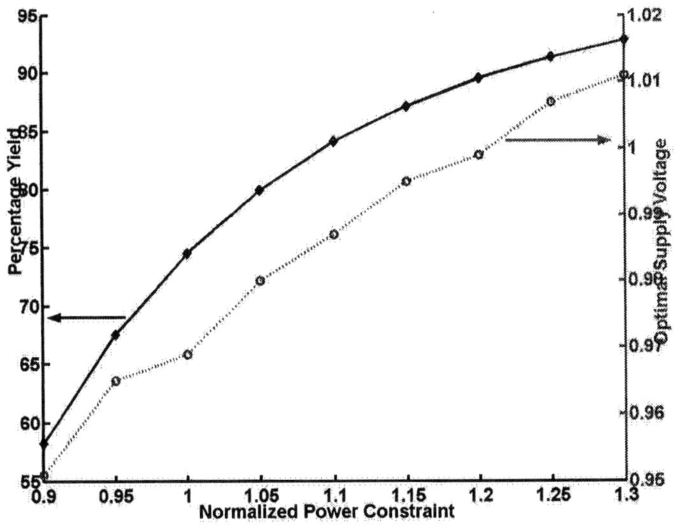

Fig. 5.17. Sensitivity of the optimized yield and the optimal supply voltage to the power constraint where the minimum frequency is constrained to be more than 0.9X the nominal frequency [116]. (©2005 IEEE)

Based on the above analysis, we see that the supply voltage has a strong influence on yield and can be used as a level to improve the yield or shipping frequency of a design. We can maximize the shipping frequency, which is defined to be the frequency of the slowest part, while ensuring a given minimum yield for the design. For a given supply voltage, the yield constraint limits the maximum gate length, which can then be used to estimate the lowest frequency of a sample, or the shipping frequency. Figure 5.18 plots this shipping frequency as the supply voltage is varied for a 20% yield in this bin. The flat regions of the plot are regions where the supply voltage is so large that the overall yield constraint cannot be met for any performance target. Also, we see that the supply voltage can be tuned for a design to maximize the shipping frequency of the design. Taking this idea a step further, we can tune the supply voltage for each sample of a design sold and this tunable supply voltage can be higher for low performance parts (which are low power) and lower for high performance parts (which are high power). Hence, using this approach we can expect to increase the yield of a design significantly. In addition, an important issue in adaptive supply designs would be to have a design flow that has been tailored to suit this approach. In this case, design should have nominal power and performance that satisfy the constraints with minimal variance. Any variation in process parameters on either side of their

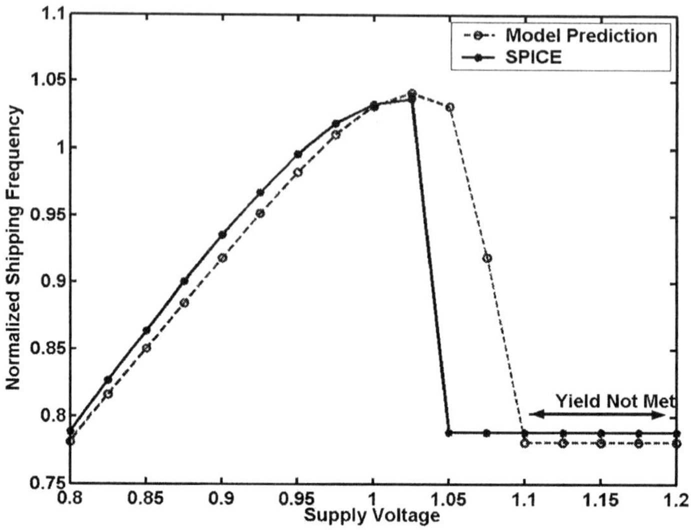

Fig. 5.18. Maximum shipping frequency as a function of the supply voltage. The desired yield is 20% with the power constrained to less than 1.2X the nominal power dissipation [116]. (©2005 IEEE)

nominal values rsulting in a variation in power and performance can then be tuned using a small range of supply voltages.

6

Statistical Optimization Techniques

To this point in the book, we have discussed the details regarding statistical analysis techniques for delay, power, and yield. In this chapter, we will use some of these analytical techniques to drive optimization methods that will be used to improve the performance of a given design. Most of the earlier work in circuit optimization has been limited to deterministic optimization using corner model based case-files. These approaches are blind to the impact of the decisions made during the optimization step on the overall parametric yield of the design, and invariably result in the formation of a timing wall as shown in Fig. 6.1. This results from the fact that the optimizer has no incentive to reduce the delay of non-critical paths. As we know, all near-critical paths can strongly affect the circuit delay distribution due to process variability; and hence, the design becomes more susceptible to process variations.

In [14] the authors proposed an uncertainty-aware heuristic approach that performed deterministic optimization using a nonlinear optimizer, while avoiding the build-up of a timing wall. This was achieved by adding an additional term to the objective function that is a function of the slack available on the primary outputs. The new objective function (to be minimized) is defined as

$$z + \sum_{i \in PO} f(d_i) \tag{6.1}$$

where z is the delay of the circuit, PO is the set of primary outputs, and f is a function of d_i, which is defined for each primary output as

$$d_i = z - AT_i - RAT_i \tag{6.2}$$

where AT and RAT represent the arrival and the required arrival times, respectively. The choice of the function f is crucial to the optimization, as it defines the level of importance given to the slack available on non-critical paths. Since our focus is to develop true statistical optimization techniques we will not go into details that need to be considered when choosing f (details can be found in [14]). The discussion is provided to emphasize that, in

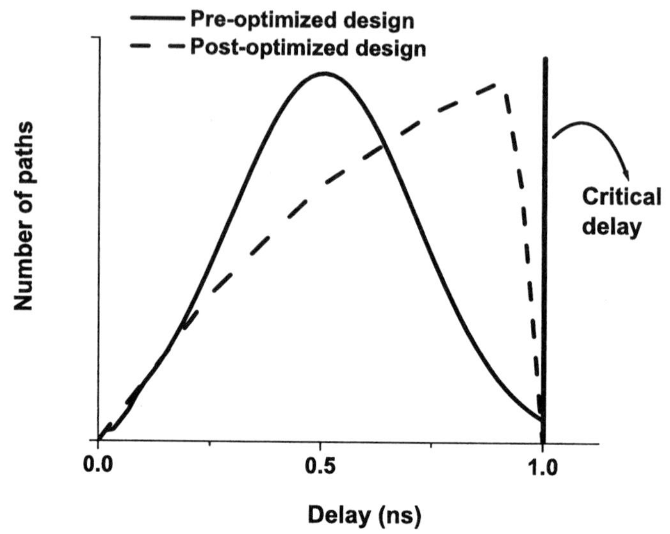

Fig. 6.1. Deterministic optimization results in the formation of a timing wall.

a statistical scenario, the definition of a critical path becomes vague and it is no longer possible to isolate a single critical path. Thus, a true statistical optimization technique will not be focused on improving the performance of a single critical path. However, the timing yield of a design can be improved by identifying paths that have a significant probability of becoming critical, given process variation. In certain situations a set of deterministic non-critical paths might have a much higher probability of becoming critical than a single deterministic critical path and in this case, efforts to improve the timing of the critical path will be futile in improving the timing yield of the design.

Another important issue that needs to be addressed before we delve into statistical optimization is the choice of objective function that should be optimized. In a deterministic scenario the objective function can be represented as the circuit delay or power. However, in a statistical scenario these performance parameters are described as probability distributions and the objective function should be based on parameters that define these distributions. Minimizing the expected value or the standard deviation or a linear combination of these parameters are some of the possible choices. In [115], the authors use the concept of utility theory to define a measure for the statistical criticality of the nodes, which is used as an objective function. The sensitivity-based approach in Sec. 6.2.5 introduces the notion of a *profit function* that directly corresponds to the revenue that can be generated. In a microprocessor setting, where companies perform speed-binning and sell higher frequency parts at a higher cost, the profit function can be approximated by a rising ramp,

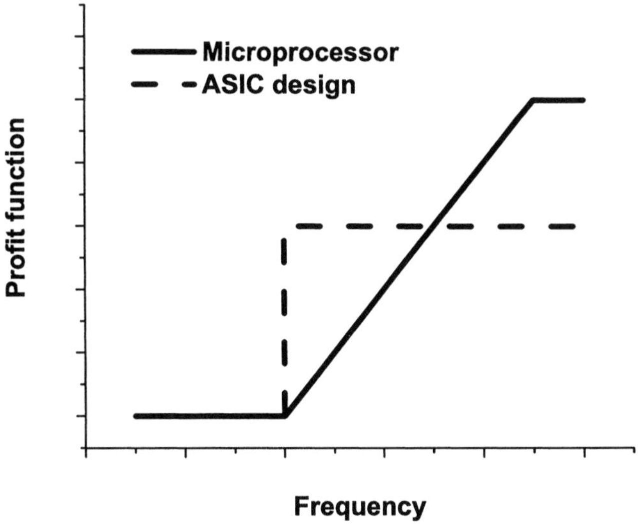

Fig. 6.2. Generic profit functions for a microprocessor and ASIC design.

which represents higher profits for high frequency components (as shown in Fig. 6.2). In the ASIC market, a chip can only be sold if it meets the specifications. Furthermore, all samples of an ASIC design are sold at the same price. Therefore, a step profit function can be used to capture this situation.

In this chapter we will first discuss a general framework that will be used to optimize values of process parameters to minimize power dissipation using a robust optimization technique. Then we will discuss statistical optimization techniques based on gate sizing to perform timing and power optimization, which has been the focus of significant recent work in this area. Finally, we will look at some of the techniques that have been proposed to perform buffer insertion and threshold voltage assignment in statistical scenarios.

6.1 Optimization of Process Parameters

Power consumption has become a top priority in modern circuit design and multiple supply and threshold voltages have been shown to be extremely effective in reducing total power dissipation. Previous implementations using multiple supply and threshold voltages have shown impressive savings in both the dynamic and leakage power of a design [143], [120]. The effectiveness of multiple supply voltage (V_{dd}) techniques was first shown to scale poorly in [58]; however, [132] showed that using multiple supply and threshold voltages (V_{th}) in conjunction is very effective in achieving large power savings in sub-1V technologies.

In this section, we will discuss the selection of appropriate values for the supply and threshold voltages to minimize power dissipation. This problem was investigated under deterministic conditions in [58], [132]. In [58], the authors proposed an approach to select optimal supply and threshold voltages in either multiple V_{dd} or multiple V_{th} designs by minimizing dynamic or static power, respectively. This approach was extended in [132] to the selection of both V_{th} and V_{dd} and considers optimization of total power in multi-V_{dd}/V_{th} systems. However, both these approaches ignore process variations and perform deterministic power optimization only.

We will discuss the robust optimization based approach developed in [131] to consider the impact of process variation while selecting the optimal V_{dd}/V_{th} values in multi-V_{dd}/V_{th} designs by minimizing power dissipation of a generic CMOS digital system. To simplify the problem, it is assumed that the network consists of non-intersecting parallel paths. A path-delay distribution of the network is assumed and is used to quantify the number of paths having a particular delay when the design is synthesized at the fastest combination of supply and threshold voltage. The goal of power optimization is to determine the fraction of each path of the network that should be assigned to a particular combination of supply and threshold voltage in order to minimize the power dissipation under a given timing yield constraint. The timing yield constraint is expressed as a minimum probability with which the delay of a path is constrained to be less than a given critical delay for the network.

Using this approach, we can estimate the achievable power savings in the available V_{dd}/V_{th} space. The V_{dd}/V_{th} point that results in the lowest power dissipation is then identified as the optimal V_{dd}/V_{th}. Note, that although we do not discuss gate leakage in this section, the approach can be easily extended to perform dual-T_{ox} allocation to minimize gate leakage power.

The high-level framework expresses both delay and power as a function of supply and threshold voltage. Since power grid variations are temporal (depending on the input vector combination) a worst-case drop can occur in any sample of a chip, and worst-case models should be used to consider V_{dd} variations. On the other hand, variations in V_{th} are process-dependent and its impact on both power and performance should be considered statistically. Threshold voltage variability is a concern due to DIBL effects as well as discrete dopant effects, which are exacerbated in highly scaled technologies. Let us define the threshold voltages, which are assumed to be normally distributed, as a sum of a nominal value and a random variable (RV), which can be expressed as

$$V_{th} = V_{TH} + v_{th} \tag{6.3}$$

where V_{TH} is the nominal value of the threshold voltage and v_{th} is a zero mean Gaussian RV with a variance equal to the variance of V_{th}. In addition, the RVs associated with different threshold voltages in a multi-V_{th} process are assumed to be mutually independent, since different threshold voltages correspond to different implantation steps in the process. However, the approach can be adopted to handle arbitrary correlations among the different threshold voltages, if the threshold voltages are jointly distributed according to a multinormal distribution. These correlations can result from variation in threshold voltage resulting from correlated variations in gate length, which will influence both threshold voltages similarly.

Consider a generic logic network that consists of a set of paths with N distinct delays. This is achieved by discretizing the path-delay distribution at N discrete delay points. Let the number of paths with delay D_i $(1 \leq i \leq N)$ be P_i, then the total number of paths in the network P is expressed as

$$P = \sum_{i=1}^{i=N} P_i. \tag{6.4}$$

Now consider one of the paths (say k) and assume V_1 and V_{th1} to be the supply voltage and threshold voltage in the initial single V_{dd}/V_{th} system. The network is assumed to be operating at the highest V_{dd} and lowest V_{th} initially, which gives the fastest design and allows for easy handling of the power optimization step. Let C_k and W_k be the total capacitance and device width of the path, respectively. The total initial dynamic power can then be expressed as

$$P_k^{dyn} = fC_k V_1^2 \tag{6.5}$$

where f is the frequency of operation. Considering the same path implemented in an $n - V_{dd}/m - V_{th}$ design, let us define C_{ij}'s and W_{ij}'s as the capacitances and device widths operating at a supply voltage V_i and threshold voltage with a nominal value V_{TH_j}. The total dynamic power dissipation can now be expressed as

$$P_k^{\mathrm{dyn}} = f \sum_{i=1}^{n} \sum_{j=1}^{m} C_{ij} V_i^2 \tag{6.6}$$

and the static power of the design can be expressed as

$$P_k^{\mathrm{static}} = I_0 \sum_{i=1}^{n} \sum_{j=1}^{m} V_i^2 W_{ij} \exp\left(\frac{-V_{thj}}{nV_T}\right) \tag{6.7}$$

where n is the subthreshold swing coefficient and $V_T = kT/q$ is the thermal voltage. Additionally the C_{ij}'s and W_{ij}'s satisfy the equation

$$\sum_{i=1}^{n} \sum_{j=1}^{m} C_{ij} = C_k \qquad \sum_{i=1}^{n} \sum_{j=1}^{m} W_{ij} = W_k. \tag{6.8}$$

The delay of the path is calculated using the alpha-power law model [123] as

$$D_k = \sum_{i=1}^{n}\sum_{j=1}^{m} \frac{C_{ij}V_{ddj}}{(V_{ddj} - V_{thk})^{\alpha}} \tag{6.9}$$

where α is the velocity saturation coefficient that varies from 1 to 2.

As shown in [58], the total capacitance and transistor width along a path are largely proportional to the path delay. Hence, the W_{ij}'s in (6.7) can be replaced by C_{ij}'s and I_0 is replaced by I_0' to absorb the change in the prefactor of (6.7) when moving from widths to capacitances. Since we have expressed both the power and delay of the circuit as a function of only one set of unknowns, the C_{ij}'s, we can formulate the power optimization problem with the C_{ij}'s as the optimization variables. In the following discussion within this section we will also refer to the solution vector of the optimization problem as $\mathbf{c}$.

6.1.1 Timing Constraint

Let us assume for now, that the required timing yield of each path in the network (ρ) is pre-determined, which allows us to express the delay constraint as

$$\mathcal{P}\left(T_{\text{critical}} - D_k \geq 0\right) \geq \rho_k. \tag{6.10}$$

This equation constrains the probability that path P_k has a delay less than the critical delay of the circuit to be more than ρ_k. We will look at an approach to estimate the ρ_k's in the Sec. 6.1.3. To simplify the timing constraint we express (6.9) as

$$D_k = \sum_{i=1}^{n}\sum_{j=1}^{m} \frac{C_{ij}V_{ddj}}{(V_{ddj} - V_{THk})^{\alpha}} \left(1 - \frac{v_{thk}}{V_{ddj} - V_{THk}}\right)^{-\alpha} \tag{6.11}$$

which can be approximated using Taylor's expansion as

$$D_k = \sum_{i=1}^{n}\sum_{j=1}^{m} \frac{C_{ij}V_{ddj}}{(V_{ddj} - V_{THk})^{\alpha}} \left(1 + \frac{\alpha v_{thk}}{V_{ddj} - V_{THk}}\right). \tag{6.12}$$

Since the variations in process parameters are typically small compared to their nominal values, the above approximation is very accurate. As an example, using nominal values of V_{th} and V_{dd} as 200 mV and 1.2 V respectively, we find an approximation error of only 1.4% for the worst-case variation (3σ point), where we have assumed a 3σ variation in V_{th} of 30%. This approximation allows us to write the delay as a linear combination of the Gaussian RVs v_{th}'s, as defined in (6.3). At this point, we can write the mean and variance of D_k as a function of $\mathbf{c}$ (the vector of C_{ij}'s), using the properties of Gaussian RVs. Now we can express (6.10) as

$$\mathcal{P}\left((T_{\text{critical}} - D_k) \sim N(\mu_k(\mathbf{c}), \sigma_k(\mathbf{c})) \geq 0\right) > \rho_k \tag{6.13}$$

which can be rewritten as

$$-\mu_k(\mathbf{c}) - \sqrt{2}\,\text{erfinv}\,(1 - 2\rho_k)\,\sigma_k\,(\mathbf{c}) < 0 \tag{6.14}$$

where *erfinv* is the inverse of the error function. Let us now show that (6.14) defines a convex set [15] under the condition that $\rho_k \geq 0.5$.

Theorem 6.1. *Let* $\mathbf{X} = (x_1, x_2, \ldots, x_n)^T$ *be a vector of random variables, whose components are jointly distributed according to a multinormal distribution. Then the variance and standard deviation of the linear combination* $y = \boldsymbol{\alpha}^T\mathbf{X}$ *are convex in* α*.*

Proof. A sufficient and necessary condition for a function to be convex is that its Hessian matrix is positive semi-definite at each point for which the function is defined. We can write the variance of y as

$$\sigma_y^2(\boldsymbol{\alpha}) = \boldsymbol{\alpha}^T\mathbf{S}\boldsymbol{\alpha} \tag{6.15}$$

where $\mathbf{S}$ is the covariance matrix of the components of $\mathbf{X}$. We know that $\mathbf{S}$ is symmetric and positive semi-definite. Since

$$\mathbf{H}_{\mathbf{u}}(\mathbf{u}^T\mathbf{A}\mathbf{u}) = \mathbf{A} + \mathbf{A}^T, \tag{6.16}$$

the Hessian matrix of (6.15) is $2\mathbf{S}$ which is positive semi-definite; and therefore, the variance of the linear combination $y = \boldsymbol{\alpha}^T\mathbf{X}$ is convex in α. To prove the convexity of standard deviation, we express the standard deviation as

$$\sigma(\mathbf{c}) = (\sum_i\sum_j c_i c_j \rho_{ij}\sigma_i\sigma_j)^{0.5}. \tag{6.17}$$

To show the convexity of (6.17), we need to show that

$$\sigma(\theta\mathbf{a} + (1-\theta)\mathbf{b}) \leq \theta\sigma(\mathbf{a}) + (1-\theta)\sigma(\mathbf{b}) \tag{6.18}$$

which is equivalent to

$$\begin{aligned}(\sum_i\sum_j(\theta a_i + (1-\theta)b_i)(\theta a_j + (1-\theta)b_j)\rho_{ij}\sigma_i\sigma_j)^{0.5} \leq \\ \theta(\sum_i\sum_j a_i a_j \rho_{ij}\sigma_i\sigma_j)^{0.5} + (1-\theta)(\sum_i\sum_j b_i b_j \rho_{ij}\sigma_i\sigma_j)^{0.5}.\end{aligned} \tag{6.19}$$

Using the standard arithmetic-geometric inequality and squaring both sides, we can rewrite the above (after some algebra) as

$$\sum_i \sum_j \sqrt{a_i a_j b_i b_j} \rho_{ij} \sigma_i \sigma_j \leq$$
$$(\sum_i \sum_j a_i a_j \rho_{ij} \sigma_i \sigma_j)^{0.5} (\sum_i \sum_j b_i b_j \rho_{ij} \sigma_i \sigma_j)^{0.5} \tag{6.20}$$

which follows from Holder's inequality [22].

Since μ_k is linear in $\mathbf{c}$, both $\mu_k(\mathbf{c})$ and $-\mu_k(\mathbf{c})$ are convex functions. $\sigma_k(\mathbf{c})$ is the standard deviation of a linear combination of mutually independent random variables, and was shown to be convex in Theorem 6.1. Also, we know that if $f(x)$ defines a convex function, then the set $\{x|f(x) < k\}$ is convex, and that non-negative weighted sums of convex functions are convex. The condition that the weights in (6.14) of a linear combination be non-negative is satisfied when

$$\sqrt{2}\,\text{erfinv}\,(1 - 2\rho_k) \leq 0, \tag{6.21}$$

which implies that $\rho_k \geq 0.5$.

Since the target yield for a given path is always much greater than 50%, this condition is easily satisfied (ρ_k is the yield of a single path; even if just a few paths have yield close to 0.5 then the overall yield of the design will be very small). It is important to note that if the ρ_k's are themselves treated as optimization variables, then the set defined by (6.14) does not remain convex, even though the inverse error function is still convex in our domain of interest. This results from the fact that (6.14), in this case, becomes a product of two convex functions, which is not convex in general [22]. Thus simultaneous optimization of total power and yield of each path is therefore not possible using the above convex formulation. To solve this, [131] separates the problems of yield allocation for each path (discussed later) and actual power minimization given the desired yield values for each path. Yield allocation is used to determine the ρ_k's values that are then used in power optimization. Let us first look at the power minimization approach itself.

6.1.2 Objective Function

The objective function of the power optimization procedure is a statistical parameter of the power dissipation of the network. We consider three different optimization functions. The objective functions are considered as a sum of the statistical parameter of dynamic and leakage power components, normalized by the nominal dynamic and leakage power of the initial design, respectively. As shown in [132], minimizing such a weighted sum can be used to minimize the total power if the weighting factor is appropriately chosen. In our case, this would allow us to minimize the appropriate statistical parameter of the total power dissipation. The weighting factors can also be chosen to result in a larger reduction in leakage or dynamic power, if so desired.

Equations (6.6) and (6.7) are used to express the dynamic and leakage power for an individual path, respectively. To calculate the total power of the network, recall that we have assumed that the amount of capacitance, and hence, the device width on a path is proportional to the initial nominal delay of the path in the single V_{dd}/V_{th} environment. Thus, the total power dissipation can be readily obtained by calculating a weighted sum of the power dissipation of all paths in the network, where the weighting factor is the initial nominal delay. Since the dynamic power does not involve any random variables, the dynamic power component can be expressed as in (6.6) for all of the following cases.

Mean Power

The leakage power, as expressed in (6.7), is a sum of lognormal (exponential of a Gaussian) RVs. If $X \sim N(\mu, \sigma)$ and $Y = exp(X)$ then the expected value of Y can be expressed as

$$E[Y] = \exp\left(\mu + \frac{\sigma^2}{2}\right). \tag{6.22}$$

Using (6.7) and (6.22), average leakage can be expressed as

$$P_{\text{mean}}^{\text{static}} = I_0' \sum_{\text{paths}} \sum_{i=1}^{n} \sum_{j=1}^{m} V_i^2 C_{ij} \exp\left(\frac{-V_{THj}}{nV_T} + \frac{1}{2}\left(\frac{\sigma_{vthj}}{nV_T}\right)^2\right). \tag{6.23}$$

Nominal Power

Nominal leakage power is the leakage power under nominal process conditions and can be expressed as

$$P_{\text{nominal}}^{\text{static}} = I_0' \sum_{\text{paths}} \sum_{i=1}^{n} \sum_{j=1}^{m} V_i^2 C_{ij} \exp\left(\frac{-V_{THj}}{nV_T}\right). \tag{6.24}$$

High Percentile Leakage Power

The variance of the lognormal RV Y, as defined above, can be written as

$$Var[Y] = \exp\left(2\left(\mu + \sigma^2\right)\right) - \exp\left(2\mu + \sigma^2\right). \tag{6.25}$$

Equations (6.7) and (6.25) can be used to express the variance of leakage power as

$$P_{\text{var}}^{\text{static}} = I_0' \sum_{j=1}^{m} \left(\sum_{\text{paths}} \sum_{i=1}^{n} \right) \cdot \tag{6.26}$$

$$\exp\left(2\left(\frac{-V_{THj}}{nV_T} + \left(\frac{\sigma_{vthj}}{nV_T}\right)^2\right) - \exp\left(\frac{-2 * V_{THj}}{nV_T} + \left(\frac{\sigma_{vthj}}{nV_T}\right)^2\right)\right).$$

It is important to consider the order of summation in (6.26), since we need to sum the capacitance (representing width) operating at a given threshold voltage across all paths before we can square the terms. A higher percentile point of the probability distribution of total power can now be expressed as

$$P_{\text{percentile}}^{\text{static}} = P_{\text{mean}}^{\text{static}} + n\sqrt{P_{\text{var}}^{\text{static}}}, \tag{6.27}$$

where n is a positive constant. Equations (6.23) and (6.24) are linear in the optimization variables C_{ij}'s, and hence represent convex functions of the components of $\mathbf{c}$. Equation (6.27) can be shown to be convex using Theorem 6.1, which shows that the standard deviation of a linear combination of RVs is convex in the weighting coefficient. Since we are interested in higher percentiles of power for which $n > 0$, we have a non-negative weighted sum of convex functions, which is convex.

Since we have represented all the objective functions as convex functions, their unique global minimum can be efficiently determined. The complete power minimization problem can now be cast as

$$\begin{aligned} &\text{Min} : f(\mathbf{c}) \\ &\text{s.t.} : \ -\mu_k(\mathbf{c}) + \beta_k \sigma_k(\mathbf{c}) \leq 0 \quad k = 1, \ldots, N \end{aligned} \tag{6.28}$$

where $f : \Re^{i*j} \rightarrow \Re$ is a convex function, β_k's are positive constant and the functions μ_k's and σ_k's are convex.

6.1.3 Yield Allocation

To perform the power optimization step discussed above, we must first identify yield constraints for all k paths in the circuit. This yield allocation problem is non-trivial as shown in the following example. The simplest yield allocation method assumes all paths in the optimized circuit will be equally likely to violate timing. In this case, if the desired parametric yield is 0.99 and there are N paths in the network, then each path must have a yield of $0.99^{1/N}$. This situation is depicted in Fig. 6.3 as the uniform yield allocation case. Assuming an initial yield distribution of the paths as shown in the figure, the optimal yield allocation, which enables the maximum reduction in power, should assign yield targets to each path that are lower than the initial yield, such that the power dissipation is minimized while the complete network has the desired timing yield. Figure 6.3 shows an expected solution of optimal yield

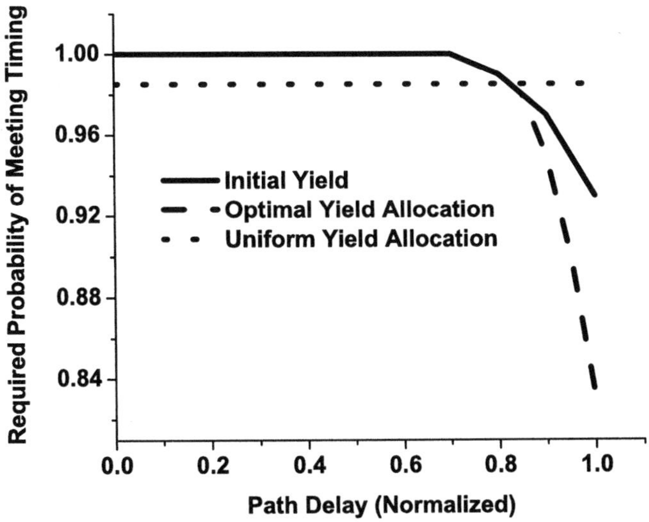

Fig. 6.3. Different yield allocation options have a significant impact on the optimization problem.

allocation which tightly constrains the fast paths and is thus able to loosely constrain the slow paths, while maintaining the overall yield of the network. On the other hand, uniform allocation can potentially constrain paths to have a post-optimized yield higher than their initial yield (this is seen in Fig. 6.3 for paths with near critical delay). In these cases, the power optimization problem becomes infeasible since there are no available means to increase the timing yield of a path. Thus, it is necessary that yield be reasonably allocated among the different paths in the network before power optimization is performed. We assumed in Sec. 6.1.1 that the yield for each path ρ_k's were pre-determined. We now set up a convex optimization problem to determine these ρ_k's in order to provide the desired timing yield for the complete network.

The global constraint on the ρ_k's can be expressed as

$$\prod_{k=1}^{N} \rho_k^{P_k} \geq Y \tag{6.29}$$

where the P_k's are as defined in (6.4) and Y is the desired timing yield of the circuit. The objective of yield allocation is to assign a yield to each path such that the objective functions defined above are minimized. We perform a gradient search to improve the yield allocation given an initial solution. Therefore, we express the objective function for yield allocation as

$$\boldsymbol{\alpha}^T \Delta \mathbf{c} = \sum_k \alpha_k \frac{\partial c_k}{\partial \rho_k} \Delta \rho_k \tag{6.30}$$

where $\boldsymbol{\alpha}$ is a constant vector representing the vector of coefficients of the C_{ij}'s in the objective function for power optimization. The gradient terms in (6.30) are estimated along the curve where (6.14) is satisfied with equality. The yield allocation problem can then be expressed as

$$\begin{aligned} \text{Min} : & \sum_k \alpha_k \frac{\partial c_k}{\partial \rho_k} \Delta \rho_k \\ \text{s.t.} : & \prod_{k=1}^{N} \rho_k^{P_k} \geq Y \\ & 0.5 \leq \rho_k \leq 1, \quad k = 1, \ldots, N. \end{aligned} \tag{6.31}$$

The above optimization problem can be easily mapped to a convex optimization problem by introducing an extra variable, and the final yield allocation problem can then be written as

$$\begin{aligned} \text{Min} : & \sum_k \alpha_k \frac{\partial c_k}{\partial \rho_k} \Delta \rho_k \\ \text{s.t.} : & \left(\prod_{k=1}^{N} \rho_k^{P_k} \right) \beta \geq 1 \\ & 0.5 \leq \rho_k \leq 1, \quad k = 1, \ldots, N \\ & 1 \leq \beta \leq 1/Y \end{aligned} \tag{6.32}$$

where β is the added variable. The equivalence of 6.31 and 6.32 can be readily established. Let us consider the convexity of (6.32). The objective function is linear in the variables of the optimization and is, therefore, convex. The last two constraints obviously define convex regions, so it needs to be established that the first constraint, which is of the form

$$\prod_i x_i^{\alpha_i} \geq 1 \tag{6.33}$$

defines a convex region. Let $\mathbf{u}$ and $\mathbf{v}$ be two vectors that satisfy (6.33). Let us form a convex combination of $\mathbf{u}$ and $\mathbf{v}$ and consider

$$\prod_i (\theta u_i + (1 - \theta) v_i)^{\alpha_i}. \tag{6.34}$$

Using the general form of the standard arithmetic-geometric inequality [22], which states that

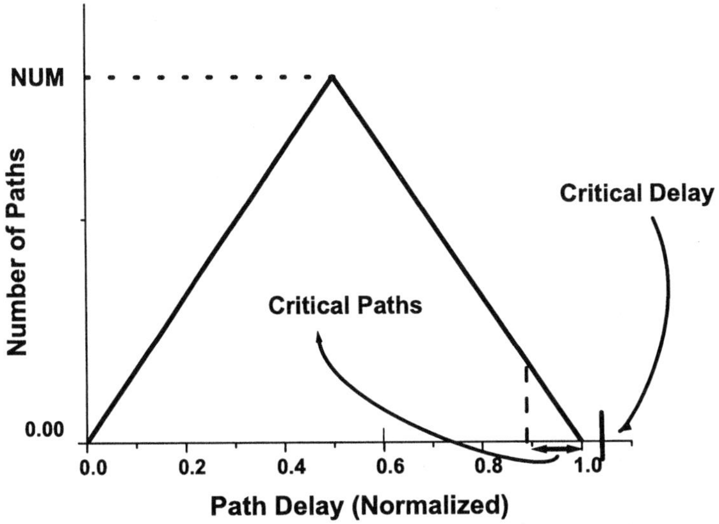

Fig. 6.4. Nominal path delay distribution showing the critical delay and a definition of critical paths.

$$a^{\theta}b^{1-\theta} \leq \theta a + (1-\theta)b \tag{6.35}$$

for $a, b \geq 0$ and $0 \leq \theta \leq 1$, we can write

$$\begin{aligned}\prod_i (\theta u_i + (1-\theta)v_i)^{\alpha_i} &\geq \prod_i \left(u_i^{\theta}\, v_i^{1-\theta}\right)^{\alpha_i} \\ &\geq \left(\prod_i u_i^{a_i}\right)^{\theta} \left(\prod_i v_i^{a_i}\right)^{1-\theta} \geq 1\end{aligned} \tag{6.36}$$

which implies that any convex combination of two points within the set defined by (6.33) belongs to the set defined in (6.33). Therefore, the constraints define a convex region and (6.32) represents a convex optimization problem.

To improve upon the initial guess, the initial yield allocation is performed by assuming the coefficients of ρ_k's in (6.32) to be the ratio of the initial delay to the initial yield. Hence, paths with lower initial timing yield (due to higher delay), have a higher weighting factor in the objective function. The yield allocation optimization then steers towards smaller values of ρ for these paths. This fulfills the goal of yield allocation, in that these initially critical paths are allowed to dominate the circuit yield, whereas non-critical paths should not be allowed to significantly degrade parametric timing yield. Yield allocation and power minimization can then be iteratively performed and the results are found to converge (within 1%) in 2-3 iterations for almost all cases.

This approach was used to analyze multi-V_{th}/V_{dd} using process and technology parameters that are typical of a 90 nm CMOS process, assuming an initial V_{dd} of $1.2\,V$ and a low V_{th} of $200\,mV$ in a dual-V_{th} process. The standard deviation of threshold voltages is considered to be 10% of its nominal value [26]. This results in higher threshold voltages having a higher standard deviation than lower threshold voltages. This may not always hold (as we saw in Chap. 2, this is true for variations originating from random dopant variations); therefore, an additional experiment where low V_{th} devices exhibit larger relative variability than their high V_{th} counterparts is considered.

The initial path delay distribution is assumed to have a symmetric triangular shape as shown in Fig. 6.4. The x-axis is the nominal path delay normalized to the nominal delay of the longest path in the network. Note that the shape of the path delay histogram can take on any form within the context of the optimization approach. Since the initial design operates at the lowest threshold voltage and highest supply voltage, both the mean and variance of all path delays are minimized. Thus, any applied optimizations will necessarily result in a lower yield, due to the introduction of a higher threshold voltage or lower supply voltage. Therefore, we must allow for a reduction in the target yield. For most of the results an initial yield of 98% is used, which is then allowed to reduce to 93%. The requirement to meet the initial yield of the circuit can be used to define the critical delay of the network. To determine the parameter NUM in Fig. 6.4, data from [53] is used, which shows that the number of critical paths in a circuit varies from 100 to 1000. A path is defined to be critical if the probability that the path can have a delay larger than the critical delay is greater than 0.5%. This information can then be used to determine the parameter NUM. For the analysis, the number of critical paths in the circuit with a 3σ V_{th} variation of 30% is 200, translating to a total of 10,000 paths in the generic logic network.

Figure 6.5 shows the mean leakage power as a function of a second threshold voltage value in a dual-V_{th} optimization (a single V_{dd} is used). The power in this and subsequent figures is shown normalized to the nominal power of the initial design operating at a single V_{dd} and V_{th}. As can be clearly seen, with an increase in the level of variability the optimal second threshold voltage reduces. For example, the difference in optimal V_{th_2} between a purely deterministic optimization and an expected 3σ level of 30% of the mean is approximately $40\,mV$. This can be understood by noting that with increasing variations, devices with higher nominal V_{th} suffer not only a larger delay penalty (due to the growing V_{th}/V_{dd} ratio), but also a larger power penalty stemming from their larger variation (consider (6.23)). Even more critical is that the achievable leakage power savings when considering process fluctuations is significantly degraded compared to the deterministic case. For the most part, this occurs since roughly half of the devices inserted for a given V_{th} will exhibit thresholds smaller than the nominal value. Due to the exponential dependency of leakage on V_{th}, the insertion of high V_{th} devices will often result in much less leakage reduction than expected from the nominal conditions

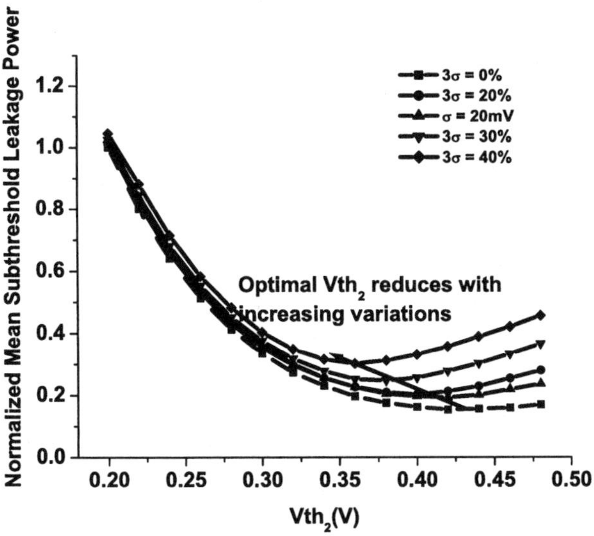

Fig. 6.5. Average power reduction as a function of the second threshold voltage.

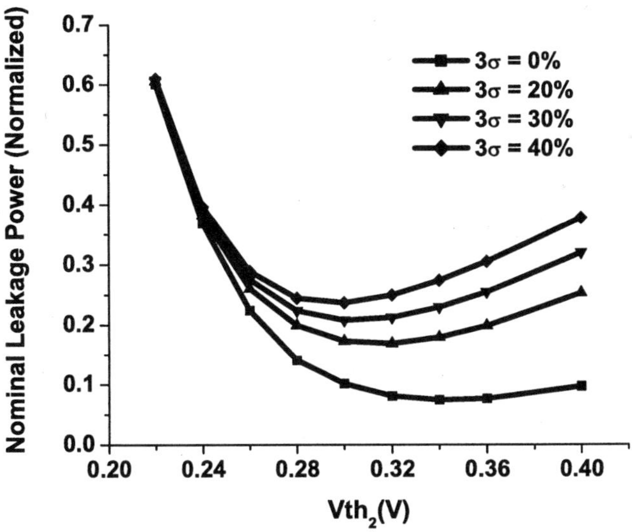

Fig. 6.6. Nominal power reduction as a function of the second threshold voltage value.

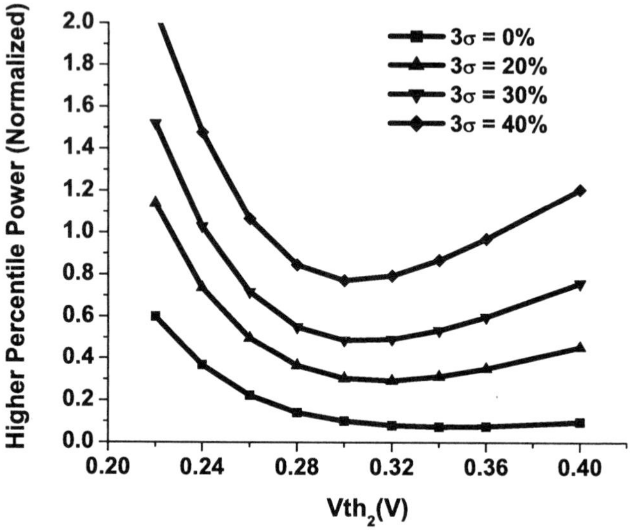

Fig. 6.7. Reduction in a higher percentile of power as a function of the second threshold voltage.

based on process spread. In Fig. 6.5, the power savings reduce from roughly 90% in the deterministic case to just over 70% under a reasonable level of V_{th} variability.

One of the curves in Fig. 6.5 is obtained by assuming a fixed level ($\sigma = 20\,mV$) of variation in V_{th}, regardless of the nominal value. This experiment attempts to investigate whether trends showing optimal second V_{th} values smaller given uncertainty are artifacts of the assumption that mean/sigma for V_{th} is assumed to be constant, therefore penalizing higher threshold values. The figure clearly shows that, even in this situation, the optimal high V_{th} in a dual-V_{th} system is lower compared to the case without considering variations. It is also interesting to note that, while determining the threshold voltages to be offered in a process, an overestimation of the process variation leads to a larger degree of performance degradation compared to an underestimation of the variability levels. This is due to the sharp increase in mean leakage for the use of threshold voltages smaller than optimal.

Figure 6.6 shows the nominal leakage power as a function of a second threshold voltage in dual-V_{th} optimization. The nominal leakage shows similar trends to the average leakage previously discussed. Also, as the second threshold voltage is reduced, the curves corresponding to different variation levels converge. As the difference between the two threshold voltages increases, the fraction of the circuit allocated to the higher V_{th} becomes strongly dependent on the magnitude of process variability (due to the constraint on timing yield)

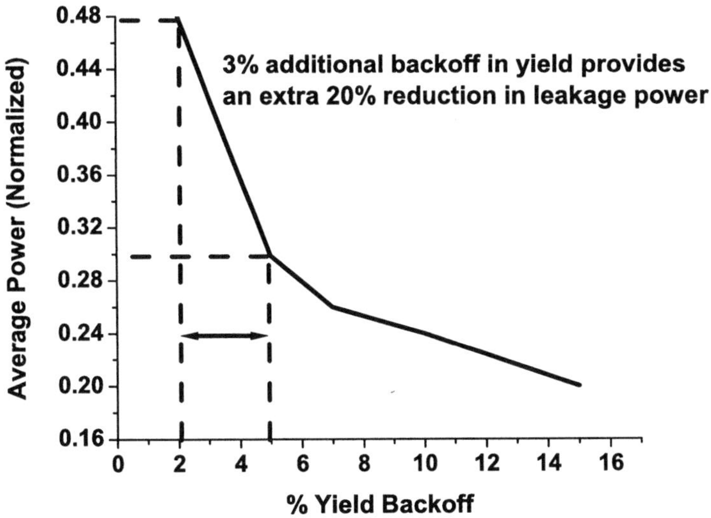

Fig. 6.8. Variation in power savings as a function of allowed timing yield degradation.

and the curves diverge. Figure 6.7 shows the dependence of a minimized higher percentile of the leakage power distribution on the second threshold voltage value. This plot corresponds to the value of leakage power that is two standard deviations away from the mean, indicating an n of 2 in (6.27). The figure shows that the curves are much more spread out here than in Fig. 6.5 and Fig. 6.6 since the power values at this point in the distribution are extremely sensitive to variation in threshold voltage. Also the trend showing that lower threshold voltages become optimal as variability rises holds true in this case as well. Thus, we can say that for power optimization, the difference between the two thresholds decreases with the increase in process variability (assuming a fixed lower bound on one of the V_{th}'s).

Figure 6.8 shows the dependence of the achievable power reduction on the yield backoff used to perform the power optimization of the network. Recall that the circuit must absorb at least a small timing yield penalty when applying low power optimization techniques such as dual-V_{th} and dual-V_{dd}. As shown in the figure, the achievable power reduction decreases rapidly as the yield loss is tightened. Specifically, reducing the backoff from 5% (our default scenario) to 2% results in an increase in power dissipation of approximately 20%. This dependence can be expected to increase for a path delay distribution in which the fraction of near-critical paths is larger than in our assumed path delay distribution.

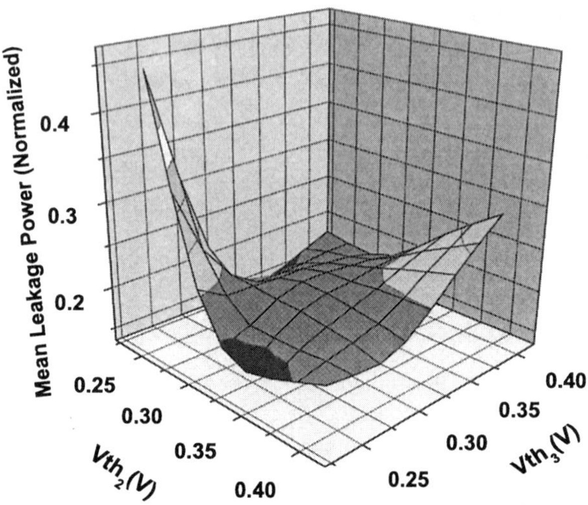

Fig. 6.9. Average leakage power reduction using three threshold voltages.

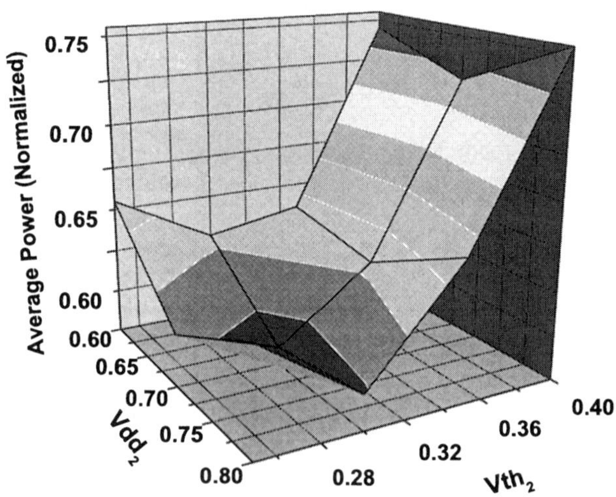

Fig. 6.10. Total power reduction using dual supply and threshold voltages.

Figure 6.9 shows the reduction in leakage power in a triple-V_{th} environment when 3σ variation is 30% of nominal. The introduction of a third V_{th} provides additional power savings of 7%, compared to dual-V_{th} alone. Keeping in mind the cost of the additional masks required for a third V_{th}, this power advantage is not overly attractive.

When dynamic power is considered during power optimization, the impact of variability is significantly reduced, since dynamic power does not vary with fluctuations in V_{th}. In real-world situations, well-controlled gate oxide thickness results in dynamic power variations that are typically very small compared to variations in components of leakage power. However, when we minimize dynamic and subthreshold leakage power using a dual-V_{dd}/V_{th} technique, we expect a change in the optimal solution. This follows from the fact that V_{th} variability leads to a significant difference in the optimal value of the second V_{th}. Hence, the optimal value of the second V_{dd} in a dual-V_{dd}/V_{th} process can also be expected to differ from the deterministic optimal value in order to maintain reasonable drive strengths for gates operating at low Vdd and high V_{th}. Figure 6.10 shows the reduction in total power (considering dynamic and subthreshold leakage) as a result of simultaneous dual-V_{dd} and dual-V_{th} power optimization without considering variations in V_{th}. The initial design at low V_{th} and high V_{dd} is assumed to have equal contribution from dynamic and leakage power. Overall the total power increases when considering variation in V_{th}. This is due to an increase in the subthreshold leakage component of the total power under process variation. However, the difference between the deterministic and statistical case is much smaller here compared to the case where only subthreshold leakage power is considered. This again results from the fact that dynamic power is insensitive to variations, and therefore, the overall impact of variations on power dissipation is considerably much smaller. The figure also shows that the optimal value of the second supply voltage shifts slightly towards a higher value, due to an increase in the optimal value of the lower threshold voltage. This is the other factor responsible for reducing the total power savings obtained by the application of a dual-V_{dd}/V_{th} process.

Thus, we see that there are two key components to the power optimization formulation: 1) we must allocate a yield budget to individual paths (or more generally, to sets of paths with similar delay characteristics) such that power can be effectively minimized, and 2) the power optimization itself which can target various points along the power distribution including high percentile points that represent problematic leaky dies for example. The yield allocation formulation is actually a more general problem, and techniques that can be used to generate a set of timing yield constraints for a lower level of hierarchy based on timing yield constraints for a higher abstraction level will be very useful for any statistical optimization technique. For example, a technique to establish yield constraints for each combinational block based on the yield constraint on the complete design will be needed to use any statistical optimization technique developed for combinational designs. This problem clearly differentiates statistical and deterministic optimization, where all combinational blocks will be required to meet a given timing constraint based on the clock cycle.

Additionally, note that the technique discussed in this section can be mapped directly to the gate sizing problem, assuming a linearized delay model, using a path-based analysis. However, the crucial step in gate sizing will be to map the yield constraints on each path to each of the timing arcs of the circuit, since the number of paths within a circuit can be exponentially large in number. The general conclusions also point to greatly degraded power savings achievable by dual-V_{th} processes when considering variability. Thus, the use of more than two threshold voltages in a process will provide substantially reduced improvements and the timing yield vs. power reduction tradeoff will allow the designer to consider interesting design decisions. In addition, the power-yield tradeoff curve in Fig. 6.8 shows that a small degradation in parametric timing yield can be seen to have large positive impact on achievable power reductions, through dual-V_{th} and other low-power design techniques.

6.2 Gate Sizing

A large amount of work has been done in the area of deterministic gate sizing. TILOS [52] was the first to show that the gate sizing problem can be expressed as a convex optimization problem. However, the convex optimization approach required an enumeration of all paths in the circuit (which increases exponentially with circuit size). Therefore, TILOS starts from a minimum sized circuit and uses a sensitivity-based heuristic to iteratively select and upsize gates that provide the maximum improvement in performance as long as the desired timing goal is not met. In [125], the authors used a novel ellipsoidal approach to solve convex optimization problems, whose runtime complexity was independent of the number of constraints, to propose an exact approach for gate sizing with polynomial complexity. This path-based formulation of a gate sizing problem (for area minimization) can be expressed as

$$\begin{aligned} \text{Min}: &\ \sum_{i}^{n} \alpha_i x_i \\ \text{s.t.}: &\ \sum_{i \in p} D_i \leq D_0 \quad \forall p \in P \\ &\ L_i \leq x_i \leq U_i \quad \forall i = 1, \ldots, n \end{aligned} \tag{6.37}$$

where x_i's are the gate sizes of an n-gate design, D_0 is the delay constraint for the design, D_i represents the delay of a gate, P is the set of paths in a circuit and L_i and U_i are the lower and upper bounds on the size of each of the gates. An equivalent problem for timing optimization can be formulated by treating D_0 as a variable and changing the objective function to minimize D_0 while enforcing a constraint on the area of the design.

The delay of a path can be shown to be a posynomial in terms of gate sizes and the optimization problem in (6.37). A posynomial, which has a general form

$$F(\mathbf{X}) = \sum_{j=1}^{m} c_j \prod_{i=1}^{n} x_i^{a_{ij}} \tag{6.38}$$

where c's are positive coefficients, can be mapped to a convex function using a simple exponential transformation. Note that in the above formulation we have a constraint for each path in the network, which might be exponential in the number of gates in a circuit. Even though the approach in [125] had polynomial complexity, the runtime performance of this approach was not found to be good in practice. The problem with the exponential number of constraints was resolved in [31], which proposed a node-based formulation for the gate sizing problem, expressed as

$$\begin{aligned}
\text{Min}: \quad & \sum_{i}^{n} \alpha_i x_i \\
\text{s.t.}: \quad & a_j \le D_0 \quad \forall j \in \text{outputs} \\
& a_j + D_i \le a_i \quad \forall i = 1, \ldots, n \text{ and } \forall j \in \text{inputs}(i) \\
& D_i \le a_i \qquad\quad \forall i \in \text{inputs} \\
& L_i \le x_i \le U_i \quad \forall i = 1, \ldots, n.
\end{aligned} \tag{6.39}$$

Here a_i's are additional variables and represent the arrival times at the respective nodes, *inputs* and *outputs* are the primary inputs and outputs of the design respectively, and *inputs*(i) refer to the inputs of the gate that feeds node i.

The above problem can be solved using *Lagrangian relaxation.* The approach proposed in [31] estimates the values of the Lagrange multipliers by solving the dual optimization problem using subgradient optimization. The values of Lagrange multipliers are then used to calculate the optimal device widths. We will discuss this technique in more detail while considering the extension of Lagrangian relaxation to statistical gate sizing in Sec. 6.2.2. The key advantage of this formulation is that the numbers of constraints are now polynomial in terms of the number of gates in the design. However, the approach is amenable only to simple convex delay models, since we need the duality gap to be zero to estimate the Lagrangian multipliers using the dual problem.

Other approaches that have been used to perform gate sizing include [137], which performs iterative slack assignment and gate sizing assignment to converge to an optimal solution. The approach presented in [37], combines an accurate dynamic timing simulator with a nonlinear optimizer, which makes the approach exact and able to handle problems associated with static timing analysis such as false paths. However, this comes at the cost of significant computational overhead. A large amount of work has also been done to tailor nonlinear optimizers to techniques that suit gate sizing, and a survey of various techniques that have been investigated can be found in [39]. Recently a number of approaches have been proposed to perform statistical gate sizing, even though the field of statistical gate sizing is much younger than deterministic gate sizing. This is still a very active area of research with new techniques being developed to reduce computational complexity and provide better performance.

A straightforward node-based formulation for statistical gate sizing can be developed by allocating a deterministic delay to each node in the circuit, which is a linear function of its mean μ and standard deviation σ, and then performing deterministic gate sizing using the standard node-based formulation approach discussed above. If a delay of $\mu+\Phi^{-1}(\eta)\sigma$, where η is the desired timing yield, is assigned to each node, then we assign a worst-case delay to each node where the worst-case is now defined by the desired yield. Therefore, we do no better than a deterministic optimization technique. However, if the value of η is chosen appropriately then the design margins assumed in worst-case design can be reduced while simplifying the optimization approach itself. However, determining the appropriate value of η is fairly complicated (in Sec. 3.4.3 we discussed an approach to estimate η for timing yield analysis).

The key difference between iterative statistical and deterministic gate sizing is that a single path cannot be identified as being critical, since a large number of paths have significant probability of being critical. Depending on the actual values of different process parameters for a given sample of the design any one of these paths can limit performance. If we seek a statistical ordering of gate delays, we need to define an ordering relationship for probability distributions instead of a deterministic ordering,. The strongest such relationship is known as *stochastic ordering*, which says that given two RVs X and Y, Y is stochastically greater than X if

$$\mathcal{P}(X > u) \geq \mathcal{P}(Y > u) \quad \forall u \in (-\infty, \infty). \tag{6.40}$$

This relationship is illustrated in Fig. 6.11. This is a very strong condition and even in simple directed acyclic graphs (DAGs), this condition is not seen to hold among a large number of delay distributions of different paths as many of the path delay distributions are found to have cross-over points. Thus, most statistical gate sizing techniques try to integrate some form of a statistical timing analyzer within the optimization and seek to identify paths that have the strongest influence on the final probability distributions of delay. This is in contrast to a deterministic approach that identifies a critical path within the network and attempts to reduce the delay of the identified critical path. Additionally, uncorrelated intra-die variations are independent across gates

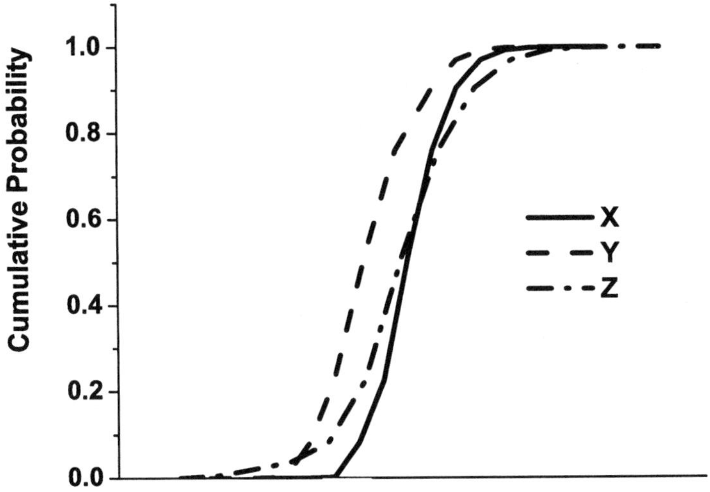

Fig. 6.11. The RV X is stochastically smaller than Y. However, no ordering can be established for Z with either X or Y in this example.

and their impact at the circuit level is much smaller due to the averaging effect, as discussed in Chap. 3. Hence, it is important that this property of intra-die variations is considered while statistically optimizing circuits.

6.2.1 Nonlinear Programming

As we saw in Chap. 3, the mean and variances of gate delay can be simply obtained by summing the mean and variance of the maximum of the input arrival times and the mean and variance of the gate delay. However, calculating the maximum of a set of probability distributions is complicated. Even under a Gaussian approximation, Clark's expressions [35] are required to estimate the probability distribution of the maximum. Using the nonlinear expression used to express the *max* function, [63] employs the node-based gate sizing formulation (6.39) to express the statistical gate sizing problem as a nonlinear optimization problem, which is solved using a large-scale nonlinear programming package, such as LANCELOT [38].

The formulation is based on a delay model of the form

$$d_i = a_{0,i} + a_{1,i} \frac{C_{\text{load},i} + \sum_{j \in \text{fanouts}(i)} C_{\text{in},j} x_j}{x_i} \tag{6.41}$$

where d_i represents the delay of gate i, a_0 represents the size independent intrinsic delay of a gate, a_1 represents the sensitivity of gate delay to output capacitive loading, C_{load} is the capacitive loading other than that associated with the input gate capacitance of the fanout gates, C_{in} represents the input gate capacitance for a gate per unit width and x represents the size of a gate. The above equation is rewritten as

$$d_i x_i = a_{0,i} x_i + a_{1,i} \left(C_{\text{load},i} + C_{\text{in}} \sum_{j \in \text{fanouts}(i)} x_j \right) \tag{6.42}$$

which results in a linear constraint in terms of gate sizes and its usefulness will be discussed later. The statistical gate sizing problem is then expressed as

$$\begin{aligned}
\text{Min}: \quad & f(\mu_o, \sigma_o^2) \\
\text{s.t.}: \quad & \mu_o = E[\max_{j \in \text{outputs}} a_j] = f_\mu(\mu_0, \ldots, \mu_j, \ldots, \sigma_0, \ldots, \sigma_j, \ldots) \\
& \sigma_o^2 = Var[\max_{j \in \text{outputs}} a_j] = f_\sigma(\mu_0, \ldots, \mu_j, \ldots, \sigma_0, \ldots, \sigma_j, \ldots) \\
& \mu_{a_i} = \mu_{d_i} + E[\max_{j \in \text{input}(i)} a_j] \quad \forall i = 1, \ldots, n \\
& \sigma_{a_i}^2 = \sigma_{d_i}^2 + Var[\max_{j \in \text{input}(i)} a_j] \quad \forall i = 1, \ldots, n \\
& \mu_{d_i} x_i = a_{0,i} x_i + a_{1,i} \left(C_{\text{load},i} + C_{\text{in}} \sum_{j \in \text{fanouts}(i)} x_j \right) \quad \forall i = 1, \ldots, n \\
& \sigma_{d_i}^2 = g(\mu_{d_i}) \quad \forall i = 1, \ldots, n \\
& L_i \le x_i \le U_i \quad \forall i = 1, \ldots, n
\end{aligned} \tag{6.43}$$

where μ_a and σ_a represent the mean and variance of delay, respectively. Similarly μ_d and σ_d represents the mean and variance of gate delay, and the variance of gate delay is assumed to be related to the mean of gate delay through the function g. The functions f_μ and f_σ are used to obtain the mean and variance of the delay of the max of a set of arrival times, respectively (the functional form is as discussed in Chap. 3). Note that the expressions for the mean and variance are only available for the max of two arrival times. Therefore, any node in the circuit that involves the max of more than two arrival times has to rewritten recursively using additional variables. Also, f is the objective function that depends on the variance and mean of the circuit delay and can have any desired form. For example, if we are interested in minimizing the average case delay, then $f = \mu_o$. However, if we are interested in maximizing FMAX (Chap. 3) for 99.8% of the samples of the design, then under the assumption that delay has a Gaussian distribution, we have $f = \mu_o + 3\sigma_o$.

To obtain efficient performance from a non-linear optimizer, it is useful to provide information regarding the first and second derivatives of the objective function and the constraints. Writing the gate delay model in the form (6.42) rather than (6.41) has the advantage that it reduces the number of nonlinear terms in the constraint, and thus simplifies the expressions for the derivatives. This was one of the first approaches that was aimed at improving the timing yield of VLSI circuits while considering within-die variations. However, this approach does not consider inter-die variations, and it is not straightforward to consider correlations in gate delays using this approach.

To handle these problems, we consider an approach that completely decouples the statistical timing analysis from the optimization step and is thus amenable to integration with any statistical timing analysis engine.

6.2.2 Lagrangian Relaxation

The nonlinear programming approach to gate sizing was based on expressing the statistical timing engine within a non-linear program used to optimize the design, and hence increased the overall complexity of the optimization. Another class of solution methodologies iteratively solves a simpler optimization problem and uses the statistical timing analyzer to perform incremental updates to the optimization problem. The approach proposed in [34] is based on this idea. It iteratively formulates the node-based gate sizing problem (6.39) while updating the timing constraints imposed on the circuit using a statistical timing analyzer. Therefore, the choice of the statistical timing analyzer in not constrained, and any of the approaches discussed in Chap. 3 can be used. The optimization problem itself is solved using the approach proposed in [31], which is based on Lagrangian relaxation.

Lagrange multipliers are used to simplify a constrained optimization problem into an unconstrained optimization problem by incorporating the constraints into the objective function. Let us consider the node-based formulation of the gate sizing problem in (6.39). The complicated constraints on the arrival times are included into the objective function using Lagrange multipliers (λ's) and the problem is restated as:

$$
\begin{aligned}
\text{Min}: \quad & L_\lambda(x,a) = \sum_i^n \alpha_i x_i + \sum_{j \in \text{inputs}} \lambda_{j0}\left(a_j - D_0\right) \\
& + \sum_{i=1}^n \sum_{j \in \text{input}(i)} \lambda_{ji}\left(a_j + D_i - a_i\right) + \sum_{i \in inputs} \lambda_{mi}\left(D_i - a_i\right) \qquad (6.44) \\
\text{s.t.}: \quad & L_i \le x_i \le U_i \quad \forall i = 1, \ldots, n.
\end{aligned}
$$

Kuhn-Tucker conditions [15] are then applied to the above problem, while minimizing the objective function (known as the Lagrangian) with respect to the variables corresponding to the arrival times. Using these conditions we obtain the relation that for any gate, the summation of the values of the Lagrange multipliers associated with the inputs is equal to the summation of the values Lagrange multipliers associated with the fanouts. This can be mathematically expressed as

$$\sum_{i \in \text{output}(k)} \lambda_{ki} = \sum_{j \in \text{input}(k)} \lambda_{jk}. \tag{6.45}$$

Plugging this condition back into the Lagrangian (6.44), we can simplify the Lagrangian such that it is independent of the arrival times. In [31] it is shown that the simplified problem can be solved exactly, using an iterative technique while considering only one gate at a time. However, to solve the problem using the iterative technique we need to determine the optimal values of Lagrange multipliers. To achieve this, a Lagrange dual problem is formulated and solved to determine these Lagrange multipliers. The Lagrange dual problem is known to be a convex optimization problem, but the objective function may not be differentiable. Hence, techniques such as subgradient optimization are used to solve this problem. An improvement to solve the Lagrangian dual problem was also suggested in [140], which combines a gradient based search technique with subgradient optimization to improve convergence.

Since a deterministic approach is not our focus here, we will not discuss the details of this approach further and the interested reader is referred to [31] for more information. Let us now discuss the technique used in [34], which iteratively updates the timing constraint in the gate sizing problem based on input from the statistical timing analyzer.

Yield Constraint

The authors in [34] perform gate sizing to minimize the area of the design while ensuring that a desired timing yield is achieved. This is achieved by iteratively solving the gate sizing problem while updating the timing constraint D_0 imposed on the circuit using information from a statistical timing analyzer. This can be understood using Fig. 6.12, which shows how the timing constraint is updated from D_0 to D_0' in going from one iteration to the next. Let us consider the case where a timing yield of 84.1% is desired. Assuming that the delay has a Gaussian distribution, the delay constraint is modified to $D_o - \sigma$, where D_0 is the initial timing constraint and σ is the standard deviation of delay obtained using statistical timing analysis. Statistical timing analysis is performed at the end of each iteration to estimate the distribution of delay of the circuit. If the variance of delay remains same and the nominal delay constraint used in the optimization correspond to the mean delay then the yield constraint is met. If the timing yield constraint is met, then we stop, otherwise the timing constraint imposed on the circuit is modified based on the variance of delay and gate sizing is performed again using the Lagrangian approach.

This approach decouples the statistical timing analysis and optimization steps and simplifies the problem. However, this makes the optimization blind

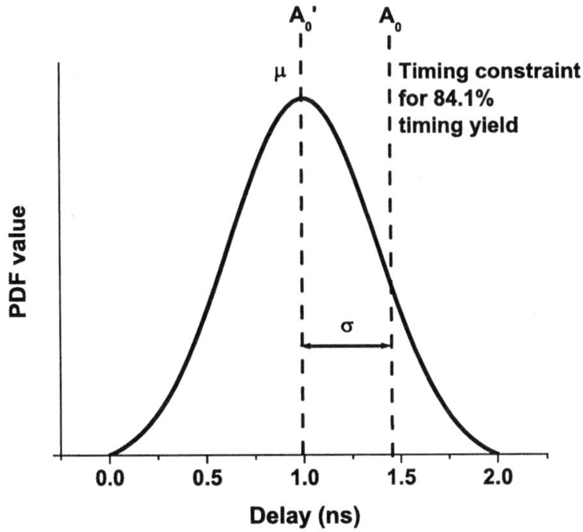

Fig. 6.12. Modifying the delay constraint to account for variations in delay.

to most of the statistical information and it is therefore not effective in providing good solutions. Additionally, the modification of the delay constraint cannot be guaranteed to converge to a final solution. Consider the case where intra-die variations have a strong influence, which results in a statistical mean delay that is significantly different from the deterministic mean delay. In this case, even if we deterministically ensure that all paths have a delay smaller than $D_0 - \sigma$, we cannot guarantee that the overall timing yield of the circuit with respect to the delay constraint of A_0 is 84.1%.

6.2.3 Utility Theory

The concepts of utility functions have been prevalent in the realm of economics, which by its nature, must be studied statistically. Utility functions are useful in situations where minimizing the expected value alone is not sufficient, and therefore a utility function is defined and one seeks to maximize the utility function or equivalently minimize a disutility function, which is defined to be the negative of the utility function.

A reasonable utility (disutility) function for circuit delay should be monotonically decreasing (increasing), since a lower delay value is always more desirable than a higher delay value. Based on the convexity or concavity we can identify two classes of utility functions. If the utility (disutility) function is concave (convex) then it defines *risk averse* behavior, otherwise the utility function is said to be *risk inclined*. Consider a game where we need to invest

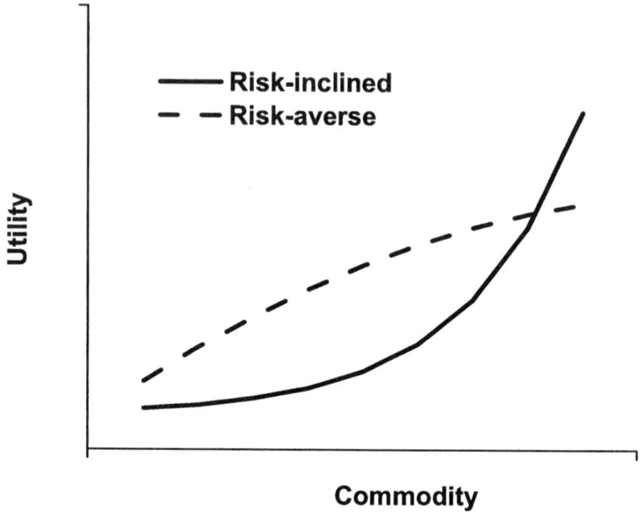

Fig. 6.13. A concave utility function represents a risk-averse preference. In addition, a concave objective function can be efficiently maximized using convex optimization techniques.

an amount (w_0) to play the game, and we either lose the money or get back double the amount we invested, both with equal probability. Consider a concave utility function, as shown in Fig. 6.13, and assume that the chances of winning and losing the game are equal. In this case, the expected value of the utility, which is the average of the utility values when we lose and win the game, is lower than before we play the game (due to the concavity of the utility function). Hence, to maximize the utility we should not invest in the game, representing a risk averse investment strategy. However, the expected utility is greater if we play the game, for a convex utility function, and we are risk inclined. The concept of utility functions was introduced in statistical gate sizing in [115] and was used to represent the potential timing yield loss associated with different paths in a circuit.

The disutility function of a path in the network is defined in [115] to be of the form

$$U_p = D_p^2 + D_p \tag{6.46}$$

where D_p and U_p are the delay and disutility of a path p, respectively. The above disutility function is convex, and hence defines a risk-averse utility function. Note that we seek to maximize (minimize) the utility (disutility) and hence standard optimization techniques can be used to efficiently optimize the problem if the utility (disutility) function is concave (convex).

Having defined the disutility function for path delay, let us define some parameters to model the delay of a path. Assume a given correlation matrix $\mathbf{C}$ for node delay with elements c_{ij}, and that each node i has a mean delay η_i and a delay variance σ_i^2. The relationship that a given path p *dominates* another path q was defined in [115] as the condition that

$$E[U_p] > E[U_q]. \tag{6.47}$$

Similarly, a subset of paths $P_U \subseteq P$ is said to be the set of *undominated* paths, if for each path $p \in P_U$ the following holds

$$E[U_p] > E[U_q] \quad \forall q \in P \backslash P_U. \tag{6.48}$$

where $P \backslash P_U$ represents the set of paths P without any paths from the set P_U. The expected value of the disutility can be expressed as

$$E[U_p] = E[D_p^2] + E[D_p]. \tag{6.49}$$

The first term in the above equation can be written as

$$E[D_p^2] = \left(\sum_{i \in p} \eta_i \right)^2 + \sum_{i \in p} \sigma_i^2 + 2 \sum_{i,j \in p} c_{ij} \sigma_i \sigma_j \tag{6.50}$$

giving

$$E[U_p] = \left(\sum_{i \in p} \eta_i \right)^2 + \sum_{i \in p} \sigma_i^2 + 2 \sum_{i,j \in p} c_{ij} \sigma_i \sigma_j + \sum_{i \in p} \eta_i. \tag{6.51}$$

Using this definition of dominance, we seek to identify the set of undominated paths from the source to the sink of the network and define the overall objective function for our optimization using the nodes that lie on these undominated paths. Let P_{si} be the set of paths from the source s to some node i within the network, and let the sink node be identified as t. The deterministic notion of domination, which implies that if a path p_1 dominates a path p_2 at the intermediate node i, then the path p_1 to t will dominate the path p_2 to t, is not valid statistically. Due to variations the path p_1 may no longer be dominated by p_2 at some arbitrary node in between i and t. This results from the fact that node delays in the fanout cone of node i may be correlated to the path delays. Therefore, if two paths $p1, p2 \in P_{si}$ satisfy

$$E\left[U_{p_1}\right] > E\left[U_{p_2}\right] \tag{6.52}$$

they are only said to imply *temporary preference*, with p_1 being preferred over p_2. *Permanent preference* is said to occur when the two paths satisfy

$$E\left[U_{p_1+p}\right] > E\left[U_{p_2+p}\right] \quad \forall p \in P_{it}. \tag{6.53}$$

This implies

$$\begin{aligned} E\left[(D_{p_1}+D_p)^2\right] + E\left[D_{p_1}+D_p\right] > \; & E\left[(D_{p_2}+D_p)^2\right] \\ & + E\left[D_{p_2}+D_p\right] \quad \forall p \in P_{it}. \end{aligned} \tag{6.54}$$

The above equation can be simplified using the following relations

$$\begin{aligned} E\left[D_{p_1}D_p\right] &= \sum_{i \in p_1}\sum_{j \in p} c_{ij}\sigma_i\sigma_j + E\left[D_{p_1}\right] + E[D_p] \\ E\left[D_{p_2}D_p\right] &= \sum_{i \in p_2}\sum_{j \in p} c_{ij}\sigma_i\sigma_j + E\left[D_{p_2}\right] + E[D_p] \end{aligned} \tag{6.55}$$

as

$$\begin{aligned} E & \left[U_{p_1}\right] - E[U_{p_2}] + 2E[D_p]\left(E[D_{p_1}] - E[D_{p_2}]\right) \\ & > 2\left(\sum_{i \in p_1, j \in p} c_{ij}\sigma_i\sigma_j - \sum_{i \in p_2, j \in p} c_{ij}\sigma_i\sigma_j\right) \quad \forall p \in P_{it}. \end{aligned} \tag{6.56}$$

Though the above expression establishes the conditions that can be used to identify the set of undominated paths in a circuit, it requires us to consider all paths P_{it} which is computationally very expensive. As we discussed before, temporary dominance does not imply permanent dominance because node delays might be correlated. To identify the set of nodes in the fanout cone of i that can be possibly correlated with the path delays p_1 and p_2, a *correlation front* is defined, denoted by $F_i(\alpha)$, and is said to be the set of nodes in the fanout cone of i whose delay distributions have correlations of at least α with any node in the fanin cone of node i. Now, let us assume α is small enough so that the path delay contributed by the nodes in the fanout cone of i other than those within the correlation front are independent of the path delays of paths p_1 and p_2. Using the above assumption, any path P_{it} can be partitioned into P_{il} and P_{lt}, such that node l lies on the correlation front. The condition for permanent preference (6.53) can then be rewritten as

$$E\left[U_{p_1+p_{il}+p_{lt}}\right] > E\left[U_{p_2+p_{il}+p_{lt}}\right] \tag{6.57}$$

which is equivalent to

$$E[U_{p_1+p_{il}}] - E[U_{p_2+p_{il}}] + 2E[D_{p_{lt}}]\left(E[D_{p_1+p_{il}}] - E[D_{p_2+p_{il}}]\right) > 0. \tag{6.58}$$

This can now be used to reduce the complexity of the steps required to check for permanent preference by only considering paths in the set $P_{il}\,\forall l \in F_i(\alpha)$, by rewriting the above condition as

$$\begin{aligned} E[U_{p_1+p_{il}}] &> E[U_{p_2+p_{il}}],\ \mathit{and} \\ E[D_{p_1+p_{il}}] - E[D_{p_2+p_{il}}] &> \frac{E[U_{p_1+p_{il}}] - E[U_{p_2+p_{il}}]}{2\max_{p\in P_{lt}} E[D_p]}\ \forall l \in F_i(\alpha). \end{aligned} \tag{6.59}$$

Now we can use a dynamic programming approach to estimate the undominated paths at the terminal node t. This is achieved by starting from node s, and pruning the set of undominated paths by establishing permanent preference among the set of paths P_{si} at each node by computing the set $F_i(\alpha)$. Even though pruning techniques can be employed, this approach has an exponential worst-case computational complexity. Also, the correlations that arise from reconvergent fanouts may be neglected as a result of the construction of the correlation front. However, it is able to capture the effect of spatial correlations if gates that are logically closer are also close together in the layout.

The goal of gate sizing optimization is to reduce the weighted sum of the disutility function of the set of nodes that lie on the set of undominated paths. For all such nodes, [115] defines a *criticality index* as the number of undominated paths that pass through the node. Additionally, the mean delay of each path is constrained to be smaller than a critical delay $T_{\max}$. The mean delay of a node i (η_i) is modeled as

$$E[d_i] = \eta_i = \tau_i + \alpha_i \frac{c\sum_{j\in\text{fanouts}(i)} s_j}{s_i} \tag{6.60}$$

where τ_i is the intrinsic gate delay independent of size s_i, c is the gate input capacitance per unit size and α_i captures the sensitivity of gate delay to load capacitance. Similarly, the variance of delay can be expressed as

$$\begin{aligned} \sigma_i^2 = {} & \left(\frac{\partial d_i}{\partial L_i}\right)^2 \sigma_{L_i}^2 + \left(\frac{\partial d_i}{\partial V_{thi}}\right)^2 \sigma_{V_{thi}}^2 + \cdots \\ & + \sum_{j\in\text{fanouts}(i)} \left(\frac{\partial d_i}{\partial L_j}\right)^2 \sigma_{L_j}^2 + \left(\frac{\partial d_i}{\partial V_{thj}}\right)^2 \sigma_{V_{thj}}^2 + \cdots \end{aligned} \tag{6.61}$$

which estimates the variance in delay based on the variation in process parameters of the gate i itself, and variations in its fanout gates. The complete gate sizing problem can now be expressed as a nonlinear optimization problem

$$\text{Min}: \sum_{i\in N} W_i E[U_i]$$

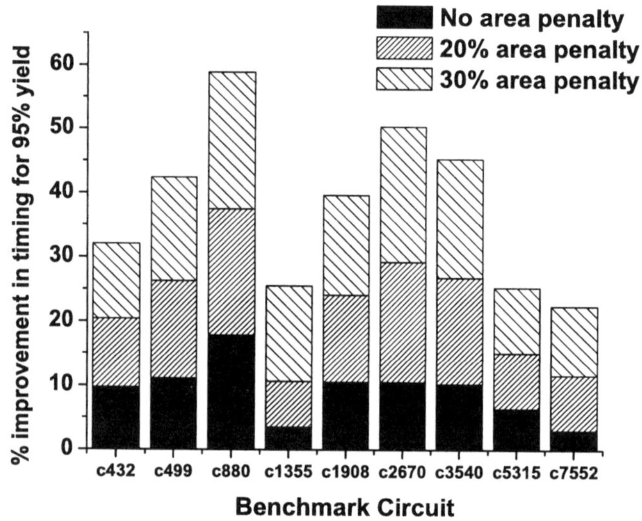

Fig. 6.14. Improvement in clock period required for 95% yield improves as a larger area penalty is allowed [115].

$$\begin{aligned} \text{s.t.}: \quad & \max_{p \in P} E[D_p] \leq T_{\max} \\ & E[U_i] = (\eta_i^2 + \sigma_i^2) + \eta_i \quad \forall i \in N \\ & l_i \leq s_i \leq u_i \\ & Area \leq A + \Delta A \end{aligned} \tag{6.62}$$

where N is the set of all nodes that lie on the set of undominated paths, η_i and σ_i are as defined in (6.60) and (6.61) respectively, and W_i is the criticality index of a node as defined above. The optimization is performed using LANCELOT and allows an area increase of ΔA while minimizing the disutility function. Results show that a large improvement in yield can be achieved with this approach compared to deterministic optimization using nominal case models even when no area penalty is incurred, as shown in Fig. 6.14. As expected, increasing the area penalty allows for further increases in timing yield as well.

Though the technique shows significant improvements, the foundation of this approach lies in path-based analysis making it unsuitable for large circuit structures since they have a very large number of paths.

6.2.4 Robust Optimization

In this section, we extend the ideas introduced in Sec. 6.1 to develop a gate sizing algorithm. This approach to gate sizing was proposed in [87]. It uses a linear gate delay model [18] to develop a robust linear programming for gate sizing. The robust optimization problem is then mapped to an equivalent second-order conic program (SOCP) problem. A SOCP involves the minimization of a linear function under a second-order conic constraint. A second-order cone constraint of dimension n is defined as

$$\| \mathbf{A_i x} + \mathbf{b_i} \| \leq \mathbf{c_i}^T \mathbf{x} + d_i \tag{6.63}$$

where $\mathbf{x}$ is the vector of optimization variables, $\mathbf{A_i}$, $\mathbf{b_i}$, $\mathbf{c_i}^T$ are arbitrary row vectors and d_i is a constant. SOCPs are a special class of convex optimization problem and can be efficiently solved using convex optimization techniques. However, specialized techniques to solve SOCP, such as primal-dual interior point methods, exist and can also be used [85]. The aim of this optimization is to minimize the sum of device sizes of a design while enforcing a timing yield constraint on the design. The sum of device sizes with different weighting factors is a measure of power dissipation, therefore the optimization problem here is the minimization of power dissipation under statistical timing constraints. We will limit the scope to power optimization and the yield allocation step as discussed in Sec. 6.1 will be assumed to have been performed a priori in an optimal manner providing the yield constraint on each path.

In a deterministic scenario the gate sizing problem is as posed in (6.37). We will write the statistical gate sizing problem as

$$\begin{aligned} \text{Min}: \quad & \sum_{i}^{n} x_i \\ \text{s.t.}: \quad & \mathcal{P}\left(\sum_{i \in p} D_i \leq D_0 \right) \geq \eta_p \quad \forall p \in P \\ & L_i \leq x_i \leq U_i \quad \forall i = 1, \ldots, n \end{aligned} \tag{6.64}$$

where P is the set of paths, D_i is the delay associated with node i (assumed to be a Gaussian RV), and the constraint enforces the condition that the timing yield of path p is η_p. The mean and variance of path delay

$$D_p = \sum_{i \in p} D_i \tag{6.65}$$

can be easily obtained using the covariance matrix $\mathbf{C}$ for the node delays and is expressed as

$$\mu_{D_p} = E[D_p] = \sum_{i \in p} E[D_i] \tag{6.66}$$

$$\sigma_{D_p}^2 = \sum_{i \in p} \sigma_{D_i}^2 + 2 \sum_{i \in p} \sum_{j \in p, j \neq i} c_{ij} \sigma_i \sigma_j.$$

Let us now assume that the RVs associated with gate delays are independent. If this is not the case, then they can be mapped to such a set of independent RVs using PCA and the following analysis remains the same. Let us rewrite the constraint in (6.64) as

$$\mathcal{P}\left(\frac{D_p - \mu_{D_p}}{\sigma_{D_p}} \leq \frac{D_0 - \mu_{D_p}}{\sigma_{D_p}}\right) \geq \eta_p. \tag{6.67}$$

Note that the RV associated with D_p has been mapped to a Gaussian RV with unit variance and zero mean. The constraint can now be rewritten as

$$\frac{D_0 - \mu_{D_p}}{\sigma_{D_p}} \geq \Phi^{-1}(\eta) \tag{6.68}$$

where Φ is the cumulative distribution function of a standard Gaussian RV. We saw in Sec. 6.1 that this condition defines a convex set if $\eta \geq 0.5$. The same assumption is made here as well.

Up until now we have neglected the dependence of gate delay on output loading, which is expressed by

$$D_i = \tau_i + \alpha_i \frac{c \sum_{j \in \text{fanouts}(i)} s_j}{s_i} \tag{6.69}$$

where τ_i is the intrinsic delay of gate i, and the second term captures the impact of loading capacitance on gate delay. Note that our optimization variables are gate sizes, and the delay model is not a linear function of the gate sizes. The problem is modeled as a robust linear problem by using a linear gate delay model [18] instead of (6.69), which sacrifices some accuracy. This linear model has the form

$$D_i = a_i - b_i s_i + c_i \sum_{j \in \text{fanouts}(i)} s_j \tag{6.70}$$

where the constants a_i, b_i and c_i are fitting coefficients. To capture the variability in delay due to variations in gate length and threshold voltage, [87] lumps the variations into coefficients in the above expression and treats them as Gaussian RVs whose parameters can be obtained using SPICE simulations. Now we can write the variance of delay as

$$Var[D_i] = s_i^2 \sigma_{b_i}^2 + \left(\sum_{j \in \text{fanouts}(i)} s_j \right)^2 \sigma_{c_i}^2. \tag{6.71}$$

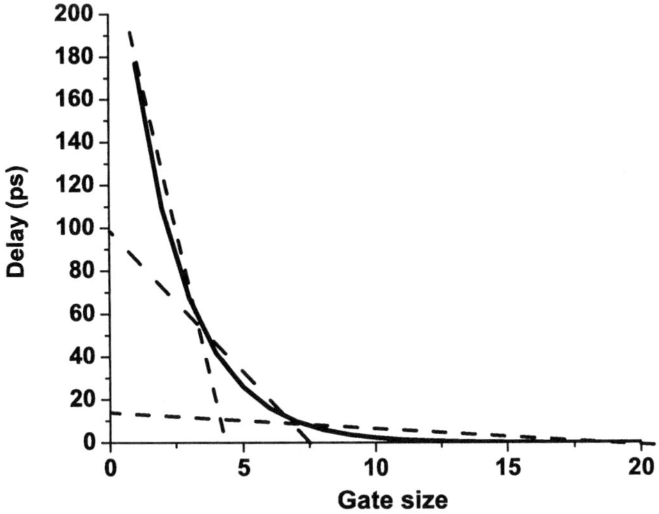

Fig. 6.15. A piecewise linear delay model can provide better accuracy at the cost of a much larger number of constraints.

The inaccuracy introduced by the linearized delay model can be reduced by using a piecewise linear delay model instead of a single linear delay model over the entire range of device widths. This delay model is defined as a maximum of a set of linearized models, which provides better accuracy for small regions of gate sizes as shown in Fig. 6.15. The constraints on delay are specified for all combinations of delay models, which automatically constrains delay to be greater than the maximum delay. The piecewise linear model provides a much better approximation for the delay of a gate. However, this improvement in accuracy is at the expense of an exponential increase in the number of constraints. If the number of logic stages in a path is n and each logic gate has a delay model with three piece-wise linear components then the number of constraints in (6.70) increases by a factor of n^3.

Using the linearized delay assumption, we can finally write down the optimization problem as

$$\text{Min}: \sum_{i}^{n} s_i$$

$$\text{s.t.}: \sum_{i \in p} \left[a_i - \mu_{b_i} s_i + \mu_{c_i} \sum_{j \in \text{fanouts}(i)} s_j \right] \tag{6.72}$$

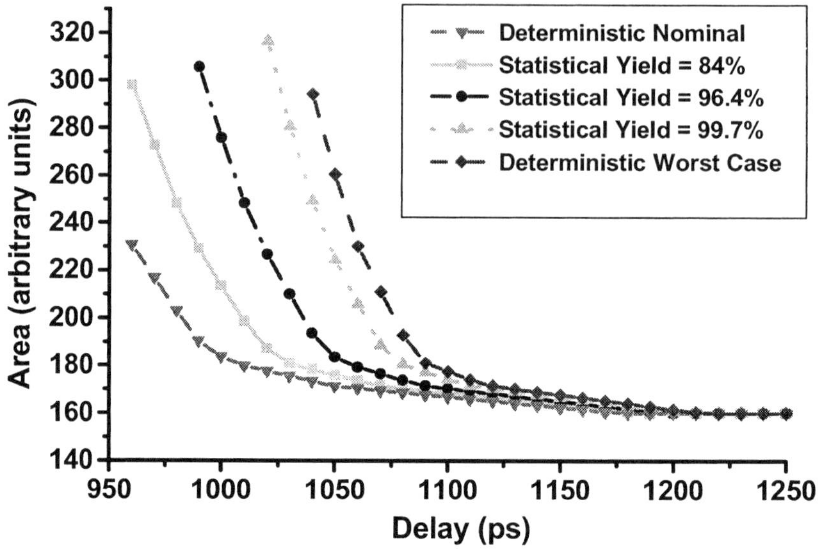

Fig. 6.16. Area-delay curves comparing optimization for different desired yields as well as deterministic optimization [87]. (©2005 IEEE)

$$+\Phi^{-1}(\eta)\left[\sum_{i\in p}\left(\sigma_{b_i}^2 s_i^2+\sigma_{c_i}^2\left(\sum_{j\in \text{fanouts}(i)} s_j\right)^2\right)\right]^{1/2} \leq D_0 \;\forall p\in P$$
$$L_i \leq x_i \leq U_i \quad \forall i=1,\ldots,n.$$

To improve the computational complexity due to the path-based nature of the problem defined above, [87] proposes to transform the above gate sizing problem (6.72) using a node-based formulation, where the delay of each gate is expressed as

$$D_i = a_i - \mu_{b_i} s_i + \mu_{c_i} + \Phi^{-1}(\eta)\left(\sigma_{b_i}^2 s_i^2+\sigma_{c_i}^2\left(\sum_{j\in \text{fanouts}(i)} s_j\right)^2\right)^{1/2}. \quad (6.73)$$

Note that using the above approximate delay expression for each gate results in a much larger variation in path delay. The expression fails to capture the averaging effect of independent intra-die variations and assumes that variations across gates are perfectly correlated. However, if intra-die variation is small then the overall impact on delay is also small, and [87] found that the above approach provided an additional average savings in area of

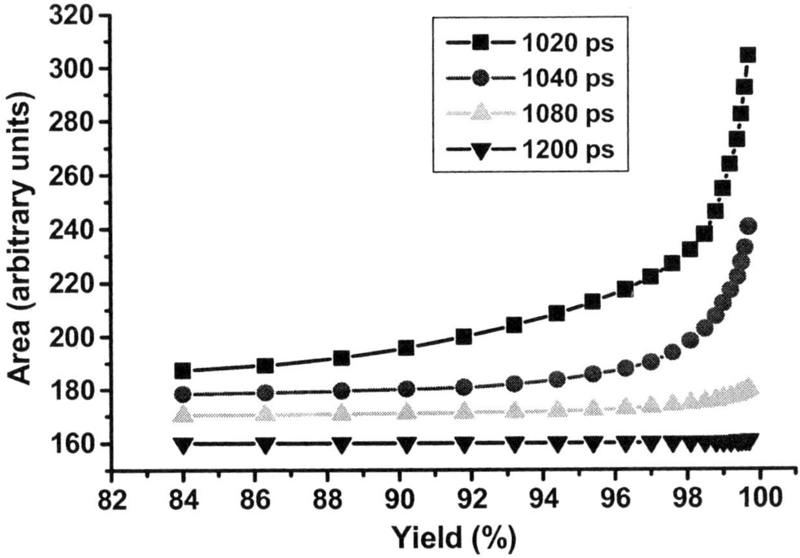

Fig. 6.17. Sensitivity of design area to desired yield for different timing constraints [87]. (©2005 IEEE)

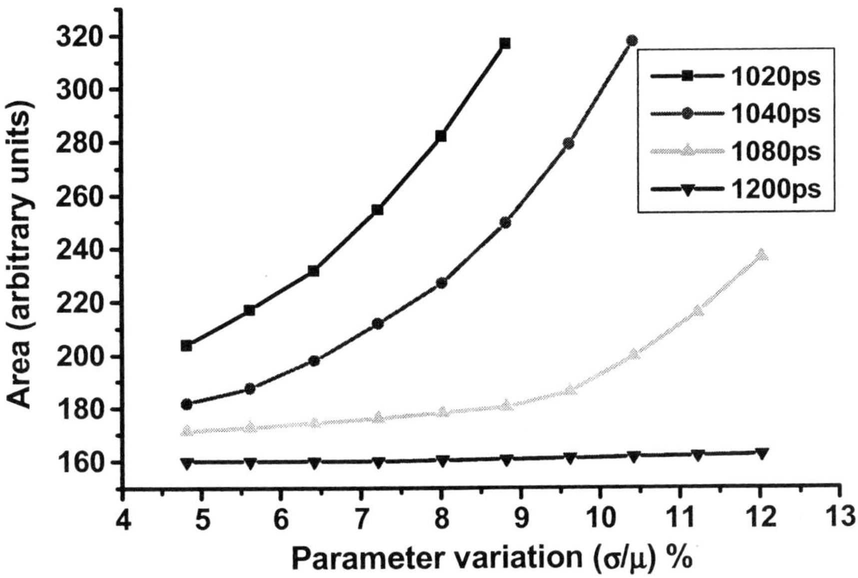

Fig. 6.18. Sensitivity of design area to variation in process parameters for different timing constraints [87]. (©2005 IEEE)

20% compared to deterministic optimization. The results also showed that the difference in the area of the final design for different yield constraints is significant for tightly constrained designs as shown in Fig. 6.16. The figure also indicates that tight timing constraints become infeasible for certain yield levels, and that designing for the nominal case results in extremely low yields. The sensitivity of area to yield for different timing constraints is also shown explicitly in Fig. 6.17, which shows that this sensitivity is very strong for tight timing constraints and grows as the desired yield increases. The amount of variation in process parameters also plays a key role in determining the area; this sensitivity is seen in Fig. 6.18. Again, we find that this sensitivity is strongest for tight timing constraints. Based on these results, we can convincingly say that statistical design techniques will be extremely important for highly performance constrained designs.

6.2.5 Sensitivity-Based Optimization

The technique proposed in [7] uses an accurate estimate of the sensitivity of the delay pdf to sizing changes of each gate. The gate that provides the maximum improvement (corresponds to the maximum sensitivity) is then sized up, and this process is repeated while the desired timing yield is not met. This work uses the discretized pdf approach for statistical timing analysis as discussed in Chap. 3, and can be used to minimize objective functions based on the profit functions discussed previously. This directly corresponds to the maximization of earned revenues. However, for ease of exposition and generality we will use the 99th percentile point of the cdf as the objective function.

The straightforward way to implement this approach is to iteratively perturb the size of each gate in the original unperturbed circuit and then perform statistical timing analysis on the perturbed circuit, which propagates the impact of upsizing the gate on gate delay to the output node. Based on the statistical timing analysis results we can estimate the sensitivity of each gate, which is defined as

$$S = \frac{\Delta T_{99}}{\Delta w} \tag{6.74}$$

where ΔT_{99} is the improvement in the 99th percentile point of the circuit delay cdf and Δw is the change in gate size. However, we soon run into runtime issues. Each upsizing move is preceded by a computation of the sensitivity for all gates. Therefore, for each upsizing move we have a runtime complexity of $O(N^2E)$. If $O(N)$ upsizing moves are performed, then the overall complexity of sizing is $O(N^3E)$. This quickly becomes untenable with increasing design size. To counter this problem, [7] introduced pruning techniques that identify gates that have a smaller sensitivity without running a full statistical static timing analysis (SSTA) run.

Consider Fig. 6.19, which shows the delay cdf corresponding to the unperturbed circuit (C). After a gate has been perturbed, the new cdf is as

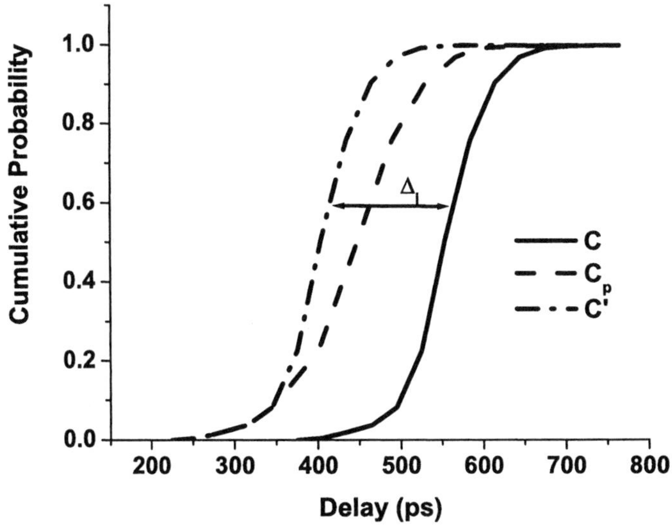

Fig. 6.19. Arrival time cdfs at the output of the candidate gate for the perturbed and unperturbed circuit and the lower bounding cdf for the perturbed circuit.

illustrated by C_p. Note that the perturbation in gate size results in both a shift and as a change in the shape of the cdf. We wish to define the perturbed cdf as simply a shifted version of the unperturbed pdf. Therefore, we define a lower bound for the perturbed cdf, which is illustrated as C' in Fig. 6.19. Note that the lower bound is defined so that the maximum difference between the unperturbed cdf and the perturbed cdf Δ_i is the same as the uniform difference between the unperturbed cdf and the lower bound of the perturbed cdf. This will be crucial in using these lower bounds to perform pruning. Note that this difference is well defined since we are dealing with discretized cdfs, and is not amenable to timing analysis techniques where continuous probability distributions are used. Now we can define C' in terms of C as

$$C'(t - \Delta_i) = C(t). \tag{6.75}$$

Also, since the shape of the two cdfs is identical, the same relation holds true for the pdfs corresponding to these two cdfs. Let us consider the impact of the basic operations of statistical timing analysis (*sum* and *max*) on the maximum difference between two cdfs of the same shape as they are propagated through a circuit. Recall that in the case of a sum operation, we convolve the arrival time cdf at the input of a node with the pdf of the node delay, to obtain the arrival time cdf at the output of the node. Let C_i and C_i' correspond to input cdfs, and C_o and C_o' correspond to the output cdfs. We can then write that

$$C_o(t) = \int_0^\infty C_i(t-\tau)p(\tau)\mathrm{d}\tau = \int_0^\infty C_i'((t-\Delta_i)-\tau)p(\tau)\mathrm{d}\tau$$
$$= C_o'(t-\Delta_i) \tag{6.76}$$

where p represents the node delay pdf. This implies that the sum operation maintains the difference between the two cdfs. Now let us consider the max operation and define two input cdfs as C_1 and C_2 and their corresponding perturbed cdfs as C_1' and C_2'. Let us assume that these cdfs are related as

$$C_1'(t-\Delta_1) = C_1(t) \quad C_2'(t-\Delta_2) = C_2(t) \tag{6.77}$$

and also assume that $\Delta_1 \geq \Delta_2$. Recall that the output cdf in the case of a max operation is the product of the input cdfs. Therefore

$$C_o(t) = C_1(t)\,C_2(t) \quad C_o'(t) = C_1'(t)\,C_2'(t). \tag{6.78}$$

We define another lower bound for C_2 that is lower than C_2' and is defined as

$$C_2''(t) = C_2'(t+\Delta_1). \tag{6.79}$$

Note that for all t, C_2'' has a larger value than C_2'. Using this we can write

$$C_o(t) = C_1'(t-\Delta_1)C_2'(t-\Delta_2) \leq C_1'(t-\Delta_1)C_2''(t-\Delta_1) = C_o'(t-\Delta_1) \tag{6.80}$$

which implies that the separation between the cdfs at the output Δ_o is less than Δ_1. This result, combined with the result obtained for the sum operation, indicates that the separation in the perturbed and unperturbed cdf is at its maximum at the perturbed node and decreases monotonically as the two cdfs are propagated through the circuit.

However, in the case of multiple-fanout nodes multiple perturbed pdfs are generated that will later recombine. Therefore, we need to define a *perturbation front*, which is the set of nodes that are at the same maximum edge distance from the perturbed node, where the maximum edge distance is defined as the length of the longest path from one node to the other. For each perturbation front, the maximum separation for the cdfs on any of the nodes is defined to be the separation between the two cdfs. This maximum separation is then used to define the sensitivity of the gate at this stage of cdf propagation.

Based on these results, we can now define a pruning criterion. The sensitivity of a node i as defined in (6.74) is bounded above by $\Delta_i/\Delta w_i$, where Δ_i is the maximum separation in the perturbed and the unperturbed cdfs at the fanout node. Moreover, we know that the separation of the cdfs decreases as we propagate the cdfs through the circuit graph. Therefore, at any stage of propagation if we find that the separation corresponds to a sensitivity

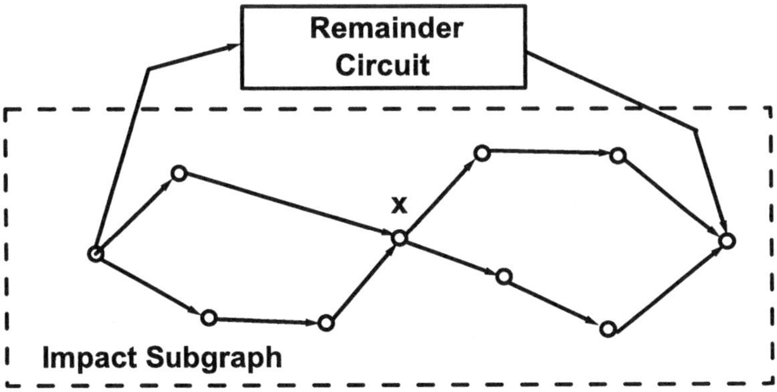

Fig. 6.20. Partitioning the circuit for efficient heuristic calculation of sensitivities.

value that is smaller than the sensitivity value of another gate (that has been previously computed), we can prune out the sensitivity calculation for this node. It is obvious that for this technique to provide good pruning we need to identify a gate with a high sensitivity early in the sensitivity computation step. A reasonable approach is to propagate the perturbed cdfs for all gates in an iterative manner. In each iteration, the perturbed cdfs for all gates are propagated towards the output node through one level. As soon as one of the propagated cdfs reaches the output, its true sensitivity is calculated and it can be used to prune out other cdfs that are still propagating to the output.

To further reduce the runtime associated with this approach, [7] uses an additional heuristic to estimate the separation in the cdfs of the perturbed and unperturbed circuits. An initial SSTA run is used to propagate the cdfs for the arrival time and required arrival times. This is achieved by partitioning the circuit as shown in Fig. 6.20 for each node x. The circuit is partitioned into a shaded region called the *impact subgraph*, which includes the fanin and fanout cone of the node x. The delay pdf of the impact subgraph can then by obtained by convolving the arrival time pdf at node x, A_{xf}, and the required arrival time pdf at node x, A_{xb}. Now, if we resize node x then the new arrival time pdf A'_{xf} can be obtained by performing a local statistical timing analysis step, where the nodes that are the immediate fanins of node x along with node x are re-analyzed. Since A_{xb} is obtained by backward traversal, resizing a node x does not impact A_{xb}. The new delay pdf for the impact subgraph can now be obtained by convolving A'_{xf} with A_{xb}. Note that the change in the pdf for the impact subgraph is not a direct a measure of the change in circuit delay. Ideally, the change in circuit delay can be obtained by convolving the delay pdfs of the perturbed impact with the delay pdf of the *remainder circuit*. However, [7] proposes a heuristic to convolve the impact

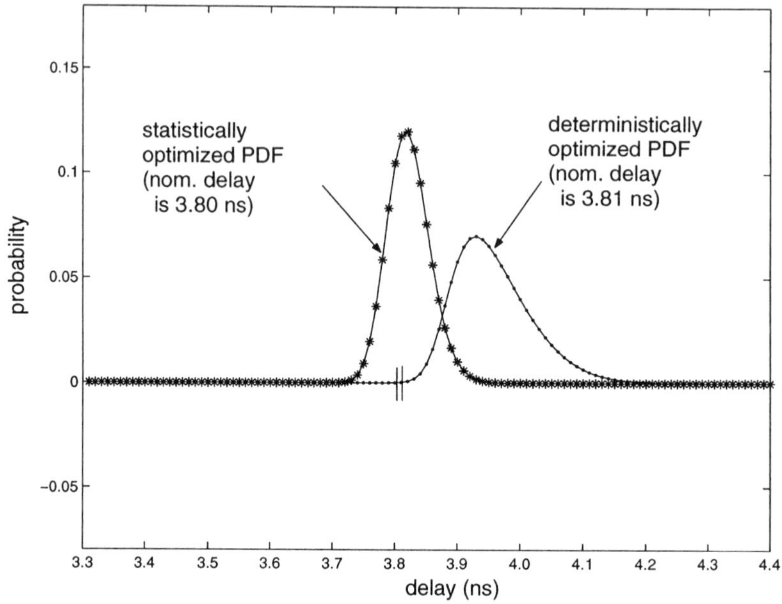

Fig. 6.21. Pre- and post-sizing gate delay pdf for c880

subgraph delay with the delay pdf of the complete circuit. This heuristic is found to work well because of the following property. Consider the case that the delay of the impact subgraph is small as compared to the delay of the complete circuit. Then small changes in the delay pdf of the impact subgraph due to perturbations in gate sizes will have a strongly diminished impact on the final circuit delay pdf, resulting in a smaller sensitivity.

The heuristic approach is inaccurate but fast, so this approach can be used to select a small fraction of the nodes that have a higher sensitivity. The pruning based accurate technique can then be used to select the node with the maximum sensitivity. This approach shows an approximate speedup of 89X while providing the same performance as the exact approach. The approach was found to provide an average improvement of 7.6% over all ISCAS'85 benchmark circuits with a maximum improvement of 16.5%.

Figure 6.21 compares the delay pdf for the ISCAS'85 benchmark circuit c880 before and after gate sizing. The sized circuit has an area penalty of 33%, but it shows a much smaller mean delay as well as a much smaller variation in delay. This improves the yield as well as the robustness of the circuit to process variability.

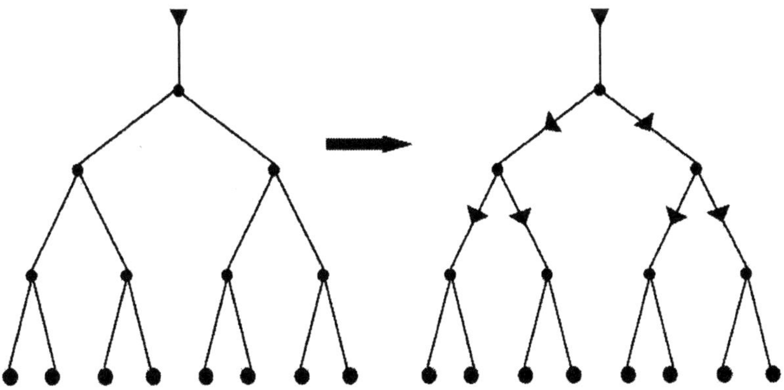

Fig. 6.22. Buffer insertion is used to reduce the delay associated with long interconnects.

6.3 Buffer Insertion

With process technology scaling, wire delay has grown to contribute a sizable fraction to total circuit delay. Buffer insertion and wire sizing are two well-known techniques that are used to reduce wire delay. The delay of a wire is proportional to the product of the resistance and capacitance of the wire. Since both the capacitance and resistance are proportional to the length of the wire, the delay of the wire exhibits a quadratic dependence on the wire length. Buffer insertion divides the wire into a number of segments and renders the wire delay linearly proportional to the length of the wire [106]. Additionally, buffers are also used to decouple large loads from the critical path of the circuit.

The buffer insertion problem can be stated formally as: given a routing tree with a single source and multiple sinks, find the set of edges of the tree to be buffered to minimize the delay of the interconnect tree (shown in Fig. 6.22). Van Ginneken [145] presented an optimal buffer insertion dynamic programming based approach. The approach has quadratic complexity in the number of possible choices for buffer locations, assuming a single buffer in the library. This approach has been the foundation for later work in the area of buffer insertion. However, nearly all these approaches neglect the impact of variations in process parameters and the lack of exact design information early in the design cycle. In this section, we provide a quick overview of Van Ginneken's approach and then discuss an extension to perform statistical buffer insertion proposed in [73], which relaxes the assumption that the exact wire lengths of the interconnects are known and instead treats them as RVs.

6.3.1 Deterministic Approach

The dynamic programming approach proposed by Van Ginneken uses the Elmore delay model while modeling a distributed RC line with the capacitor at the center of the line. The inputs to the algorithm consist of an unbuffered interconnect tree, a set of B potential locations on the tree for buffer insertion, and the delay and loading characteristics of a buffer.

The algorithm then traverses the interconnect structure from the N leaf nodes of the structure to the source while maintaining a set of solutions for each node, represented by the pair (L, D) where L is the capacitive load seen by the node and D is the delay of the subtree rooted at that node assuming that the subtree is driven by a drive with infinite strength. Suppose that a wire of length l is attached at a node n, then the solutions at the root of the attached wire can be obtained by

$$D'_n = D_n + rlL_n + \frac{rcl^2}{2} \qquad L'_n = L_n + cl \tag{6.81}$$

where r and c represent the wire resistance and capacitance per unit length, respectively. If a buffer can be inserted at node n, then an additional solution is generated. This solution is given as

$$D'_n = D_n + D_{\text{buf}} + R_{\text{buf}} L_n \qquad L'_n = C_{\text{buf}} \tag{6.82}$$

where D_{buf}, R_{buf} and C_{buf} are the intrinsic delay, drive resistance and input capacitance of the buffer, respectively. Similarly, let subtrees rooted at nodes m and n merge at node k, then the solutions at k are obtained as

$$D_k = \max(D_n, D_m) \qquad L_k = L_n + L_m. \tag{6.83}$$

Using these basic operations, solutions at the root of the tree can be generated. The solution with the minimum delay is selected and then retraced back through the tree to determine the buffering configuration that led to the minimum delay. A naive implementation of the approach generates an exponential number of solutions since the number of solutions at the root of two subtrees with $O(m)$ and $O(n)$ can be $O(mn)$, corresponding to all possible combinations of solutions. Van Ginneken's approach prunes out non-optimal solutions while traversing the tree and only generates $O(m+n)$ solutions corresponding to the above situations, and therefore runs in polynomial time (the actual complexity is $O(B^2+N)$). Van Ginneken notes that a solution is provably suboptimal if it has both a larger delay and capacitive load compared to another solution. Therefore, when we merge two subtrees (6.83) there are only $n+m$ possible delay values in the solutions, since the delay value should correspond to the delay of either of the subtrees. The minimum loading solution for each of the delay values prunes out all other solutions. Thus, a merge operation only results in a number of solutions equal to the sum of the number of solutions for each subtrees. Additionally, if a node is a possible candidate for buffer insertion, then the number of solutions is increased by only one. This results from the fact that the load capacitance for all solutions corresponding to a buffered solution is the same (C_{buf}), and the solution with the minimum delay will prune out all other solutions. Thus, the number of solutions at the root of the interconnect tree will be only $B+1$. Now we discuss an extension of this approach in a statistical scenario, which was proposed in [73].

6.3.2 Statistical Approach

The deterministic approach discussed above assumed the availability of exact wire lengths. However, the lack of information regarding the low level layout as well as variations in process parameters force us to model the wire lengths as RVs instead of deterministic numbers. If the exact wire length is not known, then an average of worst-case estimates of wire length could be used in the above approach. However, handling process variations using average values results in a large probability that the interconnect tree will fail to meet the timing constraint. On the other hand, if the worst-case estimates are used a much larger design effort is required to satisfy the timing constraints, resulting in an insertion of a large number of buffers and the corresponding adverse power implications. In [73], the delay and loading that comprise a solution in Van Ginneken's approach are represented as distributions. Thus, in a delay-capacitance space, instead of dealing with point solutions we can represent these variability-aware solutions as rectangles whose sides correspond to the interval of delay and capacitance for this solution. The solutions can again be combined using the same equations used in the deterministic scenario. However, the pruning technique, which provides polynomial complexity and optimality, cannot be applied in a statistical scenario and new pruning techniques are required.

The first pruning technique is to remove solutions that are worse both in terms of delay and capacitance. This happens when the rectangle corresponding to one of the solutions does not overlap the other rectangle in any dimension as shown in Fig. 6.23. A heuristic approach to further reduce the number of solutions was proposed in [73], which is based on a probabilistic comparison of two solutions. A probabilistic pruning measure is defined as

$$\mathcal{P}(A \Rightarrow B) \equiv \mathcal{P}(\text{A prunes B}) = \mathcal{P}(D_A \leq D_B, L_A \leq L_B) \tag{6.84}$$

which is approximated by the product of the probability that $D_A \leq D_B$ and that $L_A \leq L_B$. A symmetric expression can be written for the probability that B prunes A. The probability that both the solutions co-exist can then be written as

$$\mathcal{P}(A \sim B) \equiv \mathcal{P}(\text{A coexists with B}) = 1 - \mathcal{P}(A \Rightarrow B) - \mathcal{P}(B \Rightarrow A). \tag{6.85}$$

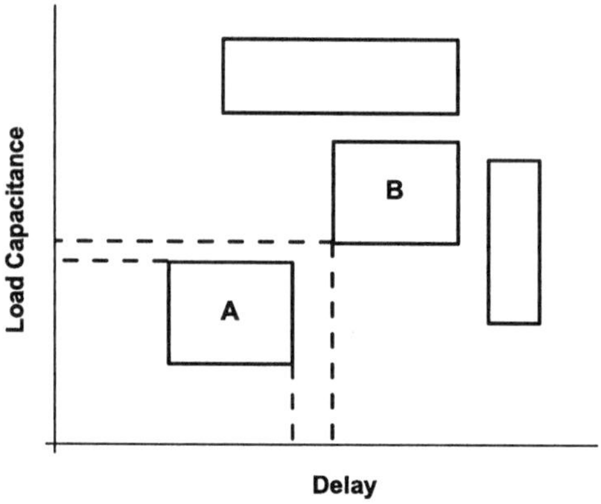

Fig. 6.23. Solution B is always worse than Solution A, and hence, can be pruned out.

The max of the three probabilities ($A \Rightarrow B$, $B \Rightarrow A and A \sim B$) defines the relation between these two solutions. The problem of selecting a set of solutions such that the maximum number of solutions can be pruned is mapped to a *DIRECTED MAXIMAL INDEPENDENT SET* problem. This is achieved by constructing a directed graph where each node corresponds to a solution, and an edge from node A to node B exists if node A prunes node B. Additionally, the cost associated with each node is defined to be the out-degree of the node and represents the number of solutions that will be pruned by selecting this node. The solution of the problem is a set of nodes such that no two nodes in the set are connected by an edge. Each node not in the set has an edge from one of the nodes in the set and can be pruned out and the sum of the costs of the nodes in the set is maximized. This problem is known to be NP-complete and is solved heuristically by iteratively assigning a node to the set from the graph that has the maximum cost and then deleting all nodes in the graph that have a directed edge from this node. This process is repeated as long as there are nodes in the graph. Though these techniques can be used to reduce the number of solutions, none of these techniques guarantee that the algorithm will run in polynomial time.

The exponential increase in the number of solutions occurs when we are not able to effectively prune out solutions while merging subtrees. Another pruning technique proposed in [73] is based on the fact that the starting time for the solutions that are generated when merging trees with m and n solutions are $m + n$. Therefore, if we select one solution for each starting point we obtain an algorithm with polynomial complexity. This can be achieved by solving the *COMPLETE R-PARTITE MAX COST CLIQUE* [12] problem on an undirected graph, which is generated as follows. Each solution corresponds to a node in the graph, and the nodes that correspond to the same

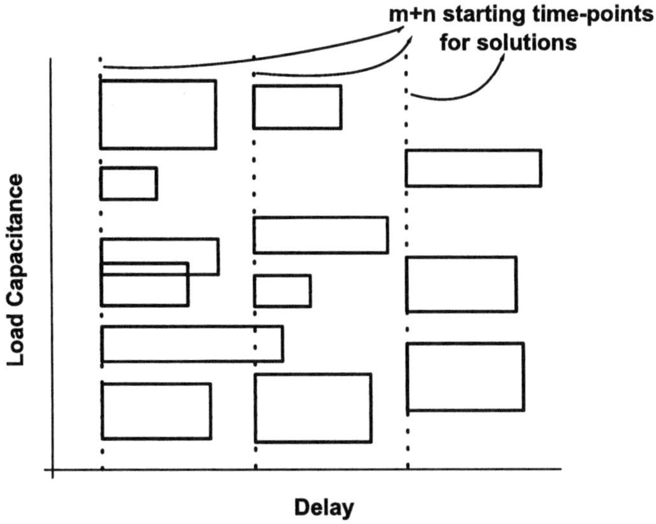

Fig. 6.24. The solutions obtained by merging subtrees with m and n solutions can be partitioned into $m + n$ sets based on starting points in time.

starting point in time are not connected by any edges. All other nodes have edges in between them, where the weight of the edge is probability of the two solutions being co-optimal (6.85). The solution of the problem then identifies a clique within this graph with the maximum cost, and therefore, has one solution for each of the $m + n$ starting time points as shown in Fig. 6.24. Moreover, the selected solutions are such that the solutions have the highest probability of co-optimality, implying that the larger solution space is covered by these solutions. However, the problem of identifying the maximal clique in a complete r-partite graph is NP-complete and needs to be solved heuristically. The graph is first levelized according to the starting time-points of the solution and is then traversed in a levelized fashion. For each node in a set, a set of cliques is formed by merging it with potential solutions from the previous levels. The solution with the maximum cost is then assigned to be the solution at this node and is stored. Similar to the deterministic case, the solutions corresponding to the addition of a buffer are actually lines, since the load capacitance is the same for all solutions. The probability that a given solution prunes out all other solutions is calculated based on (6.85), and the solution with the maximum probability is retained.

The three pruning techniques described above provide a clear trade-off in terms of runtime and optimality. The first technique prunes out the minimum number of solutions and is guaranteed to be optimal. The second pruning technique is neither optimal nor polynomial but is much more efficient in reducing the number of solutions compared to the first pruning technique. Finally, the third pruning technique has polynomial complexity but it does not guarantee optimality and is extremely aggressive in pruning out solutions. The results provided in [73] show that deterministic Van Ginneken results in a buffered interconnect tree that has a large probability of timing failure. Probabilistic optimization result in in the timing constraints being satisfied with a very high probability. Interestingly, the number of buffers inserted using deterministic and probabilistic techniques was found to be similar; however, buffer locations were different.

6.4 Threshold Voltage Assignment

Traditional leakage power optimization has been performed using assignment of non-critical sections of the circuit to a higher threshold voltage. Due to the exponential dependence of leakage power on threshold voltage a large reduction in leakage power can be achieved [147]. Since higher threshold voltages are associated with a significant delay penalty, a number of approaches have used gate sizing to increase the fraction of gates that can be assigned to the higher threshold [128][100][108][150][70]. However, these approaches do not consider variation in process parameters, and therefore have a negative impact on the parametric yield of a design.

The tremendous impact of variability on leakage currents was demonstrated in [20], which showed 20X variation in leakage power for 1.3X variations in delay between fast and slow dies. This shows that statistical information should play a key role while optimizing leakage power. Given the strong impact of variations on leakage, any optimization that seeks to optimize leakage power through techniques such as dual-V_{th} while neglecting variations will invariably result in yield loss. Figure 6.25 compares the pdfs and cdfs of a pre- and post-optimized design. The pre-optimized design refers to a design optimally sized to meet a delay target with just one threshold voltage. The design is then optimized for leakage power using an additional threshold voltage [128] while nominal delay is constrained to remain identical. Note that the post-optimized pdf exhibits many more paths at the slow-end of the distribution, which indicates a parametric yield loss. Based on this figure, we conclude that it is important to devise optimization approaches that make use of available statistical information to simultaneously maintain timing yield while improving power dissipation. Furthermore, the use of dual-V_{th} devices increases the susceptibility of the design to variations due to strong leakage variations in low V_{th} devices and a high delay sensitivity in high V_{th} devices.

6.4.1 Sensitivity-Based Optimization

The technique proposed in [133] uses a heuristic approach to extend deterministic sensitivity-based dual-V_{th} techniques to the statistical domain. The first

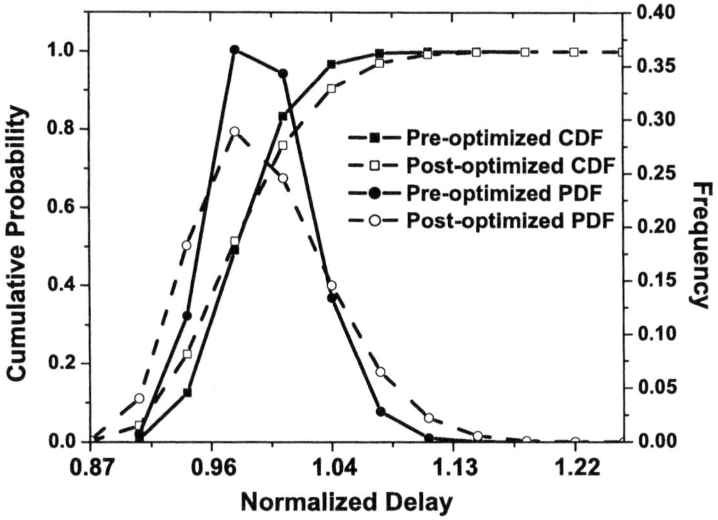

Fig. 6.25. Impact of deterministic dual-V_{th} optimization approaches on delay pdf and cdf showing timing yield loss post-optimization.

step towards statistical optimization is to use statistical analysis techniques rather than deterministic analysis techniques. This work uses the statistical timing analyzer discussed in Sec. 3.2.2 [6]. This work also introduces the concept of *statistical sensitivities*, which are used to perform sensitivity-based statistical optimization. The optimization is performed using a standard cell library, where each gate is characterized for delay and power. Leakage power is independent of capacitive loading and input transition time and is modeled by

$$P_{\text{leak}} = g\left(L_{\text{gate}}\right) = a_0 \exp\left(\frac{-L_{\text{gate}}}{a_1}\right) = \exp\left(\frac{-L_{\text{gate}}}{a_1} + \ln a_0\right) \tag{6.86}$$

where L_{gate} is gate length and a_0 and a_1 are fitting parameters that can be estimated using linear regression. Delay is modeled using a look-up table (indices are load capacitance and input transition time). For each index a delay model of the form

$$D = f\left(L_{\text{gate}}\right) = a_0 + a_1 L_{\text{gate}} + a_2 L_{\text{gate}}^2 \tag{6.87}$$

where a_0, a_1 and a_2 are fitting parameters, is used to estimate the mean and variance of gate delay, which is then fed to a statistical timing analyzer. Note that the indices need to be deterministic, and therefore, using this modeling approach requires a deterministic timing analysis run before statistical timing analysis can be performed. Let us first review a modified form of the deterministic dual-V_{th} technique [128] that will be extended to consider variations.

Deterministic Approach

The deterministic approach is a different form than the approach proposed in [128]. The initial design, which operates at the lower $\mathrm{V_{th}}$ exclusively, is first sized to meet the timing constraint using a TILOS-like optimizer [52]. A sensitivity measure is defined as

$$S_{\text{gate}}^{\text{swap}} = \frac{|\Delta P|}{|\Delta D|} Slack_{\text{gate}} \tag{6.88}$$

and evaluated for all low V_{th} gates in the circuit. ΔP and ΔD in (6.88) are the changes in power dissipation and delay of the gate when the low V_{th} gate is swapped with a high V_{th} gate (of same size and functionality). The gate with the maximum sensitivity is then swapped with a high V_{th} version of the gate. If the circuit now fails to meet timing, a new sensitivity measure is defined as

$$S_{\text{gate}}^{\text{up-size}} = \frac{1}{\Delta P} \sum_{\text{arcs}} \frac{\Delta D}{Slack_{\text{arc}} - S_{\text{min}} + K} \tag{6.89}$$

where S_{min} is the worst slack observed in the circuit and K is a small positive quantity to maintain computational stability. Equation (6.89) is then evaluated for all gates in the circuit. This form of the sensitivity metric places a higher weighting to gates lying on the critical paths of the circuit. The arcs over which the summation is taken represent the falling and rising arcs associated with each of the inputs of the gate. Thus, for a 3-input NAND gate the sensitivity measure will be obtained by summing over all six possible arcs. The ΔP and ΔD in this case are the change in power and delay when the gate is upsized to the next available size in the library. The gate with the maximum sensitivity is then upsized, and the process is repeated until either the circuit meets timing or the power dissipation increases relative to its level prior to gate G1 being set to high V_{th}. In the case of the latter event, gate G1 is set back to high V_{th} and is flagged to prevent the gate from being reconsidered for high V_{th} assignment later in the optimization. A summary of the steps in this deterministic dual-V_{th} is as follows:

deterministic dual-$\mathbf{V_{th}}$
STEP 0: Perform timing and power analysis; $Power_0$=Power
STEP 1: Calculate sensitivity (S^{swap}) of low V_{th} gates
STEP 2: Set gate with maximum S^{swap} to high V_{th}
STEP 3: Perform timing analysis
STEP 4: *if* circuit meets timing *goto* STEP0
STEP 5: Calculate sensitivity $S^{\text{up-size}}$ for all gates
STEP 6: *up-size* gate with maximum $S^{\text{up-size}}$
STEP 7: Perform timing and power analysis
STEP 8: *if* ($Power > Power_0$) undo moves and *goto* STEP 0
STEP 9: *if* timing is met *goto* STEP 0
STEP 10: *goto* STEP 5

Statistical Approach

The statistical dual-V_{th} and sizing problem can be expressed as an assignment problem that seeks to find an optimal assignment of threshold voltages (from a set of two thresholds) and drive strengths (from a set of drive strengths available in a standard cell library) for each of the gates in a given circuit network. The objective is to minimize the leakage power measured at a high percentile point of its cdf while maintaining a timing constraint imposed on the circuit. The timing constraint is also expressed as a delay target for a high percentile point of the circuit delay. These timing and power constraints can be determined based on desired yield estimates, such as 95% or 99%. This formulation serves to simplify the problem and allows traditional iterative optimization approaches to be easily adapted to statistical optimization.

Two major enhancements to the previously described deterministic approach need to be incorporated to enable statistical leakage optimization. The first enhancement requires that the timing check in STEP3 of the deterministic dual-V_{th} approach is performed using statistical timing analysis. The required percentile point on the delay cdf used to specify the constraint should now be obtained from the pdfs generated by the SSTA engine rather than a corner model case file.

A deterministic timing analyzer can be used to determine the input slope at each of the gates, which can then used along with the output capacitance as indices if look-up table based delay models are used. The mean and variance are estimated using (6.86)-(6.87), which are then passed onto the SSTA engine to evaluate the cdf of the arrival and required times at each circuit node. Note that while performing the statistical timing analysis additional dummy source and sink nodes are added to the circuit, hence the delay constraint needs to be checked at just one point within the network. As discussed in Chap. 3, using statistical delay analysis reduces the pessimism in timing since all gates cannot be expected to be simultaneously operating at their worst-case corners, an assumption that is inherently made when performing a corner-based worst-case analysis. We will see later that optimizing a circuit to meet a delay constraint using worst-case analysis results in a substantial loss in circuit performance optimality. The situation is worsened for leakage power optimization because of the exponential dependence of leakage power on threshold voltage.

The second enhancement uses the statistical information in the fitting functions of delay and power to guide the optimization by replacing the sensitivities evaluated in STEP1 and STEP4 with statistical sensitivities. These statistical sensitivities are then evaluated at a confidence point on the PDF of the sensitivity. Since generating pdfs of the sensitivity metrics themselves is fairly complicated and computationally intensive, we estimate the statistical sensitivities by evaluating the mean and standard deviation of these pdfs (i.e., we only concern ourselves with the first and second central moments of the sensitivity pdfs and not their entire shape). Also, the dependence of slack on gate length of the devices is not straightforward and we make the assumption that the slack is independent of gate length while calculating the moments of the sensitivities. The sensitivities in (6.88)-(6.89) can now be expressed as a product of two independent random variables X and Y where X is dependent on L_{gate} and Y is not. Thus, X corresponds to the ratio of the change in power and change in delay, and Y corresponds to the slack dependent terms in (6.88)-(6.89). Given two independent random variables X and Y, the expectation of their product is the same as the product of their expectation. Using this fact, we can estimate the mean and standard deviation of the sensitivities using the independence assumption made above and the following relations:

$$\begin{aligned} E[XY] &= E[X]\,E[Y] \\ Var[XY] &= E\left[(XY - E[XY])^2\right] = E\left[X^2\right]E\left[Y^2\right] - E^2[XY] \end{aligned} \tag{6.90}$$

where $E[X]$ is the expected value of X alone and $E[Y]$ is the expected value of Y alone. The mean and variance of the terms involving L_{gate} (X in (6.90)) are expressed as a function f of L_{gate} alone, using the delay and power models (6.86) and (6.87). The expected value is then written as

$$E[f(L_{\text{gate}})] = \int_{-\infty}^{\infty} f(L_{\text{gate}})\, p(L_{\text{gate}})\, \mathrm{d}L_{gate} \tag{6.91}$$

where $p(L_{\text{gate}})$ is the pdf of gate length. Applying Taylor series expansion to the above expression we can rewrite it as

$$\begin{aligned} E[f(L_{\text{gate}})] = \int_{-\infty}^{\infty} \Bigg(& f(\mu) + f'(\mu)(L_{\text{gate}} - \mu) + \cdots \\ & + f_{\mu}^{n} \frac{(L_{\text{gate}} - \mu)^n}{n!} \Bigg) p(L_{\text{gate}})\, \mathrm{d}L_{\text{gate}} \end{aligned} \tag{6.92}$$

where μ is the mean of L_{gate}. This gives

$$E\left[f\left(L_{gate}\right)\right] = f(\mu) + \frac{f''(\mu)\eta_2}{2!} + \cdots + f^{2n}(\mu)\frac{\eta_{2n}}{2n!} \tag{6.93}$$

where η_i is the ith central moment of L_{gate} and the odd central moments of L_{gate} are set to zero. This approximation can be used to obtain the mean and variance of the sensitivities. For our analysis, we found that a fourth-order approximation of $f(L_{\text{gate}})$ was sufficient for good accuracy.

The moments of the slack dependent terms (Y in Equation (6.90)) are estimated using the slack pdfs obtained from the SSTA engine. The statistical sensitivities are now redefined by evaluating at n standard deviations away from the mean. Since the shape of the sensitivity pdfs is not known, n is not known even if a known confidence point is desired. We will later look at the impact of n on the optimization results. The approach can also be easily extended to multiple delay constraints, where a set of percentile points on the delay PDF can be constrained to be less than some desired values. As an example, this flexibility is well suited to microprocessor designs where we can simultaneously constrain the 95th and 99th percentile delay to concurrently target different yields for different performance bins.

The worst-case time complexity of the algorithm can be expected to be $O(n^3)$ since the SSTA engine has a linear time complexity [6] and in the worst-case we may up-size the entire circuit each time we set a gate to high V_{th}. This would happen when we maximally size-up the circuit each time we set a gate to high V_{th} yet still fail to meet timing (this also requires the total power not to surpass the original circuit through all up-sizing moves). Note that in the worst-case the $O(n^3)$ complexity results because the total number of up-sizing moves (and reversed up-sizing moves) is $O(n^2)$ since every gate is up-sized to the maximum size available in the library whenever a gate is set to high V_{th}, and all the moves are then reversed. If the total number of up-sizing moves that are reversed is assumed to be linearly proportional to the number of gates in the circuit, the overall complexity of the algorithm reduces to $O(n^2)$ since the total number of up-sizing or cell-swapping moves now become linearly proportional to the number of gates in the circuit.

The benchmark circuits are synthesized using an industrial 130 nm standard cell library with a V_{dd} of 1.2 V and a high and low V_{th} of 0.23 V and 0.12 V, respectively. For the delay constraints, we consider two different cases where the delay is constrained at the 95th or 99th percentile. Leakage power is optimized at the same percentile point used to express the delay constraint. To make a fair comparison of the statistical and deterministic approaches, the best and worst-case corner models for the gates are developed for the same percentile point at which the delay constraint was specified for a particular experiment (95th or 99th).

Figure 6.26 shows the impact of evaluating the sensitivities at different points along their distribution (relative to the mean) on the final optimization results for two ISCAS'85 [23] benchmark circuits. The sensitivities are evaluated at a fixed number of standard deviations away from the mean, rep-

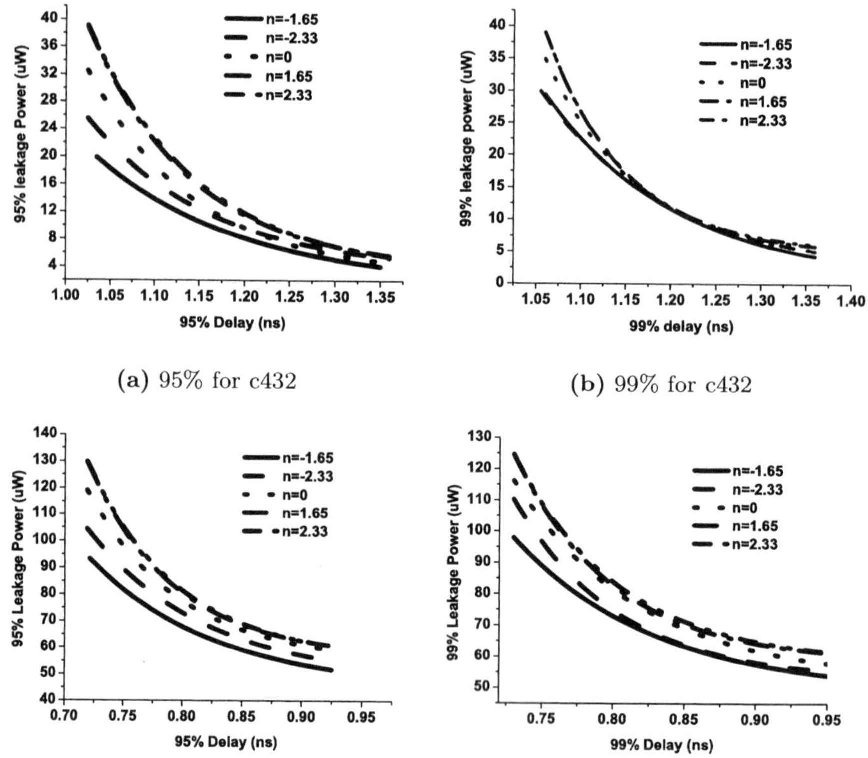

(a) 95% for c432

(b) 99% for c432

(c) 95% for c499

(d) 99% for c499

Fig. 6.26. Impact of n on statistical optimization

resented as n. The curves are obtained through multiple runs of the algorithm. Each time the algorithm is run, the delay constraint is progressively tightened to obtain a complete power-delay curve. For both 95th and 99th percentile delay constraints, we observe that considering $n = -1.63$ (corresponding to the 5th percentile point on a Gaussian) leads to the best power-delay curve characteristics. For the 99th percentile case we observe that both $n = -1.63$ and $n = -2.33$, which corresponds to the 1st percentile point in a Gaussian, perform similarly. The significant improvement over the cases where a high percentile point of the sensitivities is used to select the gate to be swapped/up-sized can be understood by noting that a low percentile point on the sensitivity distribution gives a high confidence that the sensitivity value is at least as large as the value at the decision-making point.

Figure 6.27 compares three different optimization approaches outlined above. In particular we sub-divide the statistical optimization approach into two stages – 1) *with statistical constraints*, which relies on SSTA but does

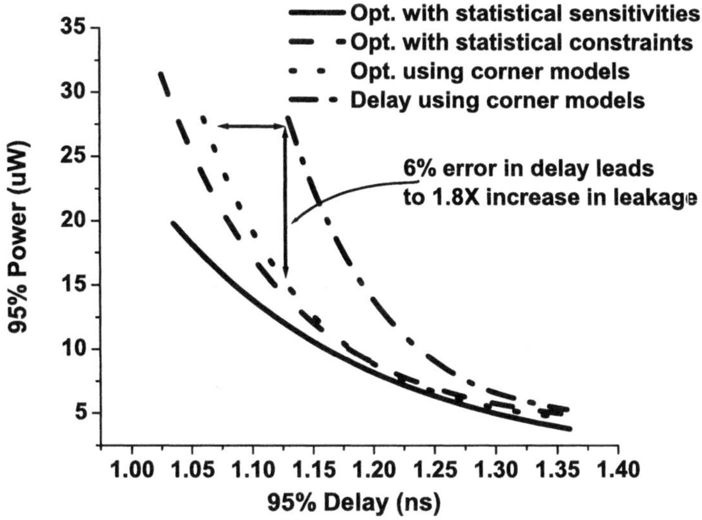

Fig. 6.27. Power-delay curves for the three optimization approaches.

not include statistical sensitivities, and 2) *with statistical sensitivities*, which includes both improvements described above. The 95% delay and 95% power are estimated using the statistical estimation techniques discussed previously for all curves except that labeled *–delay using corner models.* It is interesting to note that the incorporation of statistical sensitivities provides an additional reduction of 40% in leakage power at the tightest delay constraint compared to the case where we only use the SSTA engine to enforce the delay constraint. This indicates that, although the use of a statistical timing analysis framework is clearly important, statistically modeling the power and delay impact of change in V_{th} is equally critical. Additionally, the optimization based on corner models (using the traditional approach) is not able to meet the very tight constraints on the 95th percentile of delay that are met by optimizations that employ an SSTA engine due to the pessimism of the corner model approach.

The last curve in Fig. 6.27 labeled *delay using corner models*, shows the results for the optimization using corner models where the delay is calculated using worst-case models. The curve shows that if statistical information is not provided to the designer a small overestimation in the delay leaves large performance improvements on the table, since designs are generally optimized within a strict delay constraint. Also, high-performance circuits generally operate in a steep region of the power-delay curve and a small overestimation in delay can be expected to result in a large loss in the achievable improvements of the performance parameter being optimized. It can be seen that the

Table 6.1. Power savings for the statistical approaches compared to a corner-model based approach [133].

Circuit	Power(95%)		Power(99%)		Gate	Runtime
	OPT2	OPT3	OPT2	OPT3	count	(min)
c432	16.3%	39.3%	18.0%	35.7%	165	1
c499	4.3%	30.7%	23.9%	30.3%	519	13
c880	9.5%	13.2%	8.0%	50%	390	8
c1908	12.7%	23.6%	11.5%	35.5%	432	10
c2670	5.3%	20.3%	36.8%	45.6%	965	20
c3540	35.3%	43.5%	7.3%	15.3%	962	19
c5315	17.8%	34.3%	31.1%	39.6%	1750	68
c6288	15.0%	26.5%	22.4%	36.1%	2502	115
Average	14.5%	28.9%	19.9%	36.0%		

different optimization cases also tend to converge as the delay constraint is relaxed. This can be understood by noting that as the delay constraint is relaxed, a larger fraction of gates are assigned to high V_{th} and hence the final state becomes increasingly independent of the order in which the gates are assigned to high V_{th}.

Table 6.1 summarizes the improvement in leakage power for the ISCAS'85 benchmark circuits [23] for the statistical optimization approaches described above, compared to a deterministic approach. OPT2 and OPT3 refer to the optimization with statistical timing constraints alone, and with both statistical timing constraints and sensitivities, respectively. The results are shown for the best delay constraint that can be met using the corner models, thus the results in Table 6.1 for the 95th and 99th percentile cases correspond to different delay constraints. Average reductions in leakage power of approximately 14% and 29% can be achieved using OPT2 and OPT3, respectively, for the 95th percentile case compared to a traditional deterministic approach. A larger average improvement of approximately 20% and 36% is observed for the 99th percentile case. These delay points correspond to the high frequency bin and are most affected by leakage power dissipation. The last columns of the table list the size of the circuits and the runtime for the algorithm and we see that the runtime follows the quadratic complexity predicted above.

Figure 6.28 compares the pdf of leakage power for the three optimization approaches for both loose and tight delay constraints. These power curves are all taken with identical 95% delays, or identical performance. For loose delay constraints all three optimization approaches result in fairly similar pdfs for leakage power. This again reflects the fact that the different optimization approaches behave very similarly for loose delay constraints. The

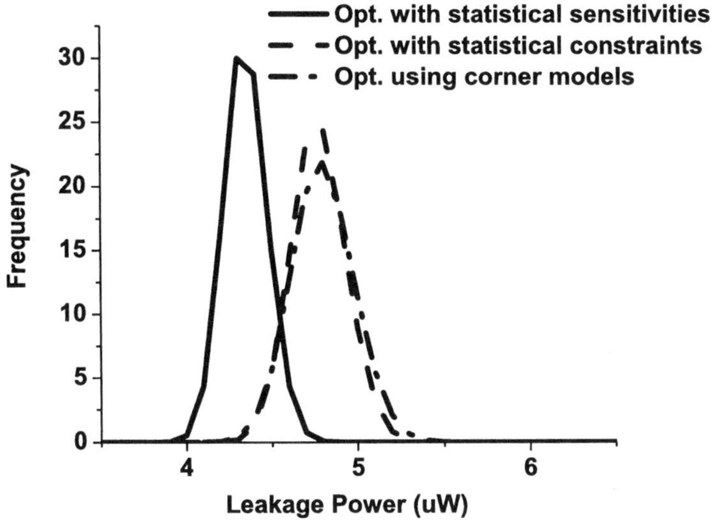

(a)

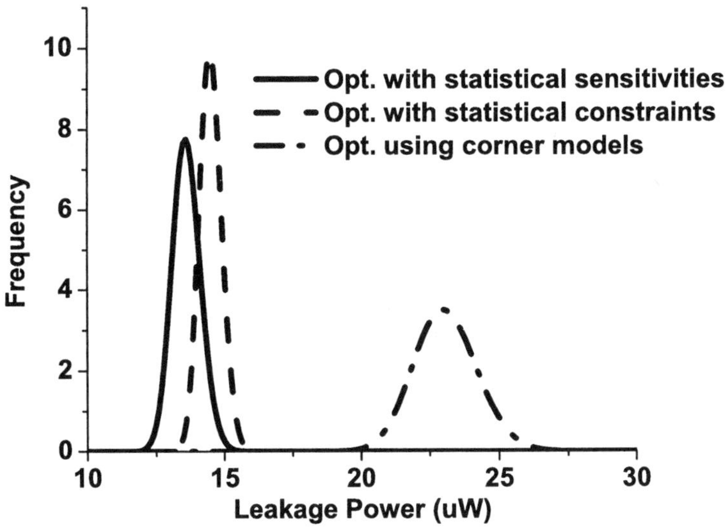

(b)

Fig. 6.28. Probability distribution functions of leakage power for a (a) loose delay constraint (b) tight delay constraint.

tighter constraints clearly separate the leakage power pdfs of the statistical and deterministic approaches. It is interesting, to note that although statistical sensitivities lead to a smaller 95th percentile leakage power compared to the other approaches, the variance is marginally larger when compared to the optimization using only statistical constraints. Fig. 6.28(b) corresponds to the highest performance parts being manufactured and using statistical optimization leads to not only a much smaller average leakage power but also reduces the spread of the distribution considerably, significantly impacting yield. Any statistical timing analyzer can be easily integrated within this approach. Additionally, correlated sources of variations can be considered using the timing and power analysis methodology, as discussed in Chap. 5, to perform gate-level yield analysis. However, note that this approach is not a true parametric yield optimization approach, since even if both the power and timing yield are 95%, the overall parametric yield of the design can be as low as 90% depending on the correlation of power and performance.

6.4.2 Dynamic Programming

Dynamic programming techniques, which are used to solve problems such as buffer insertion, was used to perform dual V_{th} assignment in [42]. Dynamic programming approaches are suitable for tree structures but suffer from the problem that it becomes very difficult to handle reconvergent fanouts, which are found commonly in DAGs used to represent combinational logic. Let us first discuss the approach, while assuming that there is no reconvergence. In this case the DAG is actually a tree. Later we will discuss heuristics presented in [42] to handle reconvergence in DAGs.

In this approach each node of a DAG is associated with a set of possible implementations. Each implementation is associated with a probability distribution for delay and a cost. The cost in dual-V_{th} assignment can be used to represent the expected value of leakage and dynamic power for that particular implementation. The nodes within the DAG are topologically ordered, and then traversed from the primary inputs to primary outputs in a topological fashion. A set of solutions, where each solution is a pair of cost and delay distribution, are obtained at each node of the DAG during the traversal. These solutions form the set of Pareto optimal solutions, and are obtained based on the set of solutions at the inputs of the gate and the different possible implementations of the gate itself.

When a single input gate is encountered the delay distribution at the output can be obtained by convolving the cdf of the delay at the input of a gate with the delay pdf of one of the possible implementations of the gate as discussed in Chap. 3. This gives the output delay cdf as

$$C_{\text{out}}(t) = \int_0^t C_{\text{in}}(t-\tau)\, p_d(\tau)\mathrm{d}\tau \tag{6.94}$$

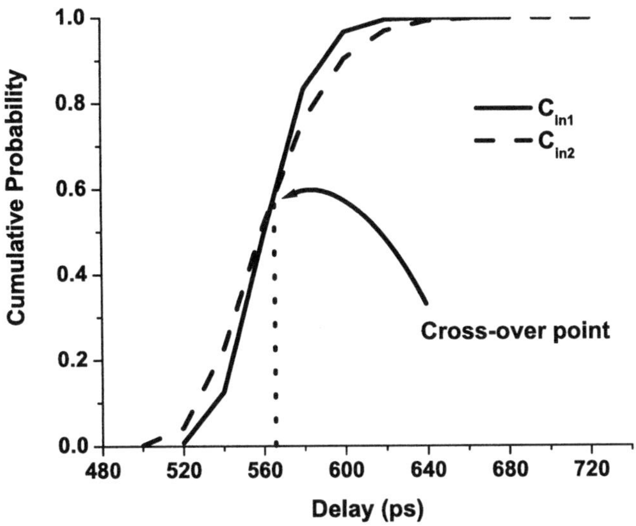

Fig. 6.29. Cross-over point before which C_{in2} has a higher value and after which C_{in1} has a higher value

where C_{out} and C_{in} are the delay cdfs at the output and input of the gate, respectively and p_d is the delay pdf of the gate. The cost at the output is the sum of the input cost and the cost associated with the gate. If the number of solutions at the input is $O(m)$ and the possible implementations for the gate is $O(n)$ then the number of possible solutions at the output of the gate is $O(mn)$. Since this becomes computationally very expensive, [42] proposes a heuristic approach to prune a large number of these solutions, as will be discussed later.

In the case of a multiple input gate, we need to perform a *max* operation before the *sum* operation (6.94) can be performed. As we saw in Chap. 3 the maximum of a set of cdfs can be obtained by simply multiplying the cdf values at the corresponding point on the time axis

$$C_{\text{max}}(t) = C_{\text{in1}}(t)\, C_{\text{in2}}(t) \tag{6.95}$$

where C_{max} represents the cdf of the max of C_{in1} and C_{in2}. The cost associated with the max operation is again the sum of the costs associated with the cdf whose max is being considered. After propagating the set of solutions to the output, a set of Pareto optimal solutions in terms of cost and the probability of timing failure are obtained.

As presented above, the number of solutions increases exponentially as we traverse the DAG. Therefore, pruning techniques need to be employed to reduce the number of solutions at a node. The pruning technique in [42] defines the concept of a cross-over point for delay cdfs. A cross-over point is a point t_0 on the time-axis such that the cdfs which that the maximum value for $t = t_0^-$ and $t = t_0^+$ are different. These cross-over points divide the time-axis into a number of partitions as shown in Fig. 6.29. Note that if cdfs A and B have $O(m)$ and $O(n)$ cross-over points, then the number of cross-over points in the cdf obtained using a max operation is $O(m+n)$. This can be understood by noting that a cross-over point in the merged max cdf must correspond to a cross-over point in either of the cdfs, since the cdf in a max operation is obtained by multiplying the cdf values of the merging cdfs. Given this, we need to retain all co-optimal solutions corresponding to the regions into which the time axis is partitioned by the cross-over points. The co-optimality results from the second part of the solution, representing the cost associated with the cdf. Therefore, if a solution in co-optimal in any of the partitions, then that particular solution is retained.

To achieve polynomial complexity for the approach [42] proposes to limit the solution at any node by the number of solutions at the fanin inputs. Thus after a max operation the solutions are pruned such that the number of solutions is equal to the summation of the number of solutions whose max is performed. Similarly, in the case of a sum operation the number of solutions is pruned to the sum of the number of solutions at the input and the number of possible implementations of the gate. This pruning is achieved by defining a *max-indicator* for each solution, which refers to the fraction of the time axis where that cdf has a value larger than any other cdf. From the possible range of max-indicator values, a set of max-indicator points are selected which are equal to the number of solutions that one wants to retain at that node. For each max-indicator value selected, the solution with the minimum cost is retained and all other solutions are pruned out. Following this pruning step only a polynomial number of solutions are retained at each node, and therefore the entire dynamic programming approach now runs in polynomial time.

We now turn to the problems associated with reconvergent fanouts that arise when using dynamic programming based approaches. The first problem appears while merging delay cdfs. The merged cdfs might correspond to different choices for the gates that are in the subtree rooted at a multiple fanout node from which the two merged cdfs originate. The second problem arises while summing costs, since summing the costs now results in the cost associated with the common subtree being counted multiple times. The second problem is solved by estimating the cost after a max operation by summing the cost of each node in the subtree for each solution generated at the output. The first problem can be solved by maintaining lists of the fanin gates and their assignment for each solution, and if a conflict is found in the assignment of a gate before merging solutions then this solution is not generated. Note that tackling these problems results in a significantly large overhead both in terms of run-time and memory requirements, and is not suitable for large designs.

Table 6.2. Comparison of cost and delay results obtained using deterministic and probabilistic constraints. PF=Probability of failure, Det.=Deterministic optimization and Stat.=Statistical optimization [42].

Benchmark	Delay Constraint	Cost Det.	Cost Stat.	PF Det.	PF Stat.	PF Constraint
c432	4485	462	496	0.87	0.24	0.3
c499	2100	3484	4843	0.36	0.14	0.2
c880	4870	668	703	0.63	0.14	0.2
c1355	2165	1149	1268	0.49	0.00	0.2
c1908	3690	812	846	0.53	0.03	0.2
x3	1610	1249	1464	0.51	0.34	0.4
x4	1600	896	2108	0.46	0.06	0.2
i2	650	420	626	0.61	0.02	0.2
i3	250	539	814	0.64	0.08	0.1
alu2	4650	740	809	0.05	0.01	0.2
dalu	6883	2522	3022	0.32	0.07	0.2
too-large	1700	624	1536	0.32	0.06	0.2
Average				0.48	0.10	

Table 6.2 compares the results obtained using deterministic and statistical dynamic programming based approaches. The second column states the delay constraint enforced on the circuit, and we can see that columns five and six show that the deterministic optimization results in a large probability of timing failure whereas the statistical approach is always able to meet the constraint on timing failure specified in the last column. Note that results are for the case where both the deterministic and statistical approaches result in similar costs, and that the cost for deterministic optimization is always smaller than the cost for statistical optimization. However, the results are generated with the expected value of the delay used as an estimate for deterministic delay, resulting in an average probability of failure of approximately 50%. Additionally, only intra-die variations are considered which have the strongest influence on the mean delay and this further penalizes the results obtained using the deterministic approach.

In this Chapter we reviewed some of the techniques that have been proposed to perform statistical optimization to date. We observed that other than sensitivity-based techniques that decouple statistical analysis from statistical optimization, most of the other optimization approaches suffer from high complexity. Mapping these techniques to a computationally efficient approach, results in a loss in achieved performance improvements. The sensitivity-based statistical techniques themselves do not guarantee optimality however, and integrating them with statistical timers that consider all components of variations is not straightforward in all cases. Current approaches that perform statistical timing or power optimization also tend to neglect the strong correlation in power and performance as discussed in the previous chapter, hence performing timing yield optimization results in loss in yield due to the power constraint and vice-versa. Therefore there remains a need to develop approaches that are truly statistical in nature and maximize yield through timing and power optimizations while considering their correlation.

References

1. A. Abu-Dayaa and N. C. Beaulieu. Outage probabilities in the presence of correlated lognormal interferers. *IEEE Transactions on Vehicular Technology*, pages 164–173, February 1994.
2. E. Acar, S. Nassif, Y. Liu, and L. Pileggi. Time-domain simulation of variational interconnect models. In *ISQED '02: Proceedings of the 2005 International Symposium on Quality of Electronic Design*, pages 419–424, 2002.
3. A. Agarwal, D. Blaauw, and V. Zolotov. Statistical timing analysis for intra-die process variations with spatial correlations. In *ICCAD '03: Proceedings of the 2003 IEEE/ACM international conference on Computer-aided design*, pages 900–907, Washington, DC, USA, 2003. IEEE Computer Society.
4. A. Agarwal, D. Blaauw, V. Zolotov, S. Sundareswaran, M. Zhou, K. Gala, and R. Panda. Statistical delay computation considering spatial correlations. In *ASP-DAC '01: Proceedings of the 2001 conference on Asia South Pacific design automation*, pages 271–276, New York, NY, USA, 2003. ACM Press.
5. A. Agarwal, F. Dartu, and D. Blaauw. Statistical gate delay model considering multiple input switching. In *DAC '04: Proceedings of the 41st annual conference on Design automation*, pages 658–663, New York, NY, USA, 2004. ACM Press.
6. A. Agarwal, V. Zolotov, and D. Blaauw. Statistical timing analysis using bounds and selective enumeration. *IEEE Transactions on Computer-aided Design of Integrated Circuits and Systems*, 22(9):1243–1260, September 2003.
7. A. Agarwal, V. Zolotov, D. Blaauw, and K. Chopra. Circuit optimization using statistical timing analysis. In *DAC '05: Proceedings of the 42nd annual conference on Design automation*, New York, NY, USA, 2005. ACM Press.
8. K. Agarwal, D. Sylvester, and D. Blaauw. Simple metrics for slew rate of rc circuits based on two circuit moments. In *DAC '03: Proceedings of the 40th conference on Design automation*, pages 950–953, New York, NY, USA, 2003. ACM Press.
9. K. Agarwal, D. Sylvester, D. Blaauw, F. Liu, S. Nassif, and S. Vrudhula. Variational delay metrics for interconnect timing analysis. In *DAC '04: Proceedings of the 41st annual conference on Design automation*, pages 381–384, New York, NY, USA, 2004. ACM Press.
10. C. J. Alpert, A. Devgan, and C. Kashyap. A two moment rc delay metric for performance optimization. In *ISPD '00: Proceedings of the 2000 international*

symposium on Physical design, pages 69–74, New York, NY, USA, 2000. ACM Press.

11. C. J. Alpert, F. Liu, C. Kashyap, and A. Devgan. Delay and slew metrics using the lognormal distribution. In *DAC '03: Proceedings of the 40th conference on Design automation*, pages 382–385, New York, NY, USA, 2003. ACM Press.
12. S. Arora. The approximability of np-hard problems. In *STOC '98: Proceedings of the thirtieth annual ACM symposium on Theory of computing*, pages 337–348, New York, NY, USA, 1998. ACM Press.
13. A. Asenov. Random dopant induced threshold voltage lowering and fluctuations in sub-0.1 μm mosfet's: a 3-d atomistic simulation study. *IEEE Transactions on Electron Devices*, 45(12):2505–2513, December 1998.
14. X. Bai, C. Visweswariah, and P. N. Strenski. Uncertainty-aware circuit optimization. In *DAC '02: Proceedings of the 39th conference on Design automation*, pages 58–63, New York, NY, USA, 2002. ACM Press.
15. M. Bazaraa, S. Hanif, and C. Shetty. *Nonlinear programming: Theory and applications.* John Wiley and sons, 1993.
16. N. Beaulieu, A. Abu-Dayya, , and P.J.McLane. Comparison of methods of computing lognormal sum distributions and outages for digital wireless applications. In *IEEE International Conference on Communications*, pages 1270–1275, 1994.
17. M. Berkelaar. Statistical delay calculation, a linear time method. In *ACM/IEEE international workshop on Timing issues in the specification and synthesis of digital systems*, pages 15–24, 1997.
18. M. R. C. M. Berkelaar and J. A. G. Jess. Gate sizing in mos digital circuits with linear programming. In *EURO-DAC '90: Proceedings of the conference on European design automation*, pages 217–221, Los Alamitos, CA, USA, 1990. IEEE Computer Society Press.
19. S. Bhardwaj, S. B. K. Vrudhula, and D. Blaauw. τau: Timing analysis under uncertainty. In *ICCAD '03: Proceedings of the 2003 IEEE/ACM international conference on Computer-aided design*, pages 615–620, Washington, DC, USA, 2003. IEEE Computer Society.
20. S. Borkar, T. Karnik, S. Narendra, J. Tschanz, A. Keshavarzi, and V. De. Parameter variations and impact on circuits and microarchitecture. In *DAC '03: Proceedings of the 40th conference on Design automation*, pages 338–342, New York, NY, USA, 2003. ACM Press.
21. K. A. Bowman, S. G. Duvall, and J. D. Miendl. Impact of die-to-die and within-die parameter fluctuations on the maximum clock frequency distribution for gigascale integration. *IEEE Journal of Solid-State Circuits*, 37(2):183–190, February 2002.
22. S. Boyd and L. Vandenberghe. *Convex optimization.* Cambridge University press, 2004.
23. F. Brglez and H. Fujiwara. A neutral netlist of 10 combinatorial benchmark circuits. In *ISCAS '85: IEEE International Symposium on Circuits and Systems*, pages 695–698, 1990.
24. J. H. Cadwell. The bivariate normal integral. *Biometrika*, pages 31–35, December 1951.
25. K. M. Cao, W. C. Lee, W. Liu, X. jin, P. Su, S. K. H. Fung, J. X. An, B. Yu, and C. Hu. Bsim4 gate leakage model including source-drain partition. In *International Electron Devices Meeting*, pages 815–818, 2000.

26. Y. Cao, P. Gupta, A. B. Kahng, D. Sylvester, and J. Yang. Design sensitivities to variability: Extrapolations and assessments in nanometer vlsi. In *IEEE International ASIC/SOC Conference*, pages 411–415, 2002.
27. M. Celik, P. Lawrence, and A. Odabasioglu. *IC interconnect analysis.* Kluwer Publishers, 2002.
28. A. Chandrakasan, W. Bowhill, and F. Fox. *Design of high-performance microprocessir circuits.* IEEE press, 2001.
29. V. Chandramouli and K. A. Sakallah. Modeling the effects of temporal proximity of input transitions on gate propagation delay and transition time. In *DAC '96: Proceedings of the 33rd annual conference on Design automation*, pages 617–622, New York, NY, USA, 1996. ACM Press.
30. H. Chang and S. S. Sapatnekar. Statistical timing analysis considering spatial correlations using a single pert-like traversal. In *ICCAD '03: Proceedings of the 2003 IEEE/ACM international conference on Computer-aided design*, pages 621–625, Washington, DC, USA, 2003. IEEE Computer Society.
31. C. P. Chen, C. C. N. Chu, and D. F. Wong. Fast and exact simultaneous gate and wire sizing by lagrangian relaxation. *IEEE Transactions on Computer-aided Design of Integrated Circuits and Systems*, 18(7):1014–1025, July 1993.
32. C.-P. Chen and D. F. Wong. Optimal wire-sizing function with fringing capacitance consideration. In *DAC '97: Proceedings of the 34th annual conference on Design automation*, pages 604–607, New York, NY, USA, 1997. ACM Press.
33. Z. Chen, M. Johnson, L. Wei, and K. Roy. Estimation of standby leakage power in cmos circuits considering accurate modeling of transistor stacks. In *ISLPED '98: Proceedings of the 1998 international symposium on Low power electronics and design*, pages 239–244, New York, NY, USA, 1998. ACM Press.
34. S. H. Choi, B. C. Paul, and K. Roy. Novel sizing algorithm for yield improvement under process variation in nanometer technology. In *DAC '04: Proceedings of the 41st annual conference on Design automation*, pages 454–459, New York, NY, USA, 2004. ACM Press.
35. C. Clark. The greates of a finite set of random variables. *Operations Resaearch*, 9:85–91, 1961.
36. J. Cohen and T. Hickey. Two algorithms for determining volumes of convex polyhedra. *Journal of the ACM*, 26:401–414, July 1979.
37. A. R. Conn, P. K. Coulman, R. A. Haring, G. L. Morill, C. Visweswarish, and C. W. Wu. Jiffytune: circuit optimization using time-domain sensitivities. *IEEE Transactions on Computer-aided Design of Integrated Circuits and Systems*, 17(12):1292–1309, December 1998.
38. A. R. Conn, N. I. M. Gould, and P. L. Toint. *LANCELOT: a FORTRAN package for large-scale nonlinear optimization, (Release A), volume 17 of Springer Series in Computational Mathematics.* Springer, 1992.
39. O. Coudert, R. Haddad, and S. Manne. New algorithms for gate sizing: a comparative study. In *DAC '96: Proceedings of the 33rd annual conference on Design automation*, pages 734–739, New York, NY, USA, 1996. ACM Press.
40. R. Cowell, A. P. David, S. L. Lauritzen, and D. J. Spiegelhalter. *Probabilistic networks and expert systems.* Springer, 1999.
41. F. Dartu and L. T. Pileggi. Teta: transistor-level engine for timing analysis. In *DAC '98: Proceedings of the 35th annual conference on Design automation*, pages 595–598, New York, NY, USA, 1998. ACM Press.

42. A. Davoodi, V. Khandelwal, and A. Srivastava. Variability inspired implementation selection problem. In *ICCAD '04: Proceedings of the 2004 IEEE/ACM international conference on Computer-aided design*, pages 423–427, Washington, DC, USA, 2004. IEEE Computer Society.
43. A. Devgan and C. Kashyap. Block-based static timing analysis with uncertainty. In *ICCAD '03: Proceedings of the 2003 IEEE/ACM international conference on Computer-aided design*, pages 607–614, Washington, DC, USA, 2003. IEEE Computer Society.
44. S. Duvall. Statistical circuit modeling and optimization. In *Workshop on statistical metrology*, pages 56–63, 2000.
45. W. C. Elmore. The transient response of damped linear networks with particular regard to wideband amplifiers. *Journal of Applied Physics*, 19:55–63, January 1948.
46. H. Elzinga. On the impact of spatial parametric variations on mos transistor mismatch. In *IEEE International Conference on Microelctronic Test Structures*, pages 173–177, 1996.
47. M. Evans, N. Hastings, and B. Peacock. *Statistical distributions.* John Wiley and sons, 1993.
48. P. Feldmann and S. W. Director. Integrated circuit quality optimization using surface integrals. *IEEE Transactions on Computer-aided Design of Integrated Circuits and Systems*, 12(12):1868–1879, December 1993.
49. E. Felt, A. Narayan, and A. Sangiovanni-Vincentelli. Measurement and modeling of mos transistor current mismatch in analog ic's. In *ICCAD '94: Proceedings of the 1994 IEEE/ACM international conference on Computer-aided design*, pages 272–277, Los Alamitos, CA, USA, 1994. IEEE Computer Society Press.
50. I. A. Ferzli and F. N. Najm. Statistical estimation of leakage-induced power grid voltage drop considering within-die process variations. In *DAC '03: Proceedings of the 40th conference on Design automation*, pages 856–859, New York, NY, USA, 2003. ACM Press.
51. I. A. Ferzli and F. N. Najm. Statistical verification of power grids considering process-induced leakage current variations. In *ICCAD '03: Proceedings of the 2003 IEEE/ACM international conference on Computer-aided design*, page 770, Washington, DC, USA, 2003. IEEE Computer Society.
52. J. Fishburn and A. Dunlop. Tilos: A posynomial programming approach to transistor sizing. In *ICCAD '85: Proceedings of the 1985 IEEE international conference on Computer-aided design*, pages 326–328, 1985.
53. D. J. Frank, P. Solomon, S. Reynolds, and J. Shin. Supply and threshold voltage optimization for low power design. In *ISLPED '97: Proceedings of the 1997 international symposium on Low power electronics and design*, pages 317–322, New York, NY, USA, 1997. ACM Press.
54. A. Gattiker, S. Nassif, R. Dinakar, and C. Long. Timing yield estimation from static timing analysis. In *ISQED '01: Proceedings of the 2001 International Symposium on Quality of Electronic Design*, pages 437–442, 2001.
55. G. Golub and C. VanLoan. *Matrix computations.* John Hopkins University press, 1996.
56. R. X. Gu and M. I. Elmasry. Power dissipation analysis and optimization of deep submicron cmos digital circuits. *IEEE Journal of Solid-State Circuits*, 31(5):707–713, May 1996.

57. J. P. Halter and F. N. Najm. A gate-level leakage power reduction method for ultra-low-power cmos circuits. In *CICC '97: Proceedings of the IEEE 1997 Custom Integrated Circuits COnference*, pages 475–478, 1997.
58. M. Hamada, Y. Ootaguro, and T. Kuroda. Utilizing surplus timing for power reduction. In *IEEE Conference on Custom Integrated Circuits*, pages 89–92, 2001.
59. C. L. Harkness and D. P. Lopresti. Interval methods for modeling uncertainty in rc timing analysis. *IEEE Transactions on Computer-aided Design of Integrated Circuits and Systems*, 11(11):1388–1401, Novemeber 1992.
60. P. Heydari and M. Pedram. Model reduction of variable-geometry interconnects using variational spectrally-weighted balanced truncation. In *ICCAD '01: Proceedings of the 2001 IEEE/ACM international conference on Computer-aided design*, pages 586–591, Piscataway, NJ, USA, 2001. IEEE Press.
61. R. Hitchcock. Timing verification and the timing analysis program. In *DAC '82: Proceedings of the annual conference on Design automation*, pages 594–604, 1982.
62. C. Hu. *Device and technology impact on low power electronics. Rabaey J. (eds).* Kluwer Publishers, 1996.
63. E. T. A. F. Jacobs and M. R. C. M. Berkelaar. Gate sizing using a statistical delay model. In *DATE '00: Proceedings of the conference on Design, automation and test in Europe*, pages 283–291, New York, NY, USA, 2000. ACM Press.
64. F. V. Jensen. *Bayesian networks and decision graphs.* Springer, 2000.
65. J. A. G. Jess, K. Kalafala, S. R. Naidu, R. H. J. M. Otten, and C. Visweswariah. Statistical timing for parametric yield prediction of digital integrated circuits. In *DAC '03: Proceedings of the 40th conference on Design automation*, pages 932–937, New York, NY, USA, 2003. ACM Press.
66. M. C. Johnson, D. Somasekhar, and K. Roy. Models and algorithms for bounds on leakage in cmos circuits. *IEEE Transactions on Computer-aided Design of Integrated Circuits and Systems*, 18(6):714–725, June 1999.
67. H. F. Jyu, S. Malik, S. Devadas, and K. Kuetzer. Statistical timing analysis of combinational logic circuits. *IEEE Transactions on Very Large Scale Integrated (VLSI) Systems*, 1(2):126–137, June 1993.
68. J. Kao, A. Chandrakasan, and D. Antoniadis. Transistor sizing issues and tool for multi-threshold cmos technology. In *DAC '97: Proceedings of the 34th annual conference on Design automation*, pages 409–414, New York, NY, USA, 1997. ACM Press.
69. J. Kao, S. Narendra, and A. Chandrakasan. Subthreshold leakage modeling and reduction techniques. In *ICCAD '02: Proceedings of the 2002 IEEE/ACM international conference on Computer-aided design*, pages 141–148, New York, NY, USA, 2002. ACM Press.
70. T. Karnik, Y. Ye, J. Tschanz, L. Wei, S. Burns, V. Govindarajulu, V. De, and S. Borkar. Total power optimization by simultaneous dual-vt allocation and device sizing in high performance microprocessors. In *DAC '02: Proceedings of the 39th conference on Design automation*, pages 486–491, New York, NY, USA, 2002. ACM Press.
71. C. V. Kashyap, C. Alpert, F. Liu, and A. Devgan. Closed-form expressions for extending step delay and slew metrics to ramp inputs for rc trees. *IEEE Transactions on Computer-aided Design of Integrated Circuits and Systems*, 23(4):509–516, April 2004.

72. K. J. Kerns and A. T. Yang. Stable and efficient reduction of large, multiport rc networks by pole analysis via congruence transformations. *IEEE Transactions on Computer-aided Design of Integrated Circuits and Systems*, 16(7):734–744, July 1997.
73. V. Khandelwal, A. Davoodi, A. Nanavati, and A. Srivastava. A probabilistic approach to buffer insertion. In *ICCAD '03: Proceedings of the 2003 IEEE/ACM international conference on Computer-aided design*, pages 560–567, Washington, DC, USA, 2003. IEEE Computer Society.
74. C. H. Kim, K. Roy, S. Hsu, A. Alvandpour, R. Krishnamurthy, and S. Borkar. A process variation compensating technique for sub-90nm dynamic circuits. In *Symposium on VLSI Circuits*, pages 205–206, 2003.
75. D. E. Knuth. *The Art of Computer Programming, volume 2: Seminumerical Algorithms.* Addison-Wesley, Reading, MA, 1997.
76. R. K. Krishnamurthy, A. Alvandpour, V. De, and S. Borkar. High-performance and low-power challenges for sub-70 nm microprocessor circuits. In *CICC '02: Proceedings of the IEEE 2002 Custom Integrated Circuits COnference*, pages 125–128, 2002.
77. J. Le, X. Li, and L. T. Pileggi. Stac: statistical timing analysis with correlation. In *DAC '04: Proceedings of the 41st annual conference on Design automation*, pages 343–348, New York, NY, USA, 2004. ACM Press.
78. D. Lee, W. Kwong, D. Blaauw, and D. Sylvester. Simultaneous subthreshold and gate-oxide tunneling leakage current analysis in nanometer cmos design. In *ISQED '03: Proceedings of the 2003 International Symposium on Quality of Electronic Design*, pages 287–292, 2003.
79. X. Li, J. Le, P. Gopalakrishnan, and L. Pileggi. Asymptotic probability extraction for non-normal distribution of circuit performance. In *ICCAD '04: Proceedings of the 2004 IEEE/ACM international conference on Computer-aided design*, pages 2–9, Washington, DC, USA, 2004. IEEE Computer Society.
80. T. Lin, E. Acar, and L. Pileggi. h-gamma: an rc delay metric based on a gamma distribution approximation of the homogeneous response. In *ICCAD '98: Proceedings of the 1998 IEEE/ACM international conference on Computer-aided design*, pages 19–25, New York, NY, USA, 1998. ACM Press.
81. J.-J. Liou, K.-T. Cheng, S. Kundu, and A. Krstic. Fast statistical timing analysis by probabilistic event propagation. In *DAC '01: Proceedings of the 38th conference on Design automation*, pages 661–666, New York, NY, USA, 2001. ACM Press.
82. F. Liu, C. Kashyap, and C. J. Alpert. A delay metric for rc circuits based on the weibull distribution. In *ICCAD '02: Proceedings of the 2002 IEEE/ACM international conference on Computer-aided design*, pages 620–624, New York, NY, USA, 2002. ACM Press.
83. Y. Liu, S. Nassif, L. Pileggi, and A. J. Strojwas. Impact of interconnect variations on the clock skew of a gigahertz microprocessor. In *DAC '00: Proceedings of the 37th Conference on Design Automation (DAC'00)*, pages 168–171, Washington, DC, USA, 2000. IEEE Computer Society.
84. Y. Liu, L. T. Pileggi, and A. J. Strojwas. Model order-reduction of rc(l) interconnect including variational analysis. In *DAC '99: Proceedings of the 36th Annual Conference on Design Automation (DAC'99)*, pages 201–206, Los Angeles, CA, USA, 1999. IEEE Computer Society.

85. M. Lobo, L. Vandenberghe, S. Boyd, and H. Lebret. Applications of second-order cone programming. *Linear Algebra and its Applications*, pages 193–228, November 1998.

86. J. Ma and R. Rutenbar. Interval valued reduced order statistical interconnect modeling. In *ICCAD '04: Proceedings of the 2004 IEEE/ACM international conference on Computer-aided design*, pages 460–467, Washington, DC, USA, 2004. IEEE Computer Society.

87. M. Mani and M. Orshansky. A new statistical optimization algorithm for gate sizing. In *ICCD '04: Proceedings of the 2005 International Conference on Computer Design*, pages 272–277, 2004.

88. N. Marlow. A normal limit theorem for power sums of independent random variables. *Bell Systems Technical Journal*, 46:2081–2090, November 1967.

89. V. Mehrotra, S. Nassif, D. Boning, and J. Chung. Modeling the effects of manufacturing variation on high-speed microprocessor interconnect performance. In *International Electron Devices Meeting*, pages 767–770, 1998.

90. C. Michael and M. Ismail. Statistical modeling of device mismatch for analog mos integrated circuits. *IEEE Journal of Solid-State Circuits*, 27(2):154–166, February 1992.

91. T. Mikolajick, V. Haublein, and H. Ryssel. The effect of random dopant fluctuations on the minimum channel length of short-channel mos transistors. *Applied Physics A Material Science and Processing*, 64:555–560, June 1997.

92. D. Montgomery and R. Myers. *Response Surface Methodology: Process and Product Optimization Using Designed Experiments.* John Wiley and sons, 2002.

93. S. Mukhopadhyay, A. Raychowdhury, and K. Roy. Accurate estimation of total leakage current in scaled cmos logic circuits based on compact current modeling. In *DAC '03: Proceedings of the 40th conference on Design automation*, pages 169–174, New York, NY, USA, 2003. ACM Press.

94. S. Mukhopadhyay and K. Roy. Modeling and estimation of total leakage current in nano-scaled cmos devices considering the effect of parameter variation. In *ISLPED '03: Proceedings of the 2003 international symposium on Low power electronics and design*, pages 172–175, New York, NY, USA, 2003. ACM Press.

95. S. Mutoh, T. Douseki, Y. Matsuya, T. Aoki, S. Shigematsu, and J. Yamada. 1-v power supply high-speed digital circuit technology with multithreshold-voltage cmos. *IEEE Journal of Solid-State Circuits*, 30(8):847–854, August 1995.

96. F. N. Najm and N. Menezes. Statistical timing analysis based on a timing yield model. In *DAC '04: Proceedings of the 41st annual conference on Design automation*, pages 460–465, New York, NY, USA, 2004. ACM Press.

97. S. Narendra, D. Antoniadis, and V. De. Impact of using adaptive body bias to compensate die-to-die vt variation on within-die vt variation. In *ISLPED '99: Proceedings of the 1999 international symposium on Low power electronics and design*, pages 229–232, New York, NY, USA, 1999. ACM Press.

98. S. Narendra, V. De, S. Borkar, D. Antoniadis, and A. Chandrakasan. Full-chip sub-threshold leakage power prediction model for sub-0.18 μm cmos. In *ISLPED '02: Proceedings of the 2002 international symposium on Low power electronics and design*, pages 19–23, New York, NY, USA, 2002. ACM Press.

99. S. R. Nassif. Modeling and analysis of manufacturing variations. In *IEEE Conference on Custom Integrated Circuits*, pages 223–228, 2001.

100. D. Nguyen, A. Davare, M. Orshansky, D. Chinnery, B. Thompson, and K. Keutzer. Minimization of dynamic and static power through joint assign-

ment of threshold voltages and sizing optimization. In *ISLPED '03: Proceedings of the 2003 international symposium on Low power electronics and design*, pages 158–163, New York, NY, USA, 2003. ACM Press.
101. A. Odabasioglu, M. Celik, and L. T. Pileggi. Prima: passive reduced-order interconnect macromodeling algorithm. In *ICCAD '97: Proceedings of the 1997 IEEE/ACM international conference on Computer-aided design*, pages 58–65, Washington, DC, USA, 1997. IEEE Computer Society.
102. K. Okada, K. Yamaoka, and H. Onodera. A statistical gate-delay model considering intra-gate variability. In *ICCAD '03: Proceedings of the 2003 IEEE/ACM international conference on Computer-aided design*, pages 908–913, Washington, DC, USA, 2003. IEEE Computer Society.
103. M. Orshansky and A. Bandyopadhyay. Fast statistical timing analysis handling arbitrary delay correlations. In *DAC '04: Proceedings of the 41st annual conference on Design automation*, pages 337–342, New York, NY, USA, 2004. ACM Press.
104. M. Orshansky, L. Milor, and C. Hu. Characterization of spatial intrafield gate cd variability, its impact on circuit performance and spatial mask-level correction. *IEEE Transactions on Semiconductor Manufacturing*, 17(1):2–11, February 2004.
105. J. M. Ortega and W. C. Rheinboldt. *Iterative solution of nonlinear equations in several variables.* SIAM, 2000.
106. R. H. J. M. Otten and R. K. Brayton. Planning for performance. In *DAC '98: Proceedings of the 35th annual conference on Design automation*, pages 122–127, New York, NY, USA, 1998. ACM Press.
107. P. Pant, V. De, and A. Chatterjee. Device-circuit optimization for minimal energy and power consumption in cmos random logic networks. In *DAC '97: Proceedings of the 34th annual conference on Design automation*, pages 403–408, New York, NY, USA, 1997. ACM Press.
108. P. Pant, R. Roy, and A. Chatterjee. Dual-threshold voltage assignment with transistor sizing for low power cmos circuits. *IEEE Transactions on Very Large Scale Integrated (VLSI) Systems*, 9(2):390–394, April 2001.
109. A. Papoulis. *Probability, random variables and dtochastic processes.* McGraw Hill, NY, 1965.
110. J. Pearl. *Probabilistic reasoning in intelligent systems: networks of plausible inference.* Morgan Kaufmann, 1988.
111. M. J. M. Pelgrom, A. C. J. Duinmaijer, and A. P. G. Welbers. Dual-threshold voltage assignment with transistor sizing for low power cmos circuits. *IEEE Journal of Sold-State Circuits*, 24(5):1433–1439, October 1989.
112. L. Pileggi. Timing metrics for physical design of deep submicron technologies. In *ISPD '98: Proceedings of the 1998 international symposium on Physical design*, pages 28–33, New York, NY, USA, 1998. ACM Press.
113. L. Pillage and R. A. Rohrer. Asymptotic waveform evaluation for timing analysis. *IEEE Transactions on Computer-aided Design of Integrated Circuits and Systems*, 9(4):352–366, April 1990.
114. W. H. Press, B. P. Flannery, S. A. Teukolsky, and W. T. Vetterling. *Numerical recipies in C.* Cambridge University Press, 1992.
115. S. Raj, S. B. K. Vrudhula, and J. Wang. A methodology to improve timing yield in the presence of process variations. In *DAC '04: Proceedings of the 41st annual conference on Design automation*, pages 448–453, New York, NY, USA, 2004. ACM Press.

116. R. Rao, K. Agarwal, A. Devgan, K. Nowka, D. Sylvester, and R. Brown. Parametric yield analysis and constrained-based supply voltage optimization. In *ISQED '05: Proceedings of the 2005 International Symposium on Quality of Electronic Design*, pages 284–290, 2005.
117. R. Rao, A. Srivastava, D. Blaauw, and D. Sylvester. Statistical analysis of subthreshold leakage current for vlsi circuits. *IEEE Trans. Very Large Scale Integr. Syst.*, 12(2):131–139, 2004.
118. R. R. Rao, A. Devgan, D. Blaauw, and D. Sylvester. Parametric yield estimation considering leakage variability. In *DAC '04: Proceedings of the 41st annual conference on Design automation*, pages 442–447, New York, NY, USA, 2004. ACM Press.
119. C. L. Ratzlaff, N. Gopal, and L. T. Pillage. Rice: Rapid interconnect circuit evaluator. In *DAC '91: Proceedings of the 28th conference on ACM/IEEE design automation*, pages 555–560, New York, NY, USA, 1991. ACM Press.
120. N. Rohrer and et al. A 480 mhz risc microprocessor in a 0.12 m leff cmos technology with copper interconnects. In *IEEE International Solid-State Circuits Conference*, pages 5–7, 1998.
121. K. Roy, S. Mukhopadhyay, and H. Mahmoodi-Meimand. Leakage current mechanisms and leakage reduction techniques in deep-submicrometer cmos circuits. *Proceedings of the IEEE*, 91(2):305–327, February 2003.
122. J. Rubenstein, P. Penfield, and M. A. Horowitz. Signal delay in rc tree networks. *IEEE Transactions on Computer-aided Design of Integrated Circuits and Systems*, 2(3):202–211, July 1983.
123. T. Sakurai and A. R. Newton. Alpha-power law mosfet model and its application to cmos inverter delay and other formulas. *IEEE Journal of Solid-State Circuits*, 25(2):584–594, April 1990.
124. S. B. Samaan. The impact of device parameter variation on the frequency and performance of vlsi chips. In *ICCAD '04: Proceedings of the 2004 IEEE/ACM international conference on Computer-aided design*, pages 343–346, Washington, DC, USA, 2004. IEEE Computer Society.
125. S. S. Sapatnekar, V. B. Rao, P. M. Vaidya, and S. M. Kang. An exact solution to the transistor sizing problem for cmos circuits using convex optimization. *IEEE Transactions on Computer-aided Design of Integrated Circuits and Systems*, 12(11):1621–1634, November 1993.
126. L. Scheffer. The count of monte carlo. In *ACM/IEEE international workshop on Timing issues in the specification and synthesis of digital systems*, 2004.
127. S. C. Schwartz and Y. S. Yeh. On the distribution function and moments of power sums with lognormal components. *Bell Systems Technical Journal*, 61:1441–1462, September 1982.
128. S. Sirichotiyakul, T. Edwards, C. Oh, J. Zuo, A. Dharchoudhury, R. Panda, and D. Blaauw. Stand-by power minimization through simultaneous threshold voltage selection and circuit sizing. In *DAC '99: Proceedings of the 36th ACM/IEEE conference on Design automation*, pages 436–441, New York, NY, USA, 1999. ACM Press.
129. A. Srivastava, R. Bai, D. Blaauw, and D. Sylvester. Modeling and analysis of leakage power considering within-die process variations. In *ISLPED '02: Proceedings of the 2002 international symposium on Low power electronics and design*, pages 64–67, New York, NY, USA, 2002. ACM Press.

130. A. Srivastava, S. Shah, K. Agarwal, D. Sylvester, D. Blaauw, and S. Director. Accurate and efficient parametric yield estimation considering correlated variations in leakage power and performance. In *DAC '05: Proceedings of the 42nd annual conference on Design automation*, New York, NY, USA, 2005. ACM Press.
131. A. Srivastava and D. Sylvester. A general framework for low-power design space exploration considering process variation. In *ICCAD '04: Proceedings of the 2004 IEEE/ACM international conference on Computer-aided design*, pages 808–813, New York, NY, USA, 2004. ACM Press.
132. A. Srivastava and D. Sylvester. Minimizing total power by simultaneous vdd and vth assignment. *IEEE Transactions on Computer-aided Design of Integrated Circuits and Systems*, 23(5):665–677, May 2004.
133. A. Srivastava, D. Sylvester, and D. Blaauw. Statistical optimization of leakage power considering process variations using dual-vth and sizing. In *DAC '04: Proceedings of the 41st annual conference on Design automation*, pages 773–778, New York, NY, USA, 2004. ACM Press.
134. J. Stolfi and L. H. de Figueiredo. *Self validated numerical methods and applications.* Brazilian Mathematics Colloquium Monograph, Rio De Janeiro, Brazil, 1997.
135. H. Su, E. Acar, and S. R. Nassif. Power grid reduction based on algebraic multigrid principles. In *DAC '03: Proceedings of the 40th conference on Design automation*, pages 109–112, New York, NY, USA, 2003. ACM Press.
136. H. Su, F. Liu, A. Devgan, E. Acar, and S. Nassif. Full chip leakage estimation considering power supply and temperature variations. In *ISLPED '03: Proceedings of the 2003 international symposium on Low power electronics and design*, pages 78–83, New York, NY, USA, 2003. ACM Press.
137. V. Sundararajan, S. S. Sapatnekar, and K. K. Parhi. Minflotransit: Min cost flow based transistor sizing tool. In *DAC '00: Proceedings of the 37th Annual Conference on Design Automation (DAC'00)*, pages 150–155, Washington, DC, USA, 1999. IEEE Computer Society.
138. K. Takeuchi, T. Tatsumi, and A. Furukaea. Channel engineering for the reduction of random-dopant-placement-induced threshold voltage fluctuation. In *International Electron Devices Meeting*, pages 841–844, 1997.
139. K. Takeuchi, T. Tatsumi, and A. Furukaea. A critical examination of the mechanics of dynamic nbti for pmosfets. In *International Electron Devices Meeting*, pages 14.4.1–14.4.4, 2003.
140. H. Tennakoon and C. Sechen. Gate sizing using lagrangian relaxation combined with a fast gradient-based pre-processing step. In *ICCAD '02: Proceedings of the 2002 IEEE/ACM international conference on Computer-aided design*, pages 395–402, New York, NY, USA, 2002. ACM Press.
141. S. Tsukiyama, M. Tanaka, and M. Fukui. A statistical static timing analysis considering correlations between delays. In *ASP-DAC '01: Proceedings of the 2001 conference on Asia South Pacific design automation*, pages 353–358, New York, NY, USA, 2001. ACM Press.
142. S. Tyagi and et al. A 130 nm generation logic technology featuring 70 nm transistors, dual vt transistors and 6 layers of cu interconnects. In *International Electron Devices Meeting*, pages 567–570, 2000.
143. K. Usami, M. Igarashi, F. Minami, T. Ishikawa, M. Kanzawa, M. Ichida, and K. Nogami. Automated low-power technique exploiting multiple supply volt-

ages applied to a media processor. *IEEE Journal of Solid-State Circuits*, 33(3):463–472, March 1998.

144. L. Vandenberghe, S. Boyd, and S. P. Wu. Determinant maximization with linear matrix inequality constraints. *SIAM Journal on Matrix analysis and Applications*, 19:499–533, 1998.
145. L. P. P. P. VanGinneken. Buffer placement in distributed rc-tree networks for minimal elmore delay. In *ISCAS '90: IEEE International Symposium on Circuits and Systems*, pages 865–868, 1990.
146. C. Visweswariah, K. Ravindran, K. Kalafala, S. G. Walker, and S. Narayan. First-order incremental block-based statistical timing analysis. In *DAC '04: Proceedings of the 41st annual conference on Design automation*, pages 331–336, New York, NY, USA, 2004. ACM Press.
147. Q. Wang and S. Vrudhula. Algorithms for minimizing standby power in deep submicrometer, dual-vt cmos circuits. *IEEE Transactions on Computer-aided Design of Integrated Circuits and Systems*, 21(3):306–318, March 2002.
148. Q. Wang and S. B. K. Vrudhula. Static power optimization of deep submicron cmos circuits for dual vt technology. In *ICCAD '98: Proceedings of the 1998 IEEE/ACM international conference on Computer-aided design*, pages 490–496, New York, NY, USA, 1998. ACM Press.
149. L. Wei, Z. Chen, K. Roy, Y. Ye, and V. De. Mixed- vth (mvt) cmos circuit design methodology for low power applications. In *DAC '99: Proceedings of the 36th Annual Conference on Design Automation (DAC'99)*, pages 430–435, Washington, DC, USA, 1999. IEEE Computer Society.
150. L. Wei, K. Roy, and C. Koh. Power minimization by simutaneous dual-v_{th} assignment and gate sizing. In *CICC '00: Proceedings of the IEEE 2000 Custom Integrated Circuits COnference*, pages 413–416, 2000.
151. Y. Ye, S. Borkar, and V. De. A new technique for standby leakage reduction in high-performance circuits. In *Symposium on VLSI Circuits*, pages 11–13, 1998.
152. S. H. Yen, D. H. Du, and S. Ghanta. Efficient algorithms for extracting the k most critical paths in timing analysis. In *DAC '89: Proceedings of the 26th ACM/IEEE conference on Design automation*, pages 649–654, New York, NY, USA, 1989. ACM Press.
153. G. Yoh and F. Najm. A statistical model for electromigration failures. In *ISQED '00: Proceedings of the 2000 International Symposium on Quality of Electronic Design*, pages 45–50, 2000.
154. J. Zhang and M. Styblinski. *Yield and variability optimziation of integrated circuits*. Kluwer Publishers, 1995.
155. L. Zhang, W. Chen, Y. Hu, and C. C. Chen. Statistical timing analysis with extended pseudo-canonical timing model. In *DATE '05: Proceedings of the conference on Design, Automation and Test in Europe*, pages 952–957, Washington, DC, USA, 2005. IEEE Computer Society.
156. S. Zhang, V. Wason, and K. Banerjee. A probabilistic framework to estimate full-chips subthreshold leakage power distribution considering within-die and die-to-die p-t-v variations. In *ISLPED '04: Proceedings of the 2004 international symposium on Low power electronics and design*, pages 156–161, New York, NY, USA, 2004. ACM Press.
157. Y. Zhang and L. Gao. On numerical solution of the maximum ellipsoid problem. Technical Report TR01-15, Rice University, Department of Computational and Applied Mathematics, August 2001.

158. P. Zuchowski, P. A. Habitz, J. D. Hayes, and J. H. Oppold. Process and environmental variation impacts on asic timing. In *ICCAD '04: Proceedings of the 2004 IEEE/ACM international conference on Computer-aided design*, pages 336–342, Washington, DC, USA, 2004. IEEE Computer Society.

Index